Introduction to Linear Circuit Analysis and Modelling

From DC to RF

To our families

Introduction to Linear Circuit Analysis and Modelling

From DC to RF

Luis Moura
University of Algarve, Portugal

Izzat Darwazeh
University College London, UK

AMSTERDAM · BOSTON · HEIDELBERG · LONDON · NEW YORK · OXFORD
PARIS · SAN DIEGO · SAN FRANCISCO · SINGAPORE · SYDNEY · TOKYO

Newnes is an imprint of Elsevier

Newnes
An imprint of Elsevier
Linacre House, Jordan Hill, Oxford OX2 8DP
30 Corporate Drive, Burlington, MA 01803

First published 2005

Copyright © Luis Moura and Izzat Darwazeh, 2005. All rights reserved.

The right of Luis Moura and Izzat Darwazeh to be identified as the authors of this work has been asserted in accordance with the Copyright, Designs and Patents Act 1988.

No part of this publication may be reproduced in any material form (including photocopying or storing in any medium by electronic means and whether or not transiently or incidentally to some other use of this publication) without the written permission of the copyright holders except in accordance with the provisions of the Copyright, Design and Patents Act 1988 or under the terms of a license issued by the Copyright Licensing Agency Ltd, 90 Tottenham Court Road, London, England W1P 4LP. Applications for the copyright holders' written permission to reproduce any part of this publication should be addressed to the publishers.

British Library Cataloguing in Publication Data
A catalogue record for this book is available from the British Library

ISBN 07506 59327

Working together to grow
libraries in developing countries

www.elsevier.com | www.bookaid.org | www.sabre.org

ELSEVIER   BOOK AID International   Sabre Foundation

Contents

Preface   ix

Acknowledgements   xi

1 Elementary electrical circuit analysis   1
    1.1 Introduction 1
    1.2 Voltage and current 1
       1.2.1 Voltage sources 2
       1.2.2 Current sources 3
    1.3 Electrical passive elements 5
       1.3.1 Resistance and conductance 5
       1.3.2 Capacitance 8
       1.3.3 Inductance 9
    1.4 Kirchhoff's laws 10
       1.4.1 Series and parallel combinations of passive elements . 11
       1.4.2 Other types of circuit element connections 17
       1.4.3 Electrical network analysis – Nodal analysis 18
       1.4.4 Resistive voltage and current dividers 20
       1.4.5 Controlled sources 21
    1.5 Thévenin's theorem 22
    1.6 Norton's theorem 23
    1.7 Superposition theorem 25
    1.8 Bibliography 26
    1.9 Problems 27

2 Complex numbers: An introduction   32
    2.1 Introduction 32
    2.2 Definition 32
    2.3 Elementary algebra 34
       2.3.1 Addition 34
       2.3.2 Subtraction 34
       2.3.3 Multiplication 35
       2.3.4 Division 37
       2.3.5 Complex equations 38
       2.3.6 Quadratic equations 38
    2.4 Polar representation 39

		2.4.1	Multiplication and division	40
	2.5	The exponential form		41
		2.5.1	Trigonometric functions and the exponential form	42
	2.6	Powers and roots		43
	2.7	Bibliography		46
	2.8	Problems		46

3 Frequency domain electrical signal and circuit analysis — 48

- 3.1 Introduction — 48
- 3.2 Sinusoidal AC electrical analysis — 48
  - 3.2.1 Effective electrical values — 49
  - 3.2.2 I–V characteristics for passive elements — 50
  - 3.2.3 Phasor analysis — 54
  - 3.2.4 The generalised impedance — 56
  - 3.2.5 Maximum power transfer — 64
- 3.3 Generalised frequency domain analysis — 65
  - 3.3.1 The Fourier series — 65
  - 3.3.2 Fourier coefficients, phasors and line spectra — 72
  - 3.3.3 Electrical signal and circuit bandwidths — 73
  - 3.3.4 Linear distortion — 79
  - 3.3.5 Bode plots — 80
  - 3.3.6 The Fourier transform — 88
  - 3.3.7 Transfer function and impulse response — 98
  - 3.3.8 The convolution operation — 100
- 3.4 Bibliography — 106
- 3.5 Problems — 106

4 Natural and forced responses circuit analysis — 109

- 4.1 Introduction — 109
- 4.2 Time domain analysis — 109
- 4.3 Transient analysis using Fourier transforms — 113
  - 4.3.1 Differentiation theorem — 113
  - 4.3.2 Integration theorem — 114
  - 4.3.3 I–V characteristics for passive elements — 114
- 4.4 The Laplace transform — 117
  - 4.4.1 Theorems of the Laplace transform — 119
  - 4.4.2 Partial-fraction expansion — 123
- 4.5 Analysis using Laplace transforms — 127
  - 4.5.1 Solving differential equations — 127
  - 4.5.2 I–V characteristics for passive elements — 128
  - 4.5.3 Natural response — 130
  - 4.5.4 Response to the step function — 136
- 4.6 Bibliography — 147
- 4.7 Problems — 147

5 Electrical two-port network analysis — 150
- 5.1 Introduction — 150
- 5.2 Electrical representations — 150
  - 5.2.1 Electrical impedance representation — 150
  - 5.2.2 Electrical admittance representation — 153
  - 5.2.3 Electrical chain representation — 157
  - 5.2.4 Conversion between electrical representations — 159
  - 5.2.5 Miller's theorem — 161
- 5.3 Computer-aided electrical analysis — 163
- 5.4 Bibliography — 167
- 5.5 Problems — 167

6 Basic electronic amplifier building blocks — 169
- 6.1 Introduction — 169
- 6.2 Modelling the amplification process — 169
- 6.3 Operational amplifiers — 177
  - 6.3.1 Open-loop and feedback concepts — 177
  - 6.3.2 Other examples and applications — 179
- 6.4 Active devices — 183
  - 6.4.1 The junction or p–n diode — 183
  - 6.4.2 The bipolar junction transistor — 183
  - 6.4.3 The insulated gate field-effect transistor — 192
  - 6.4.4 The common-emitter amplifier — 196
  - 6.4.5 The differential pair amplifier — 215
- 6.5 Bibliography — 221
- 6.6 Problems — 221

7 RF circuit analysis techniques — 224
- 7.1 Introduction — 224
- 7.2 Lumped versus distributed — 224
- 7.3 Electrical model for ideal transmission lines — 227
  - 7.3.1 Voltage Standing Wave Ratio – VSWR — 234
  - 7.3.2 The $\lambda/4$ transformer — 239
  - 7.3.3 Lossy transmission lines — 242
  - 7.3.4 Microstrip transmission lines — 247
- 7.4 Scattering parameters — 248
  - 7.4.1 S-parameters and power waves — 254
  - 7.4.2 Power waves and generalised S-parameters — 258
  - 7.4.3 Conversions between different two-port parameters — 263
- 7.5 The Smith chart — 264
  - 7.5.1 The impedance and the reflection coefficient planes — 264
  - 7.5.2 Representation of impedances — 266
  - 7.5.3 Introduction to impedance matching — 272
- 7.6 Bibliography — 276
- 7.7 Problems — 276

8 Noise in electronic circuits — 279
- 8.1 Introduction — 279
- 8.2 Random variables — 279
  - 8.2.1 Moments of a random variable — 285
  - 8.2.2 The characteristic function — 288
  - 8.2.3 The central limit theorem — 290
  - 8.2.4 Bivariate Gaussian distributions — 293
- 8.3 Stochastic processes — 295
  - 8.3.1 Ensemble averages — 295
  - 8.3.2 Stationary random processes — 296
  - 8.3.3 Ergodic random processes — 296
  - 8.3.4 Power spectrum — 298
  - 8.3.5 Cross-power spectrum — 301
  - 8.3.6 Gaussian random processes — 302
  - 8.3.7 Filtered random signals — 303
- 8.4 Noise in electronic circuits — 305
  - 8.4.1 Thermal noise — 305
  - 8.4.2 Electronic shot-noise — 306
  - 8.4.3 $1/f$ noise — 306
  - 8.4.4 Noise models for passive devices — 307
  - 8.4.5 Noise models for active devices — 308
  - 8.4.6 The equivalent input noise sources — 310
  - 8.4.7 The noise figure — 319
- 8.5 Computer-aided noise modelling and analysis — 323
  - 8.5.1 Noise representations — 323
  - 8.5.2 Calculation of the correlation matrices — 325
  - 8.5.3 Elementary two-port interconnections — 327
  - 8.5.4 Transformation matrices — 328
- 8.6 Bibliography — 339
- 8.7 Problems — 340

A Mathematical formulae for electrical engineering — 342

B Elementary matrix algebra — 350

C Two-port electrical parameters — 352

Index — 356

Preface

The mathematical representation and analysis of circuits, signals and noise are key tools for electronic engineers. These tools have changed dramatically in recent years but the theoretical basis remain unchanged. Nowadays, the most complicated circuits can be analysed quickly using computer-based simulation. However, good appreciation of the fundamentals on which simulation tools are based is essential to make the best use of them.

In this book we address the theoretical basis of circuit analysis across a broad spectrum of applications encountered in today's electronic systems, especially for communications. Throughout the book we follow a mathematical-based approach to explain the different concepts using plenty of examples to illustrate these concepts.

This book is aimed at engineering and sciences students and other professionals who want solid grounding in circuit analysis. The basics covered in the first four chapters are suitable for first year undergraduates. The material covered in Chapters five and six is more specialist and provides a good background at an intermediate level, especially for those aiming to learn about electronic circuits and their building blocks. The last two chapters are more advanced and require good grounding in the concepts covered earlier in the book. These two chapters are suited to students in the final year of their engineering degree and to post-graduates.

In the first chapter we begin by reviewing the fundamental laws and theorems applicable to electrical circuits. In Chapter two we include a review of complex numbers, crucial for dealing with AC signals and circuits.

The varied and complex nature of signals and electronic systems require a thorough understanding of the mathematical description of signals and the circuits that process them. Frequency domain circuit and signal analysis, based on the application of Fourier techniques, are discussed in Chapter three with special emphasis on the use of these techniques in the context of circuits. Chapter four then considers time domain analysis and Laplace techniques, again with similar emphasis.

Chapter five covers the analysis techniques used in two-port circuits and also covers various circuit representations and parameters. These techniques are important for computer-based analysis of linear electronic circuits. In this chapter the treatment is frequency-domain-based.

Chapter six introduces basic electronic amplifier building blocks and describes frequency-domain-based analysis techniques for common circuits and circuit topologies. We deal with bipolar and field-effect transistor circuits as well as operational amplifiers.

Radio-frequency and microwave circuit analysis techniques are presented in Chapter seven where we cover transmission lines, S-parameters and the concept and application of the Smith chart. Chapter eight discusses the mathematical representation of noise and its origins, analysis and effects in electronic circuits. The analysis techniques outlined in this chapter also provide the basis for an efficient computer-aided analysis method.

Appendix A and B provide a synopsis of frequently used mathematical formulae and a review of matrix algebra. Appendix C gives a summary of various two-port circuit parameters and their conversion formulae.

In writing this book we have strived to make it suitable for teaching and self-study. Concepts are illustrated using examples and the reader's acquired knowledge can be tested using the problems at the end of each chapter. The examples provided are worked in detail throughout the book and the problems are solved in the solution manual provided as a web resource. For both examples and problems we guide the reader through the solution steps to facilitate understanding.

Luis Moura
Izzat Darwazeh

Acknowledgements

Work on this book was carried out at the University of Algarve, in Portugal and at University College London, in the UK. We should like to thank our colleagues at both universities for their support and enlightening discussions. We are particularly grateful to Dr Mike Brozel for his perceptive comments on the content and style of the book and for his expert and patient editing of the manuscript.

We also thank Elsevier staff for their help throughout the book writing and production processes.

The text of the book was set using LaTeX and the figures were produced using Xfig and Octave/Gnuplot running on a Debian Linux platform. We are grateful to the Free Software community for providing such reliable tools.

Lastly, we are indebted to our wives, Guida and Rachel for their love, patience and support to 'keep on going'.

1 Elementary electrical circuit analysis

1.1 Introduction

Analogue electronic circuits deal with signal processing techniques such as amplification and filtering of electrical and electronic[1] signals. Such signals are voltages or currents. In order to understand how these signals can be processed we need to appreciate the basic relationships associated with electrical currents and voltages in each electrical component as well as in any combination that make the complete electrical circuit. We start by defining the basic electrical quantities – voltage and current – and by presenting the main passive electrical devices; resistors, capacitors and inductors.

The fundamental tools for electrical circuit analysis – Kirchhoff's laws – are discussed in section 1.4. Then, three very important electrical network theorems; Thévenin's theorem, Norton's theorem and the superposition theorem are presented.

The unit system used in this book is the *International System of Units* (SI) [1]. The relevant units of this system will be mentioned as the physical quanti-

femto-	pico-	nano-	micro-	milli-	kilo-	mega-	giga	tera
(f)	(p)	(n)	(μ)	(m)	(k)	(M)	(G)	(T)
10^{-15}	10^{-12}	10^{-9}	10^{-6}	10^{-3}	10^{3}	10^{6}	10^{9}	10^{12}

Table 1.1: Powers of ten.

ties are introduced. In this book detailed definition of the different units is not provided as this can be found in other sources, for example [2, 3], which address the physical and electromagnetic nature of circuit elements. At this stage it is relevant to mention that the SI system incorporates the decimal prefix to relate larger and smaller units to the basic units using these prefixes to indicate the various powers of ten. Table 1.1 shows the powers of ten most frequently encountered in circuit analysis.

1.2 Voltage and current

By definition electrical current is the rate of flow (with time) of electrical charges passing a given point of an electrical circuit. This definition can be expressed as follows:

$$i(t) = \frac{d\,q(t)}{d\,t} \qquad (1.1)$$

---

[1] The term 'electronic signals' is sometimes used to describe low-power signals. In this book, the terms 'electrical' and 'electronic' signals are used interchangeably to describe signals processed by a circuit.

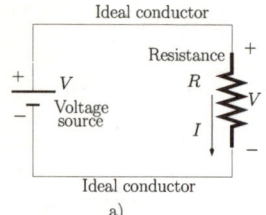

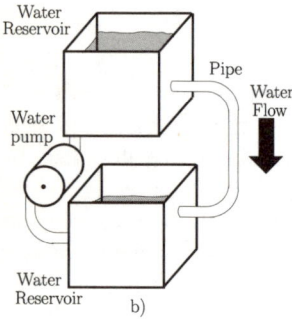

Figure 1.1: *a) Voltage source driving a resistance. b) Hydraulic equivalent system.*

where $i(t)$ represents the electrical current as a function of time represented by t. The unit for the current is the ampere (A). $q(t)$ represents the quantity of flowing electrical charge as a function of time and its unit is the coulomb (C). The elementary electrical charge is the charge of the electron which is equal to 1.6×10^{-19} C.

At this stage it is relevant to mention that in this chapter we represent constant quantities by uppercase letters while quantities that vary with time are represented by the lower case. Hence, a constant electrical current is represented by I while an electrical current varying with time is represented by $i(t)$.

Electrical current has a a very intuitive hydraulic analogue; water flow. Figure 1.1 a) shows a voltage source which is connected to a resistance, R, creating a current flow, I, in this circuit. Figure 1.1 b) shows an hydraulic equivalent system. The water pump *together* with the water reservoirs maintain a constant water pressure across the ends of the pipe. This pressure is equivalent to the voltage potential difference at the resistance terminals generated by the voltage source. The water flowing through the pipe is a consequence of the pressure difference. It is common sense that the narrower the pipe the greater the water resistance and the lower the water flow through it. Similarly, the larger the electrical resistance the smaller the electrical current flowing through the resistance. Hence, it is clear that the equivalent to the pipe water resistance is the electrical resistance R.

The electrical current, I, is related to the potential difference, or voltage V, and to the resistance R by Ohm's law:

$$I = \frac{V}{R} \qquad (1.2)$$

The unit for resistance is the ohm (also represented by the Greek symbol Ω). The unit for the potential difference is the volt (or simply V)[2]. Ohm's law states that the current that flows through a resistor is inversely proportional to the value of that resistance and directly proportional to the voltage across the resistance. This law is of fundamental importance for electrical and electronic circuit analysis.

Now we discuss voltage and current sources. The main purpose of each of these sources is to provide power and energy to the circuit to which the source is connected.

1.2.1 Voltage sources

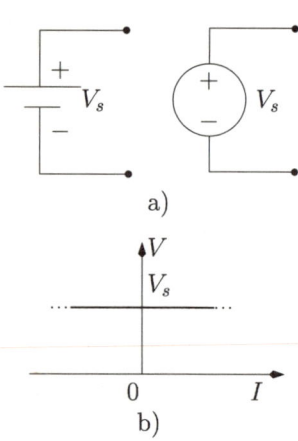

Figure 1.2: *Ideal voltage source. a) Symbols. b) V–I characteristic.*

Figure 1.2 a) shows the symbols used to represent voltage sources. The plus sign, the anode terminal, indicates the higher potential and the minus sign, the cathode terminal, indicates the lower potential. The positive flow of current supplied by a voltage source is from the anode, through the exterior circuit, such as the resistance in figure 1.1 a), to the cathode. Note that the positive current flow is conventionally taken to be in the opposite direction to the flow of electrons. An ideal constant voltage source has a voltage–current, V–I,

---

[2]It is common practice to use the letter V to represent the voltage, as well as its unit. This practice is followed in this book.

characteristic like that illustrated in figure 1.2 b). From this figure we observe that an ideal voltage source is able to maintain a constant voltage V across its terminals regardless of the value of the current supplied to (positive current) or the current absorbed from (negative current) an electrical circuit.

When a voltage source, such as that shown in figure 1.1 a), provides a constant voltage at its terminals it is called a direct current (or DC) voltage source. No practical DC voltage source is able to maintain the same voltage across its terminals when the current increases. A typical V–I characteristic of a practical voltage source is as shown in figure 1.3 a). From this figure we observe that as the current I increases up to a value I_x the voltage drops from V_s to V_x in a linear manner. A practical voltage source can be modelled according to the circuit of figure 1.3 b) which consists of an ideal voltage source and a resistance R_s whose value is given by:

$$R_s = \frac{V_s - V_x}{I_x} \qquad (1.3)$$

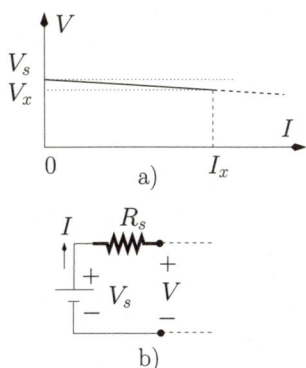

Figure 1.3: *Practical voltage source. a) V–I characteristic. b) Electrical model.*

This resistance is called the 'source output resistance'. Examples of DC voltage sources are the batteries used in radios, in cellular phones and automobiles.

An alternating (AC) voltage source provides a time varying voltage at its terminals which is usually described by a sine function as follows:

$$v_s(t) = V_s \sin(\omega t) \qquad (1.4)$$

where V_s is the amplitude and ω is the angular frequency in radians per second. An ideal AC voltage source has a V–I characteristic similar to that of the ideal DC voltage source in the sense that it is able to maintain the AC voltage regardless of the amount of current supplied or absorbed from a circuit. In practice AC voltage sources have a non-zero output resistance. An example of an AC voltage source is the domestic mains supply.

Example 1.2.1 Determine the output resistance of a voltage source with $V_s = 12$ V, $V_x = 11.2$ V and $I_x = 34$ A.

Solution: The output resistance is calculated according to:

$$\begin{aligned} R &= \frac{V_s - V_x}{I_x} \\ &= 0.024\,\Omega \\ &= 24\text{ m}\Omega \end{aligned}$$

1.2.2 Current sources

Figure 1.4 a) shows the symbol for the ideal current source[3]. The arrow indicates the positive flow of the current. Figure 1.4 b) shows the current–voltage,

---

[3] Although the symbol of a current source is shown with its terminals in an open-circuit situation, the practical operation of a current source requires an electrical path between its terminals or the output voltage will become infinite.

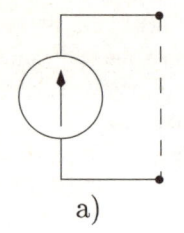

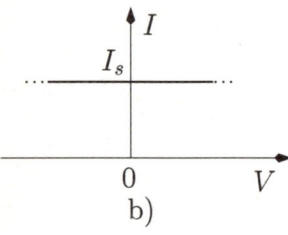

Figure 1.4: *Ideal current source. a) Symbol. b) I–V characteristic.*

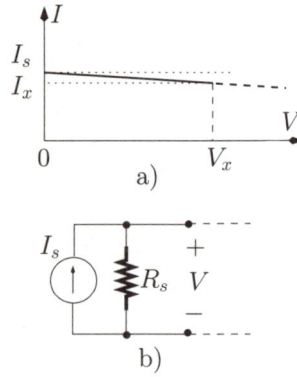

Figure 1.5: *Practical current source. a) I–V characteristic. b) Equivalent circuit.*

I–V, characteristic of an ideal current source. From this figure it is clear that an ideal current source is able to provide a given current regardless of the voltage at its terminals. Practical current sources have an I–V characteristic like that represented in figure 1.5 a). As the voltage across the current source increases up to a value V_x the current tends to decrease in a linear fashion. Figure 1.5 b) shows the equivalent circuit for a practical current source including a resistance R_s which is once again called the 'source output resistance'. The value of this resistance is:

$$R_s = \frac{V_x}{I_s - I_x} \quad (1.5)$$

Examples of simple current sources are difficult to provide at this stage. Most current sources are implemented using active devices such as transistors. Active devices are studied in detail in Chapter 6 where it is shown that, for example, the field-effect transistor, in specific configurations, displays current source behaviour.

Example 1.2.2 Determine the output resistance of a current source whose output current falls from 2 A to 1.99 A when its output voltage increases from 0 to 100 V.

Solution: The output resistance is calculated according to:

$$\begin{aligned} R_s &= \frac{V_x}{I_s - I_x} \\ &= 10^4 \, \Omega \\ &= 10 \text{ k}\Omega \end{aligned}$$

Power supplied by a source

As mentioned previously, the main purpose of a voltage or current source is to provide power to a circuit. The instantaneous power delivered by either source is given by the product of the current supplied with the voltage at its terminals, that is,

$$p_s(t) = v(t)\,i(t) \quad (1.6)$$

The unit for power is the watt (W) when the voltage is expressed in volts (V) and the current is expressed in amperes (A). If the voltage and current are constant then eqn 1.6 can be written as:

$$P_s = VI \quad (1.7)$$

Often it is of interest to calculate the *average* power, P_{AV_s}, supplied by a source during a period of time T. This average power can be calculated by the successive addition of all values of the instantaneous power, $p_s(t)$, during the time interval T and then dividing the outcome by the time interval T. That is, P_{AV_s} can be calculated as follows[4]:

$$\begin{aligned} P_{AV_s} &= \frac{1}{T} \int_{t_o}^{t_o+T} p_s(t)\, dt \\ &= \frac{1}{T} \int_{t_o}^{t_o+T} v(t)\, i(t)\, dt \end{aligned} \quad (1.8)$$

where t_o is a chosen instant of time. For a periodic signal (voltage or current) T is usually chosen as the period of the signal.

Example 1.2.3 A 12 volt DC source supplies a transistor circuit with periodic current of the form; $i(t) = 3 + 2\cos(2\pi 100 t)$ mA. Plot the instantaneous power and the average power supplied by this source in the time period $0 < t < 0.01$ s.

Solution: The instantaneous power is calculated using eqn 1.6:

$$\begin{aligned} p_s(t) &= 12 \times [3 + 2\cos(2\pi 100 t)]\, 10^{-3} \\ &= 36 + 24\cos(2\pi 100 t) \text{ (mW)} \end{aligned}$$

This is plotted in figure 1.6. The average power is calculated according to eqn 1.8:

$$\begin{aligned} P_{AV_s} &= \frac{1}{0.01} \int_0^{0.01} 12 \times [3 + 2\cos(2\pi 100 t)]\, 10^{-3}\, dt \\ &= 100 \times 12 \times 10^{-3} \left[3t + \frac{2}{2\pi 100}\sin(2\pi 100 t) \right]_0^{0.01} \\ &= 36 \text{ mW} \end{aligned}$$

Note that the same average power will be obtained if eqn 1.8 is applied over any time interval T as long as T is a multiple of the period of the waveform.

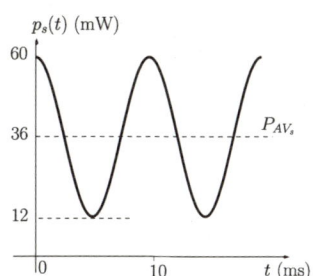

Figure 1.6: *Instantaneous and average power.*

1.3 Electrical passive elements

The main passive electrical elements are the resistor, the capacitor and the inductor. For each of these elements we study the voltage–current relationship and we also present hydraulic analogies as suggested by Wilmshurst [4].

1.3.1 Resistance and conductance

The resistance[5] has been presented in the previous section. Ohm's law relates the voltage at the terminals of a resistor with the current which flows through it according to eqn 1.2. The hydraulic analogue for a resistance has also been presented above in figure 1.1. It is worth mentioning that if the voltage varies

---

[4] Recall that the integral operation is basically an addition operation.

[5] Strictly speaking, the suffix *-or* designates the name of the element (like resist*or*) while the suffix *-ance* designates the element property (like resist*ance*). Often these two are used interchangeably.

with time then the current varies with time in exactly the same manner, as illustrated in figure 1.7. Therefore, the resistance appears as a scaling factor which relates the amplitude of the two electrical quantities; current and voltage. So, we can generalise eqn 1.2 as follows:

$$i(t) = \frac{v(t)}{R} \quad (1.9)$$

and

$$v(t) = R\,i(t) \quad (1.10)$$

Figure 1.7 illustrates this concept.

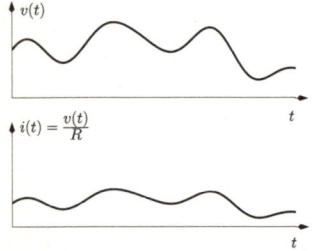

Figure 1.7: *Voltage and current in a resistance.*

Example 1.3.1 Consider a current $i(t) = 0.5\sin(\omega t)$ A flowing through a resistor of 10 Ω. Determine an expression for the voltage across the resistor.

<u>Solution</u>: Using eqn 1.10 we obtain the voltage $v(t)$ as

$$\begin{aligned} v(t) &= R\,i(t) \\ &= 5\sin(\omega t) \text{ V} \end{aligned}$$

Often it is useful to express Ohm's law as follows:

$$I = GV \quad (1.11)$$

where $G = R^{-1}$ is known as the 'conductance'. The unit of the conductance is the siemen (S) and is equal to $(1 \text{ ohm })^{-1}$.

A resistance dissipates power and generates heat. When a resistance is driven by a DC source this power dissipation, P_R, is given by:

$$P_R = VI \quad (1.12)$$

where V represents the voltage across the resistance terminals and I is the current that flows through it. Using Ohm's law we can express eqn 1.12 as follows:

$$\begin{aligned} P_R &= RI^2 & (1.13) \\ &= \frac{V^2}{R} & (1.14) \end{aligned}$$

These two eqns (1.13 and 1.14) appear to be contradictory in terms of the role the resistance plays in determining the level of power dissipation. Does the dissipated power increase with increasing the resistance (eqn 1.13) or does it decrease (eqn 1.14)? The answer to this question relates to the way we view the circuit and to what quantity we measure across the resistor. Let us consider the case where a resistor is connected across the terminals of an ideal voltage source. Here, the stimulus is the voltage that results in a current through the resistor. In this situation the larger the resistance the smaller the current is and, according to eqn 1.12, there is less power dissipation in the resistance. Note that, in the extreme situation of $R \to \infty$ the resistance behaves as an open circuit (the current is zero) and there is no power dissipation. On the other hand, if the resistance is driven by an ideal current source, then the voltage across the resistor is the resulting effect. Hence, the larger the resistance, the larger the voltage developed across its terminals and, according to eqn 1.12, the power dissipation increases.

For time-varying sources the instantaneous power dissipated in a resistor is given by:

$$p_R(t) = v(t)\,i(t) \qquad (1.15)$$
$$= R\,i^2(t) \qquad (1.16)$$
$$= \frac{v^2(t)}{R} \qquad (1.17)$$

where $v(t)$ represents the voltage across the resistance and $i(t)$ is the current that flows through it. The average power dissipated in a period of time T can be expressed as follows:

$$P_{AV_R} = \frac{1}{T}\int_{t_o}^{t_o+T} p_R(t)\,dt \qquad (1.18)$$
$$= \frac{R}{T}\int_{t_o}^{t_o+T} i^2(t)\,dt \qquad (1.19)$$
$$= \frac{1}{RT}\int_{t_o}^{t_o+T} v^2(t)\,dt \qquad (1.20)$$

where t_o is a chosen instant of time. When the resistor is driven by a periodic signal, T is normally chosen to be its period.

Example 1.3.2 An AC voltage $v(t) = A\sin(\omega t)$, with $A = 10$ V, is applied to a resistance $R = 50\,\Omega$. Determine the average power dissipated.

Solution: According to eqn 1.20 we can write:

$$P_{AV_R} = \frac{1}{TR}\int_0^T V_A^2 \sin^2(\omega t)\,dt$$
$$= \frac{1}{T}\frac{V_A^2}{R}\int_0^T \frac{1-\cos(2\omega t)}{2}\,dt$$
$$= \frac{V_A^2}{2TR}\left[t - \frac{1}{2\omega}\sin(2\omega t)\right]_0^T$$

Since the period of the AC waveform is $T = 2\pi/\omega$, the last eqn can be written as:

$$P_{AV_R} = \frac{1}{R}\frac{V_A^2}{2} \qquad (1.21)$$
$$= 1\text{ W}$$

1.3.2 Capacitance

Figure 1.8 a) shows a capacitor connected to a voltage source. A capacitor is usually implemented using two metal plates separated by an insulator. This means that the capacitor does not allow the passage of direct current. However, when a voltage is applied across the terminals of an uncharged capacitor electronic charge can be added to one of the metal plates and removed from the other. This charge is proportional to the applied voltage and defines the capacitance according to the following eqn:

$$q(t) = C\,v(t) \tag{1.22}$$

where C is the capacitance and $v(t)$ is the voltage applied across its terminals. The capacitance can be seen to have the ability to accumulate charge. The unit for the capacitance is the farad (F) when, in eqn 1.22, $v(t)$ is expressed in volts and $q(t)$ is expressed in coulombs. In other words, a capacitor of one farad will store one coulomb of charge if a potential difference of one volt exists across its plates. From eqns 1.1 and 1.22 we can relate the current through a capacitor with the voltage across its terminals according to the following eqn:

$$i(t) = C\,\frac{d\,v(t)}{d\,t} \tag{1.23}$$

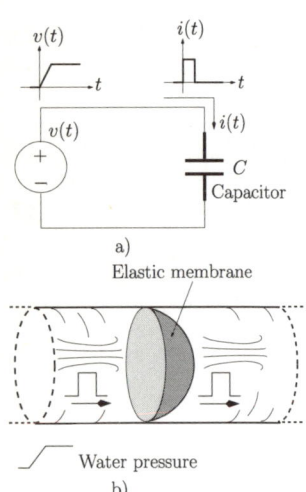

Figure 1.8: *The capacitor. a) In an electrical circuit. b) Hydraulic analogue.*

Figure 1.8 b) shows a hydraulic analogue for the capacitor which is an elastic membrane covering the section of a water pipe. In this analogy the voltage becomes the water pressure, the capacitor becomes the membrane elasticity, and the charge becomes the water volume displaced by the membrane. Let us consider that there is water on both sides of the elastic membrane. This membrane does not allow the direct crossing of water between the two sides of the pipe. However, if there is a pulsed increase of the water pressure in one side of the membrane, as illustrated in fig 1.8 b), the stretching of this membrane causes an effective travelling of the pulse from one side of the membrane to the other (with no water passing through the membrane!). This process is conceptually similar to the flow of charges (current) in the metal plates of a capacitor. Note that if there is too much pressure on the membrane it will eventually breakdown and the same can happen to a capacitor if too much voltage is applied to it.

Integrating eqn 1.23 we obtain an expression for the voltage in terms of the current, that is:

$$v(t) = \frac{1}{C}\int_0^t i(t)\,dt + V_{co} \tag{1.24}$$

where V_{co} represents the initial voltage across the terminals of the capacitor at $t=0$.

Unlike the resistor the capacitor does not dissipate any power. In fact because this circuit element is able to accumulate charge on its metal plates it stores energy. To illustrate this we return to the hydraulic analogue; if two shutters are inserted across each side of the pipe in order to stop the stretched

1. Elementary electrical circuit analysis

membrane relaxing to its original state this is equivalent to the disconnection of a charged capacitor from the circuit. The energy stored in a capacitor can be written as

$$E_C = \frac{1}{2} C V^2 \quad \text{(joule)} \tag{1.25}$$

where V is the voltage across the capacitor terminals.

1.3.3 Inductance

Figure 1.9 a) shows an inductor connected to a current source. The inductor is usually formed by a coil of a metal wire. The passage of current, $i(t)$, in a metal wire induces a magnetic flux which in turn (according to Faraday's law) results in a voltage, $v(t)$, developing across the terminals, such that:

$$v(t) = L \frac{d\,i(t)}{d\,t} \tag{1.26}$$

where L is the inductance of the wire. The unit for the inductance is the henry (H) when, in eqn 1.26, $v(t)$ is expressed in volts and the current rate of change, $di(t)/dt$, is expressed in amperes per second. Thus, an inductance of one henry will have a potential difference of one volt across its terminals when the current passing through it is changing at a rate of one ampere per second. Figure 1.9 b) shows a hydraulic analogue for the inductor which consists of paddles connected to a flywheel. In this analogy the voltage becomes the water pressure and the current becomes the rate of water flow. The inductance becomes the flywheel moment of inertia. The flywheel requires water pressure to change the speed of the paddles which, in turn, change the rate of the water flow. In figure 1.9 b) we illustrate the situation where the application of a pulse of water pressure causes an increase in the speed of the water flow which will be maintained constant by the flywheel inertia until a different level of water pressure is applied. In electrical terms this means that the voltage difference at the terminals of an inductor is proportional to the rate of variation of the current that flows through it.

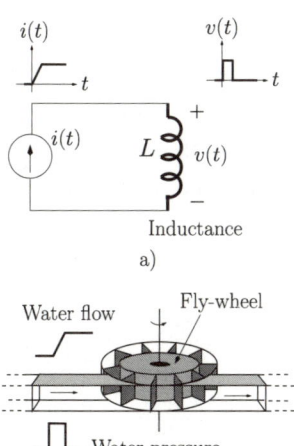

Figure 1.9: *The inductance.*
a) In an electrical circuit.
b) Hydraulic analogue.

Integrating eqn 1.26 we obtain an expression for the current that flows in the inductor in terms of the voltage at its terminals, that is:

$$i(t) = \frac{1}{L} \int_0^t v(t)\,dt + I_{lo} \tag{1.27}$$

where I_{lo} represents the initial current in the inductor at $t = 0$.

Like the capacitor, the inductor does not dissipate energy and is capable of storing energy. However, the energy is now stored in terms of the magnetic flux created by the current. This energy can be written as

$$E_L = \frac{1}{2} L I^2 \quad \text{(joule)} \tag{1.28}$$

where I is the current through the inductor.

1.4 Kirchhoff's laws

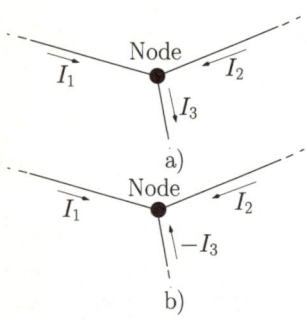

Figure 1.10: *Kirchhoff's current law. a) Illustration of the current law. b) Equivalent representation.*

Kirchhoff's laws provide the basis of all circuit analysis techniques as long as such circuits can be described by lumped elements such as resistors, capacitors, etc. There are two Kirchhoff's laws: the current law and the voltage law. These two laws are quite simple in terms of concept. However, the application of these laws requires careful attention to the algebraic sign conventions of the current and voltage.

The current law

The current law states that the sum of currents entering a node is equal to the sum of the currents leaving the node. A node is a point at which two or more electrical elements have a common connection. Figure 1.10 a) shows an example of a node where the currents I_1 and I_2 are entering the node while the current I_3 is leaving the node. According to the current law, we can write:

$$I_1 + I_2 = I_3$$

which can also be expressed as:

$$I_1 + I_2 + (-I_3) = 0$$

This is equivalent to reversing the direction of the current I_3, as shown in figure 1.10 b). Hence, the current law can also be stated as follows: the sum of all currents flowing *into* a node, taking into account their algebraic signs, is zero.

Example 1.4.1 Consider the circuit of figure 1.11. Determine the currents I_1, I_2 and I_3.

Solution: From figure 1.11 we can write the following eqns for nodes X, Y and Z, respectively:

$$\begin{aligned} 6 + 7 + I_1 &= 3 + 4 + 5 \\ 3 + 4 &= I_2 \\ 5 + I_2 &= I_3 \end{aligned}$$

Solving to obtain I_1, I_2 and I_3 we obtain:

$$\begin{aligned} I_1 &= -1 \text{ A} \\ I_2 &= 7 \text{ A} \\ I_3 &= 12 \text{ A} \end{aligned}$$

Note that the current I_1 is negative which means that the direction of the current is the opposite of that shown in the figure.

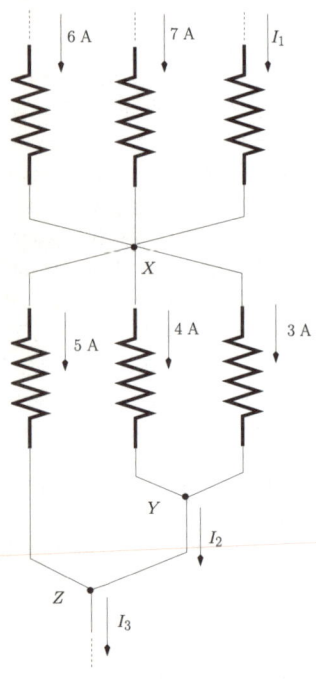

Figure 1.11: *Circuit for the application of the current law.*

The voltage law

The voltage law states that the sum of all voltages around any closed electrical loop, taking into account polarities[6], is zero. Figure 1.12 shows a circuit with

---
[6]In this book we use curved arrows to indicate the potential difference between two points in a circuit, with the arrow head pointing to the lower potential.

1. Elementary electrical circuit analysis

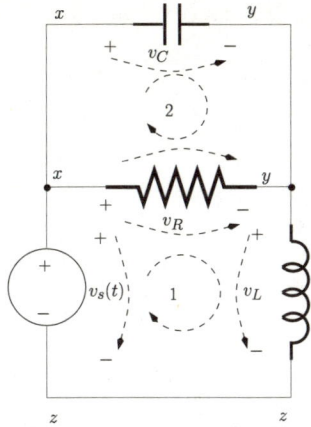

Figure 1.12: *Kirchhoff's voltage law.*

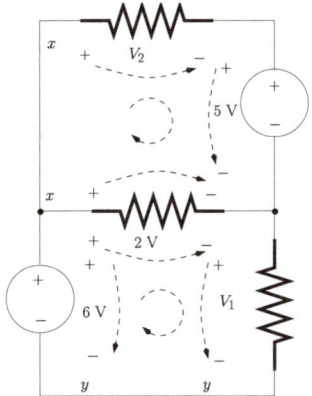

Figure 1.13: *Circuit for the application of the voltage law.*

two closed loops. By applying the voltage law we can write the following equations for loop 1 and loop 2, respectively:

$$v_R(t) + v_L(t) - v_s(t) = 0$$
$$v_C(t) - v_R(t) = 0$$

It should be noted that since the connections of the elements are assumed as ideal (zero resistance), voltage differences are observed only across the various elements. Hence, for example, node z is, from an electrical point-of-view, the same at the low end of the voltage source and at the low-end of the inductor.

Example 1.4.2 Consider the circuit of figure 1.13. Determine the voltages V_1 and V_2.

<u>Solution</u>: From figure 1.13 and starting from point x we can apply the voltage law to the upper and lower loop, respectively, as indicated below:

$$V_2 + 5 - 2 = 0$$
$$2 + V_1 - 6 = 0$$

Solving we get:

$$V_2 = -3 \text{ V}$$
$$V_1 = 4 \text{ V}$$

We observe that V_2 is negative meaning that the polarity of the voltage drop is opposite to that chosen originally.

1.4.1 Series and parallel combinations of passive elements

Two connected elements are said to be in series if the same current flows through each and in parallel if they share the same voltage across their terminals. We study now the series and parallel combinations of resistors, capacitors and inductors.

Resistance

Figure 1.14 a) shows two resistors in a *series* connection. These two resistors can be replaced by a resistor with an equivalent resistance, R_{eq}, as shown in figure 1.14 b). Hence, R_{eq} must draw the same amount of current I as the series combination. Applying Kirchhoff's voltage law to the circuit of figure 1.14 a) we obtain:

$$-V_s + V_1 + V_2 = 0 \tag{1.29}$$

that is

$$V_s = V_1 + V_2 \tag{1.30}$$

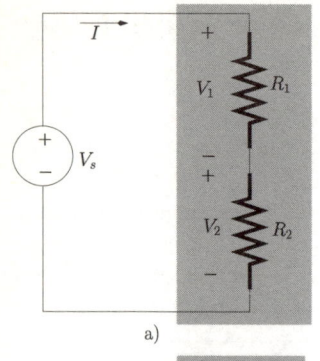

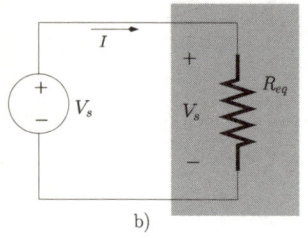

Figure 1.14: *a) Series combination of two resistors. b) Equivalent resistance.*

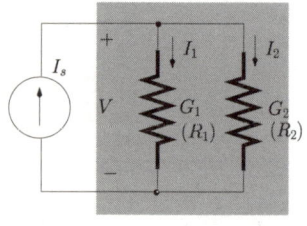

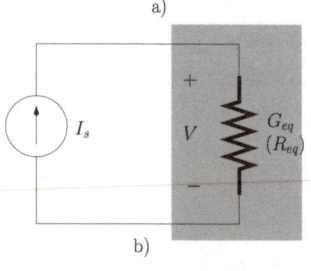

Figure 1.15: *a) Parallel combination of two resistors. b) Equivalent resistance (conductance).*

Since the current that flows through both resistors is the same we can write the last eqn as follows:

$$\begin{aligned} V_s &= R_1 I + R_2 I \\ &= (R_1 + R_2) I \end{aligned} \quad (1.31)$$

Applying Ohm's law to the circuit of figure 1.14 b) we obtain:

$$V_s = R_{eq} I \quad (1.32)$$

therefore

$$R_{eq} = R_1 + R_2 \quad (1.33)$$

It can be shown (see problem 1.4) that the above can be generalised for the series combination of N resistors as follows:

$$R_{eq} = \sum_{k=1}^{N} R_k \quad (1.34)$$

that is, the equivalent resistance is obtained by the addition of all resistances that make the series connection.

Figure 1.15 a) shows two resistors in a *parallel* connection. Each resistor R_k can be expressed as a conductance $G_k = R_k^{-1}$. Applying Kirchhoff's current law to the circuit of figure 1.15 a) we obtain:

$$I_s = I_1 + I_2 \quad (1.35)$$

Since the voltage across each conductance is the same we can write the last eqn as follows:

$$\begin{aligned} I_s &= G_1 V + G_2 V \\ &= (G_1 + G_2) V \end{aligned} \quad (1.36)$$

Applying Ohm's law to the circuit of figure 1.15 b) we obtain:

$$I_s = G_{eq} V \quad (1.37)$$

so that

$$G_{eq} = G_1 + G_2 \quad (1.38)$$

It can be shown (see problem 1.5) that this result can be generalised for the parallel combination of N resistors as follows:

$$G_{eq} = \sum_{k=1}^{N} G_k \quad (1.39)$$

that is, the equivalent conductance is the addition of each conductance that composes the parallel connection.

1. Elementary electrical circuit analysis

The equivalent resistance, R_{eq}, for the parallel combination of two resistances can be obtained by re-writing eqn 1.38 as follows:

$$\frac{1}{R_{eq}} = \frac{1}{R_1} + \frac{1}{R_2} \qquad (1.40)$$

or,

$$R_{eq} = \frac{R_1 R_2}{R_1 + R_2} \qquad (1.41)$$

Often, we use the notation $\|$ to indicate the parallel connection of resistances, that is, $R_1\|R_2$ means R_1 in parallel with R_2.

Example 1.4.3 Consider two resistances $R_1 = 3$ kΩ and $R_2 = 200$ Ω.

1. Determine the equivalent resistance, R_{eq}, for the series connection of R_1 and R_2.

2. Determine the equivalent resistance, R_{eq}, for the parallel connection of R_1 and R_2.

Solution:

1. According to eqn 1.33 the equivalent resistance is $R_{eq} = 3.2$ kΩ.

2. According to eqn 1.41 the equivalent resistance is $R_{eq} = 188$ Ω.

It should be remembered that for the series connection of resistances the equivalent resistance is larger than the largest resistance in the series chain whilst for the parallel connection the equivalent resistance is smaller than the smallest resistance in the connection.

Capacitance

Figure 1.16 a) shows two capacitors in a *series* connection. These can be replaced by a capacitor with an equivalent capacitance, C_{eq}, as shown in figure 1.16 b). Applying Kirchhoff's voltage law to the circuit of figure 1.16 a) we obtain:

$$v_s(t) = v_1(t) + v_2(t) \qquad (1.42)$$

Since the current that flows through both capacitors is the same then, according to eqn 1.24, we can rewrite the above eqn as follows:

$$v_s(t) = \frac{1}{C_1} \int_0^t i(t)\,dt + \frac{1}{C_2} \int_0^t i(t)\,dt$$

$$= \left(\frac{1}{C_1} + \frac{1}{C_2}\right) \int_0^t i(t)\,dt \qquad (1.43)$$

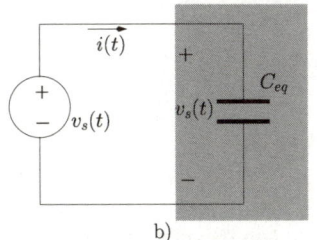

Figure 1.16: *a) Series combination of two capacitors. b) Equivalent capacitor.*

Here we assume that the initial voltage across each capacitor is zero. Applying eqn 1.24 to the capacitor of the circuit of figure 1.16 b) we obtain:

$$v_s(t) = \frac{1}{C_{eq}} \int_0^t i(t)\,dt \qquad (1.44)$$

From eqns 1.44 and 1.43 we conclude that

$$\frac{1}{C_{eq}} = \frac{1}{C_1} + \frac{1}{C_2} \qquad (1.45)$$

that is:

$$C_{eq} = \frac{C_1 C_2}{C_1 + C_2} \qquad (1.46)$$

It can be shown (see problem 1.7) that eqn 1.45 can be generalised for the series combination of N capacitors as follows:

$$\frac{1}{C_{eq}} = \sum_{k=1}^{N} \frac{1}{C_k} \qquad (1.47)$$

Figure 1.17 a) shows two capacitors in a *parallel* connection. Applying Kirchhoff's current law to the circuit of figure 1.17 a) we obtain:

$$i_s(t) = i_1(t) + i_2(t) \qquad (1.48)$$

Since the voltage across each capacitor is the same then, using eqn 1.23, we can write the above eqn as follows:

$$i_s(t) = C_1 \frac{dv(t)}{dt} + C_2 \frac{dv(t)}{dt}$$
$$= (C_1 + C_2) \frac{dv(t)}{dt} \qquad (1.49)$$

Applying eqn 1.23 to the capacitor of the circuit of figure 1.15 b) we obtain:

$$i_s(t) = C_{eq} \frac{dv(t)}{dt} \qquad (1.50)$$

so that

$$C_{eq} = C_1 + C_2 \qquad (1.51)$$

It can be shown (see problem 1.8) that the above eqn can be generalised for the parallel combination of N capacitors as follows:

$$C_{eq} = \sum_{k=1}^{N} C_k \qquad (1.52)$$

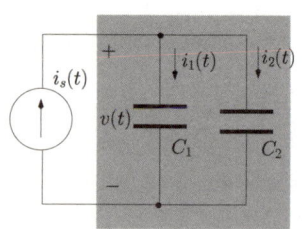

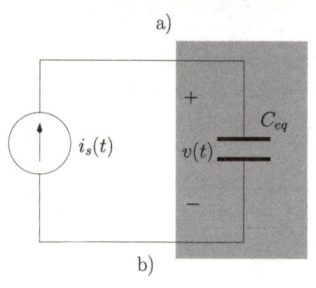

Figure 1.17: a) Parallel combination of two capacitors. b) Equivalent capacitor.

1. Elementary electrical circuit analysis   15

Example 1.4.4 Consider two capacitances $C_1 = 0.3$ μF and $C_2 = 1$ μF.

1. Determine the equivalent capacitance, C_{eq}, for the series connection of C_1 and C_2.

2. Determine the equivalent capacitance, C_{eq}, for the parallel connection of C_1 and C_2.

<u>Solution</u>:

1. According to eqn 1.46 the equivalent capacitance is $C_{eq} = 0.2$ μF.

2. According to eqn 1.51 the equivalent capacitance is $C_{eq} = 1.3$ μF.

It should be noted that, for capacitors, the series connection results in a decreased value for the equivalent capacitance while the parallel connection results in an increased value for the equivalent capacitance.

Inductance

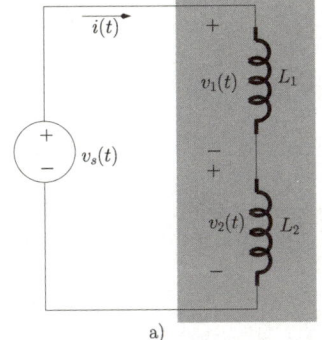

Figure 1.18 a) shows two inductors[7] in a *series* connection. These two inductors can be replaced by a single inductor with an equivalent inductance, L_{eq}, as shown in figure 1.18 b). Applying Kirchhoff's voltage law to the circuit of figure 1.18 a) we obtain:

$$v_s(t) = v_1(t) + v_2(t) \tag{1.53}$$

Since the current that flows through both inductors is the same then, according to eqn 1.26, we can rewrite the last eqn as follows:

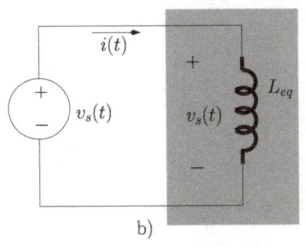

$$\begin{aligned} v_s(t) &= L_1 \frac{d\,i(t)}{d\,t} + L_2 \frac{d\,i(t)}{d\,t} \\ &= (L_1 + L_2) \frac{d\,i(t)}{d\,t} \end{aligned} \tag{1.54}$$

Applying eqn 1.26 to the equivalent inductor in figure 1.14 b) we obtain:

$$v_s(t) = L_{eq} \frac{d\,i(t)}{d\,t} \tag{1.55}$$

Figure 1.18: *a) Series combination of two inductors. b) Equivalent inductor.*

Comparing this eqn with eqn 1.54 we conclude that

$$L_{eq} = L_1 + L_2 \tag{1.56}$$

It can be shown (see problem 1.10) that the last eqn can be generalised for the series combination of N inductors as follows:

$$L_{eq} = \sum_{k=1}^{N} L_k \tag{1.57}$$

---

[7] We assume the inductors to be uncoupled, that is they are sufficiently far apart or mounted in such a way so that their magnetic fluxes do not interact.

Figure 1.19 a) shows two inductors in a *parallel* connection. Applying Kirchhoff's current law to the circuit of figure 1.19 a) we obtain:

$$i_s(t) = i_1(t) + i_2(t) \tag{1.58}$$

Since the voltage across each inductor is the same, we can apply eqn 1.27 to eqn 1.58 as follows:

$$\begin{aligned} i_s(t) &= \frac{1}{L_1}\int_0^t v(t)\,dt + \frac{1}{L_2}\int_0^t v(t)\,dt \\ &= \left(\frac{1}{L_1} + \frac{1}{L_2}\right)\int_0^t v(t)\,dt \end{aligned} \tag{1.59}$$

where we assume that the initial current in each inductor is zero. Applying eqn 1.27 to the equivalent inductor of figure 1.15 b) we obtain:

$$i_s(t) = \frac{1}{L_{eq}}\int_0^t v(t)\,dt \tag{1.60}$$

Comparing the last eqn with eqn 1.59 we conclude that

$$\frac{1}{L_{eq}} = \frac{1}{L_1} + \frac{1}{L_2} \tag{1.61}$$

The equivalent inductance, L_{eq}, for the parallel combination of two inductances can be obtained from eqn 1.61 as follows:

$$L_{eq} = \frac{L_1 L_2}{L_1 + L_2} \tag{1.62}$$

It can be shown (see problem 1.11) that the above eqn can be generalised for the parallel combination of N inductors as follows:

$$\frac{1}{L_{eq}} = \sum_{k=1}^{N} \frac{1}{L_k} \tag{1.63}$$

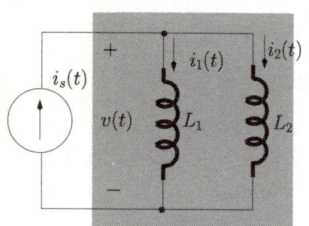

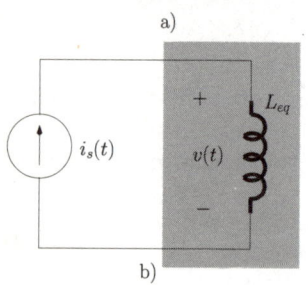

Figure 1.19: *a) Parallel combination of two inductors. b) Equivalent inductor.*

Example 1.4.5 Consider two inductances $L_1 = 1.5$ mH and $L_2 = 1$ mH.

1. Determine the equivalent inductance, L_{eq}, for the series connection of L_1 and L_2.

2. Determine the equivalent capacitance, L_{eq}, for the parallel connection of L_1 and L_2.

Solution:

1. According to eqn 1.56 the equivalent inductance is $C_{eq} = 2.5$ mH.

2. According to eqn 1.62 the equivalent inductance is $C_{eq} = 0.6$ mH.

It is interesting to note that like resistors (but unlike capacitors) the series connection of inductors increases the value of the equivalent inductance while the parallel connection decreases it.

1.4.2 Other types of circuit element connections

Circuit elements can be connected in combinations which are neither parallel nor series. For example, take the circuit of figure 1.20 a) for which we want to determine the equivalent resistance, R_{eq}, between terminals A and B. In

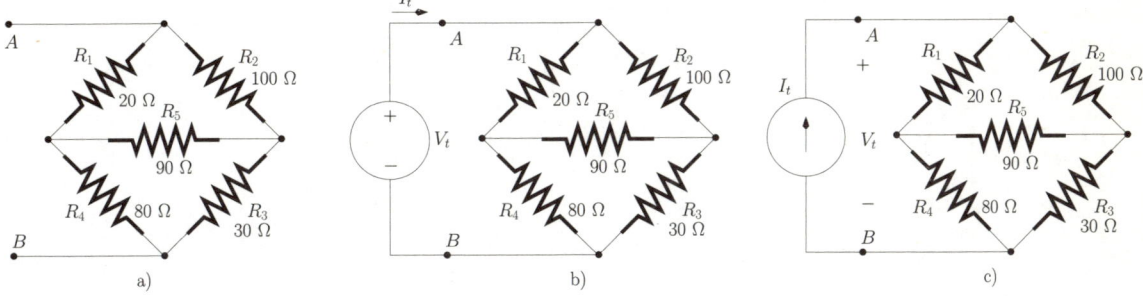

Figure 1.20: *a) Resistive circuit. b) Calculation of the equivalent resistance, R_{eq}, by applying a test voltage source, V_t. c) Calculation of the equivalent resistance, R_{eq}, by applying a test current source, I_t.*

this circuit there is not a single combination of two resistances which share the same current through or the same voltage across their terminals and, therefore, there is not a single parallel or series connection. This means that we cannot directly apply the rules discussed previously, for parallel and series connections of resistances, to determine R_{eq}. However, the calculation of the equivalent resistance can be done by applying a test voltage source, V_t, to the terminals of the circuit as shown in figure 1.20 b). Then, we determine the current I_t supplied by this source. Finally, by calculating the ratio V_t/I_t we can find the equivalent resistance, R_{eq}, effectively applying Ohm's law. Alternatively, we can apply a test current source, I_t, to the circuit as illustrated in figure 1.20 c). Again, by calculating the ratio V_t/I_t we obtain R_{eq}. In general, this procedure can be applied to any circuit.

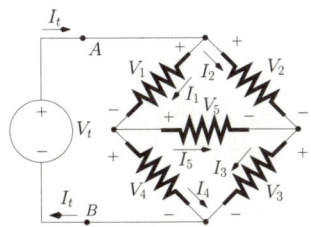

Figure 1.21: *Calculation of R_{eq}.*

Example 1.4.6 Determine the equivalent resistance, R_{eq}, of the circuit of figure 1.20 between terminals A and B.

<u>Solution</u>: We apply a test voltage source to the circuit as shown in figure 1.21. This figure also shows the definition of the voltages across and currents through each resistance. Applying Kirchhoff's current law we can write:

$$\begin{cases} I_t = I_1 + I_2 \\ I_t = I_4 + I_3 \\ I_1 = I_4 + I_5 \\ I_3 = I_2 + I_5 \end{cases} \quad (1.64)$$

and applying Kirchhoff's voltage law we can write

$$\begin{cases} V_t = V_1 + V_4 \\ V_t = V_2 + V_3 \end{cases} \quad (1.65)$$

These two sets of eqns can be rewritten as:

$$\begin{cases} I_t = \dfrac{V_1}{R_1} + \dfrac{V_2}{R_2} \\ I_t = \dfrac{V_4}{R_4} + \dfrac{V_3}{R_3} \\ \dfrac{V_1}{R_1} = \dfrac{V_4}{R_4} + \dfrac{V_5}{R_5} \\ \dfrac{V_3}{R_3} = \dfrac{V_2}{R_2} + \dfrac{V_5}{R_5} \\ V_t = V_1 + V_4 \\ V_t = V_2 + V_3 \end{cases} \quad (1.66)$$

Solving, to obtain $V_t/I_t = R_{eq}$, we get:

$$\begin{aligned} \frac{V_t}{I_t} &= \frac{R_2 R_5 (R_1 + R_4) + R_1 R_2 (R_3 + R_4)}{R_2(R_5 + R_3 + R_4) + R_5(R_4 + R_1 + R_3) + R_1(R_3 + R_4)} \\ &+ \frac{R_3 R_4 (R_2 + R_5) + R_1 R_3 (R_4 + R_5)}{R_2(R_5 + R_3 + R_4) + R_5(R_4 + R_1 + R_3) + R_1(R_3 + R_4)} \\ &= 49.5 \, \Omega \end{aligned}$$

1.4.3 Electrical network analysis – Nodal analysis

There are various electrical analysis methods, all derived from Kirchhoff's laws, to analyse electrical circuits. One of most effective and computationally efficient is the Nodal analysis method. Therefore, we now illustrate the application of this method to resistive electrical networks.

Figure 1.22 a) shows a circuit for which we want to determine the current I_A. Since the resistance R_2 is short-circuited the voltage across this resistance is zero and, according to Ohm's law, the current that flows through R_2 is zero. Hence, the circuit of figure 1.22 a) can be replaced by its equivalent represented in figure 1.22 b). First, we indicate the voltages at each node. These voltages indicate the potential difference between the node being considered and a reference node which can be chosen arbitrarily. This node is traditionally called 'node zero' (0) or the 'ground terminal' and is often chosen as the node with the highest number of attached electrical elements. For this circuit there are three nodes (X, Y, Z) plus the reference node zero, as shown in figure 1.22 b). Then, we consider, in an arbitrary manner, the current direction in each branch,

1. Elementary electrical circuit analysis

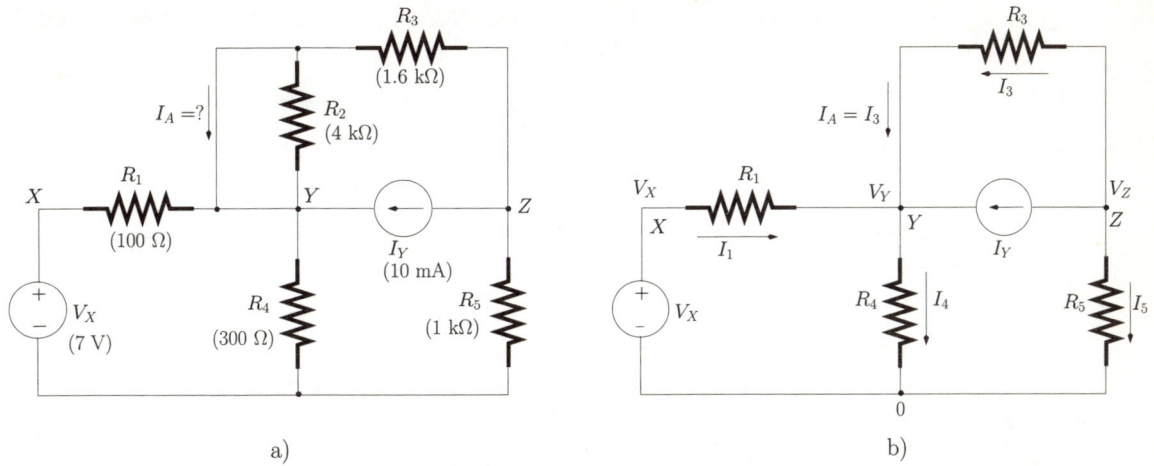

Figure 1.22: *a) Resistive electrical network. b) Equivalent circuit.*

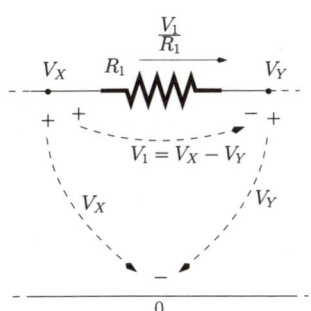

Figure 1.23: *Expressing the current across R_1 in terms of $(V_X - V_Y)/R_1$.*

as indicated in figure 1.22 b). The current that flows through each resistance can be expressed, according to Ohm's law, as the ratio of the voltage across that resistance and the resistance value. Figure 1.23 illustrates this procedure for the current I_1 that flows through R_1; since the voltages V_X and V_Y have been defined, referenced to ground, then by applying Kirchhoff's voltage law we can express the voltage across R_1, V_1, as the difference between V_X and V_Y. Hence we can write:

$$I_1 = \frac{V_X - V_Y}{R_1} \tag{1.67}$$

Applying this technique to the remaining currents, as defined in the circuit of figure 1.22 b), we can write:

$$I_3 = \frac{V_Z - V_Y}{R_3} \tag{1.68}$$

$$I_4 = \frac{V_Y}{R_4} \tag{1.69}$$

$$I_5 = \frac{V_Z}{R_5} \tag{1.70}$$

Applying Kirchhoff's current law to nodes Y and Z we can write:

$$I_1 + I_Y + I_3 = I_4 \tag{1.71}$$

$$I_Y + I_3 + I_5 = 0 \tag{1.72}$$

These two eqns can be expressed, using eqns 1.67–1.70, as follows:

$$\frac{V_X - V_Y}{R_1} + I_Y + \frac{V_Z - V_Y}{R_3} = \frac{V_Y}{R_4} \tag{1.73}$$

$$I_Y + \frac{V_Z - V_Y}{R_3} + \frac{V_Z}{R_5} = 0 \tag{1.74}$$

Note that the only two unknown quantities are V_Y and V_Z. Solving the last two eqns in order to obtain V_Y and V_Z we get:

$$V_Y = R_4 \frac{R_3 V_X + I_Y R_3 R_1 + V_X R_5}{R_3 R_4 + R_4 R_5 + R_1 R_4 + R_1 R_3 + R_1 R_5} \quad (1.75)$$
$$= 5.6 \text{ V}$$
$$V_Z = R_5 \frac{R_4 V_X - R_3 R_4 I_Y - I_Y R_3 R_1}{R_3 R_4 + R_4 R_5 + R_1 R_4 + R_1 R_3 + R_1 R_5} \quad (1.76)$$
$$= -4.0 \text{ V}$$

Since $I_A = I_3$, eqn 1.68 gives us $I_A = -6$ mA.

1.4.4 Resistive voltage and current dividers

Resistive voltage and current dividers are simple yet very important circuits which allow us to obtain fractions of a source voltage or current, respectively. In addition, these circuits play a major role in the calculation of voltage and current gains in electronic amplifier analysis.

Resistive voltage divider

Figure 1.24 a) shows the resistive voltage divider formed by resistances R_1 and R_2. For this circuit we observe that the current flowing through R_1 and R_2 is the same. Hence, we can write:

$$\frac{V_s - V_o}{R_1} = \frac{V_o}{R_2} \quad (1.77)$$

Solving this eqn in order to obtain V_o we get:

$$V_o = V_s \frac{R_2}{R_1 + R_2} \quad (1.78)$$

We observe that if $R_1 = R_2$ then $V_o = 0.5 V_s$. Also, if $R_1 \gg R_2$ then V_o tends to zero. On the other hand, if $R_1 \ll R_2$ then V_o tends to V_s.

Resistive current divider

Figure 1.24 b) shows the resistive current divider formed by resistances R_1 and R_2. For this circuit the voltage across each resistor is the same. Hence, we can write the following set of eqns:

$$I_s = \frac{V}{R_1} + \frac{V}{R_2} \quad (1.79)$$
$$I_o = \frac{V}{R_2} \quad (1.80)$$

Rearranging, we obtain I_o:

$$I_o = I_s \frac{R_1}{R_1 + R_2} \quad (1.81)$$

We can conclude that if $R_1 = R_2$ then $I_o = 0.5 I_s$. Also, if $R_1 \gg R_2$ then I_o tends to I_s. On the other hand, if $R_1 \ll R_2$ then I_o tends to zero.

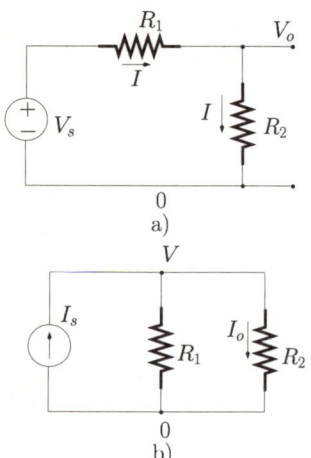

Figure 1.24: a) Resistive voltage divider. b) Resistive current divider.

1. Elementary electrical circuit analysis

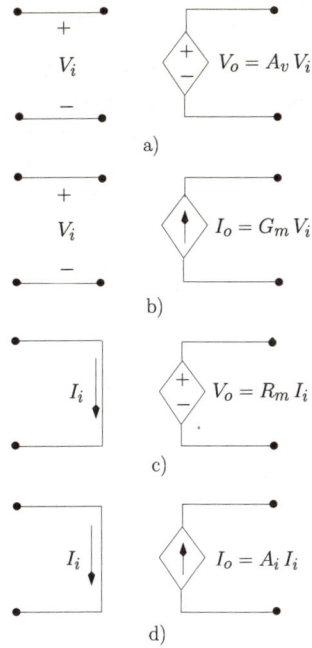

Figure 1.25: *a) Voltage-controlled voltage source. b) Voltage-controlled current source. c) Current-controlled voltage source. d) Current-controlled current source.*

1.4.5 Controlled sources

The voltage and the current sources presented in section 1.2 are called *independent* sources. Other types can be controlled by either a voltage or current existing elsewhere in the circuit. These controlled (or dependent) sources are often used to model the gain of transistors operating in their linear region and also to model the gain of linear electronic amplifiers, as will be discussed in Chapter 6. There are four types of controlled sources, drawn as diamond shapes, as illustrated in figure 1.25.

- Voltage-controlled voltage sources: the output of the source is a voltage, V_o, and the quantity that controls it is also a voltage, V_i. A_v is the ratio V_o/V_i and is called the 'voltage gain' of the source. A_v is dimensionless;

- Voltage-controlled current sources: the output of the source is a current, I_o, but the quantity that controls it is a voltage, V_i. G_m is the ratio I_o/V_i and is called the 'transconductance gain' of the source. The dimension of G_m is the siemen;

- Current-controlled voltage sources: the output of the source is a voltage, V_o, but the controlling quantity is a current I_i. R_m is the ratio V_o/I_i and is called the 'transresistance gain' of the source. The dimension of R_m is the ohm;

- Current-controlled current sources: the output of the source is a current, I_o, and the controlling quantity is also a current, I_i. A_i is the ratio I_o/I_i and is called the 'current gain' of the source. A_i is dimensionless.

It is important to note that the output characteristics of each of the dependent sources is exactly the same as those of its corresponding independent source (see figures 1.2 b) and 1.4 b)) but with I and V values being controlled by a quantity occurring somewhere else in the circuit.

Example 1.4.7 Consider the circuit of figure 1.26 containing a voltage-controlled current source. Determine the voltage across the resistance R_2.

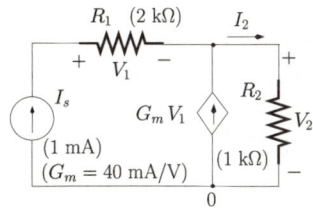

Figure 1.26: *Circuit containing a voltage-controlled current source.*

Solution: According to Kirchhoff's current law the current that flows through R_2 is the sum of I_s with the current supplied by the voltage-controlled current source, $G_m V_1$, that is:

$$I_2 = I_s + G_m V_1$$

Since $V_1 = R_1 I_s$ we can write the last eqn as follows:

$$I_2 = I_s (1 + G_m R_1)$$

Finally;

$$\begin{aligned} V_2 &= R_2 I_2 \\ &= R_2 I_s (1 + G_m R_1) \\ &= 81 \text{ V} \end{aligned}$$

1.5 Thévenin's theorem

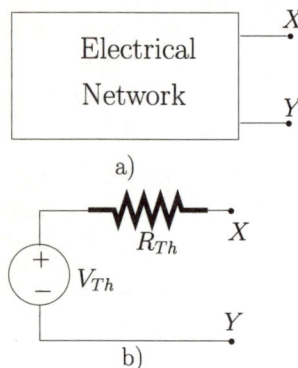

Figure 1.27: a) Generic DC electrical network. b) Thévenin equivalent circuit.

Thévenin's theorem states that any two-terminal electrical network, such as that depicted in figure 1.27 a) consisting of resistances and independent sources (voltage or current or both) can be replaced by an equivalent ideal voltage source in series with an equivalent resistance, as shown in figure 1.27 b). The value of the equivalent voltage source, V_{Th}, known as the Thévenin voltage, is the open-circuit voltage at the output terminals. R_{Th} is the Thévenin resistance 'looking into' the terminals of the network when all independent voltage sources are replaced by short-circuits and all independent current sources are replaced by open-circuits.

Example 1.5.1 Determine the Thévenin equivalent circuit for the circuit of figure 1.28 a) where the two terminals to be considered are X and Y.

Solution: The calculation of the Thévenin voltage can be performed by analysis of the circuit of figure 1.28 b). Note that V_{Th} is equal to V_A. We can write the following set of eqns:

$$I_r = I_2 + I_1$$
$$V_s = V_B - V_A$$
$$I_1 = I_3$$

These can be rewritten as follows:

$$I_r = \frac{V_C}{R_2} + \frac{V_C - V_B}{R_1}$$
$$V_s = V_B - V_A$$
$$\frac{V_A}{R_3} = \frac{V_C - V_B}{R_1}$$

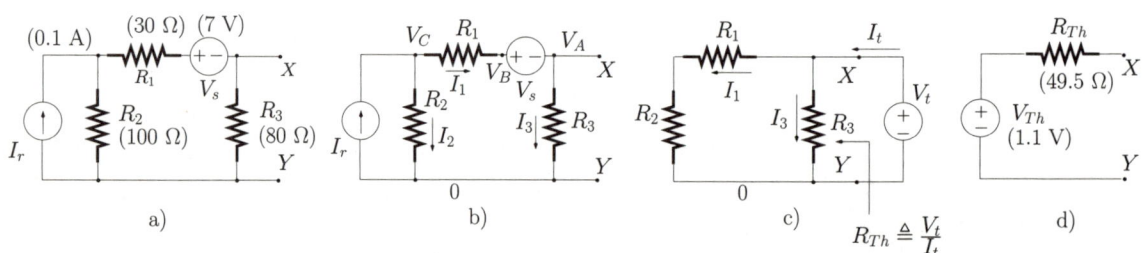

Figure 1.28: a) Electrical network. b) Calculation of the Thévenin voltage. c) Calculation of the Thévenin resistance. d) Thévenin equivalent circuit.

Solving, in order to obtain the voltages V_A, V_B and V_C we get:

$$V_A = R_3 \frac{I_r R_2 - V_s}{R_3 + R_1 + R_2} \quad (1.82)$$
$$= 1.1 \text{ V}$$

1. Elementary electrical circuit analysis

$$V_B = \frac{I_r R_2 R_3 + R_1 V_s + R_2 V_s}{R_3 + R_1 + R_2} \tag{1.83}$$
$$= 8.1 \text{ V}$$
$$V_C = R_2 \frac{I_r R_3 + V_s + I_r R_1}{R_3 + R_1 + R_2} \tag{1.84}$$
$$= 8.6 \text{ V}$$

Since the Thévenin voltage is the voltage across points X and Y it is the same as V_A, that is, $V_{Th} = 1.1$ V. The calculation of the Thévenin equivalent resistance can be carried out by analysing the circuit of figure 1.28 c) where it can be seen that the voltage source, V_s, has been replaced by a short-circuit and the current source, I_r, has been replaced by an open-circuit. The calculation of the Thévenin resistance can be calculated by applying a test voltage, V_t, between terminals X and Y, the terminals where the resistance is to be determined. The ratio between V_t and the current supplied by this test voltage source, I_t, is the required resistance. Applying Kirchhoff's current law to the circuit of figure 1.28 c) we can write:

$$I_t = I_1 + I_3 \tag{1.85}$$

From this figure we can also observe that the voltage V_t is applied to R_3 and also to the series combination of R_1 and R_2. Hence we can write:

$$I_t = \frac{V_t}{R_1 + R_2} + \frac{V_t}{R_3} \tag{1.86}$$

Solving for V_t/I_t we obtain:

$$R_{Th} = \frac{R_3(R_1 + R_2)}{R_1 + R_2 + R_3} \tag{1.87}$$
$$= 49.5 \text{ }\Omega$$

It should be noted that this resistance could also be calculated by close inspection of figure 1.28 c) after recognising that the resistance R_{Th} is the parallel combination of R_3 with the series combination of R_1 and R_2.

1.6 Norton's theorem

Norton's theorem states that any two-terminal electrical network, such as that depicted in figure 1.29 a) consisting of resistances and independent sources (voltage or current or both) can be replaced by an equivalent independent current source in parallel with an equivalent resistance, as shown in figure 1.29 b). The value of the current source is the current flowing from X to Y when X to Y are short-circuited. The equivalent resistance (Norton resistance), R_{Nt} is the resistance 'looking into' the terminals of the network when the independent voltage sources are replaced by short-circuits and the independent current sources are replaced by open-circuits. Norton's theorem is the dual of Thévenin's since the equivalent voltage source, V_{Th}, is replaced by an equivalent current source, I_{Nt}, and the series resistance, R_{Th}, is replaced by a parallel resistance, R_{Nt}.

Example 1.6.1 Determine the Norton equivalent circuit for the circuit of figure 1.28 a) where the two terminals to be considered are X and Y.

Solution: The calculation of the Norton short-circuit current, I_{Nt}, can be performed by analysing the circuit of figure 1.30 a) for which we can write the following set of eqns:

$$I_{Nt} = I_1 \tag{1.88}$$
$$I_r = I_1 + I_2 \tag{1.89}$$
$$V_B = V_s \tag{1.90}$$

Note that since R_3 is short-circuited, the voltage across its terminals is zero and there is no current flowing through this resistance. Eqn 1.89 can be written as follows:

$$I_r = \frac{V_C - V_B}{R_1} + \frac{V_C}{R_2} \tag{1.91}$$

Since $V_B = V_s$ we have that:

$$I_r = \frac{V_C - V_s}{R_1} + \frac{V_C}{R_2} \tag{1.92}$$

Solving in order to obtain V_C we can write:

$$\begin{aligned} V_C &= \frac{R_1 R_2}{R_1 + R_2} I_r + \frac{R_2}{R_1 + R_2} V_s \\ &= 7.7 \text{ V} \end{aligned} \tag{1.93}$$

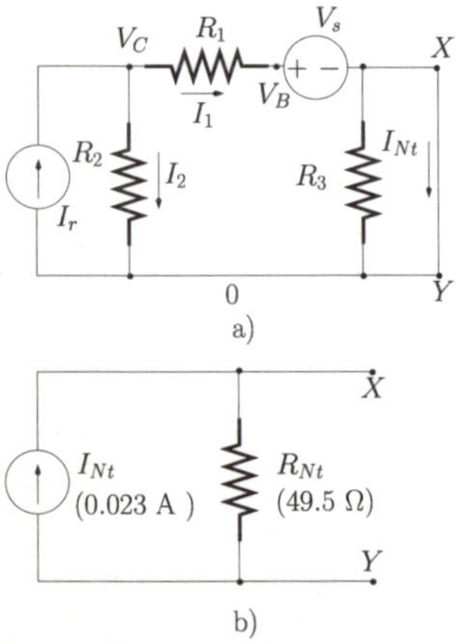

The Norton equivalent short-circuit current is,

$$\begin{aligned} I_{Nt} &= \frac{V_C - V_B}{R_1} \\ &= 0.023 \text{ A} \end{aligned} \tag{1.94}$$

The Norton equivalent resistance is determined in a fashion similar to that used to calculate the Thévenin resistance in example 1.5.1. Consequently, the Norton equivalent circuit for the circuit of 1.28 a) is as shown in figure 1.30 b).

It is useful to note the straightforward equivalence between the Thévenin and Norton theorems. If, for example, the Thévenin equivalent circuit is known then the equivalent Norton circuit can be obtained as shown in figures 1.31 a) and 1.31 b), that is:

$$I_{Nt} = \frac{V_{Th}}{R_{Th}} \tag{1.95}$$
$$R_{Nt} = R_{Th} \tag{1.96}$$

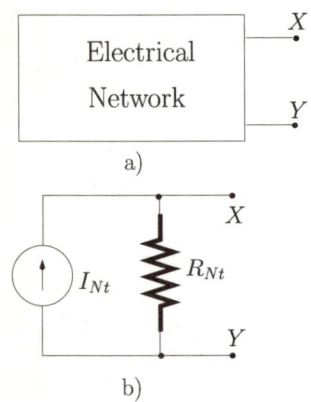

Figure 1.29: *a) Generic electrical network. b) Norton equivalent circuit.*

Figure 1.30: *a) Equivalent circuit for the calculation of the short-circuit current $I_{Nt} = I_1$. b) Norton equivalent circuit.*

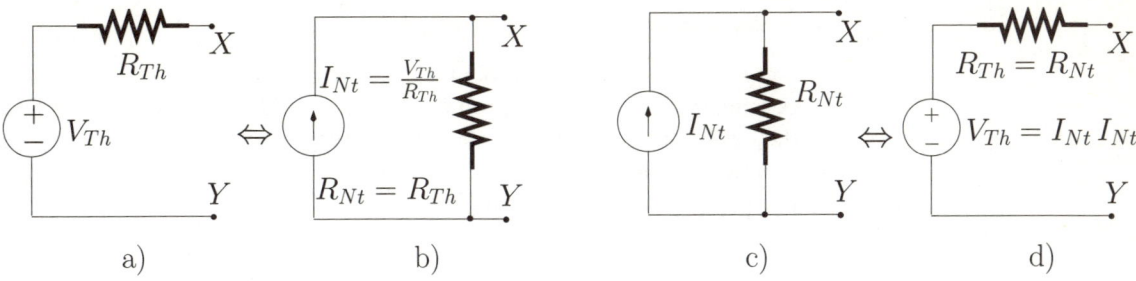

Figure 1.31: *a) Thévenin circuit. b) Equivalent Norton circuit. c) Norton circuit. d) Equivalent Thévenin circuit.*

On the other hand, if the Norton equivalent circuit is known then the Thévenin equivalent circuit can be obtained as shown in figures 1.31 c) and 1.31 d), that is:

$$V_{Th} = I_{Nt} R_{Nt} \quad (1.97)$$
$$R_{Th} = R_{Nt} \quad (1.98)$$

It is left to the reader to prove these equivalences.

We emphasise that the discussion of Thévenin and Norton theorems presented above applies to circuit networks containing only independent sources (voltage and current). However, there are algebraic techniques which allow us to obtain Thévenin and Norton equivalent circuits when the networks include dependent sources [5].

1.7 Superposition theorem

The superposition theorem is of considerable importance since it can provide useful insight into the relative contribution of a given independent source to the current flowing through or the voltage across a given circuit element. The superposition theorem also plays a major role in the frequency domain circuit analysis, discussed in Chapter 3, and in the noise analysis of linear electronic circuits presented in Chapter 8.

The superposition theorem applies to linear circuits and it can be stated as follows: "In a network containing several current and/or voltage sources, the voltage across (or the current flowing through) any circuit element can be obtained from the algebraic sum of the voltages (currents) caused by each *independent* source considered individually with all other *independent* voltage sources considered as short-circuits and all other *independent* current sources considered as open-circuits". As stated above, the superposition theorem does not allow for the substitution of controlled voltage sources and controlled current sources by short-circuits and open-circuits, respectively. In this respect the superposition theorem applies only to independent sources[8].

Example 1.7.1 Apply the superposition theorem to determine the equivalent Thévenin voltage of the circuit of figure 1.28 a) where the two terminals to be

---

[8]Using mathematical manipulation techniques it possible to apply the superposition theorem to dependent sources [5].

considered are the X and Y terminals.

Solution: Figure 1.32 a) shows the equivalent circuit for the calculation of the contribution of V_s to the Thévenin voltage across the terminals X and Y. Note that the positive current supplied by the voltage source flows as indicated in this figure. Since all resistances are in series the current can be determined, according to Ohm's law, as follows:

$$\begin{aligned} I &= \frac{V_s}{R_1 + R_2 + R_3} \\ &= 33.3 \text{ mA} \end{aligned}$$

The contribution to the Thévenin voltage is the voltage across R_3 and can be expressed as follows:

$$\begin{aligned} V_{Th_a} &= -I\,R_3 \\ &= -\frac{V_s\,R_3}{R_1 + R_2 + R_3} \\ &= -2.7 \text{ V} \end{aligned}$$

Figure 1.32 b) shows the equivalent circuit derived to calculate the contribution of I_r to the Thévenin voltage. The current that flows through R_3 can be calculated using the current divider concept. Hence, since R_2 is in a series connection with R_3 we can write:

$$\begin{aligned} I_1 &= I_r \frac{R_2}{R_1 + R_2 + R_3} \\ &= 47.6 \text{ mA} \end{aligned}$$

and this contribution to the Thévenin voltage, the voltage across R_3, can be expressed as follows:

$$\begin{aligned} V_{Th_b} &= I_1\,R_3 \\ &= I_r \frac{R_2\,R_3}{R_1 + R_2 + R_3} \\ &= 3.8 \text{ V} \end{aligned}$$

Adding V_{Th_a} and V_{Th_b} we obtain the Thévenin voltage as:

$$V_{Th} = 1.1 \text{ V}$$

Note that the superposition theorem allows us to identify the contribution of each independent source in a clear manner. While the contribution of V_s to the Thévenin voltage is negative (-2.7 V), the contribution of I_r to V_{Th} is positive (3.8 V). This results in a net voltage V_{Th} of 1.1 V.

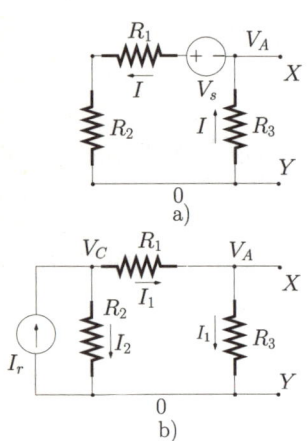

Figure 1.32: *Application of the superposition theorem to the circuit of figure 1.28 a). a) The contribution from V_s. b) The contribution from I_r.*

1.8 Bibliography

1. K.C.A. Smith and R.E. Alley, *Electrical Circuits, an Introduction*, 1992 (Cambridge University Press).

2. J.D. Kraus, *Electromagnetics with applications*, 1999 (McGraw-Hill) 5th edition.

1. Elementary electrical circuit analysis 27

3. A.J. Compton, *Basic Electromagnetism and its Applications*, 1990 (Chapman and Hall).

4. T.H. Wilmshurst, *Analog Circuit Techniques, with Digital Interfacing*, 2001 (Newnes).

5. A.M. Davies, *Some fundamental topics in introductory circuit analysis: A critique*, IEEE Transactions on Education, Aug. 2000, Vol. 43, No. 3, pp. 330–335.

1.9 Problems

1.1 A voltage $v(t) = 10\,\sin(2\pi\,100\,t + \pi/4)$ volts is applied across the terminals of a 1 μF capacitor. Sketch the current through the capacitor as a function of time from $t = 0$ to $t = 20$ ms.

1.2 A current $i(t) = 20\,\cos(2\pi\,5000\,t)$ mA flows through a 3 mH inductor. Sketch the voltage across the inductor as a function of time from $t = 0$ to $t = 500$ μs.

1.3 Find the current through and the voltage across each resistance for the circuits of figure 1.33. Take $V_1 = 2$ V, $V_2 = 3$ V, $I_1 = 0.2$ A and $I_2 = 0.5$ A.

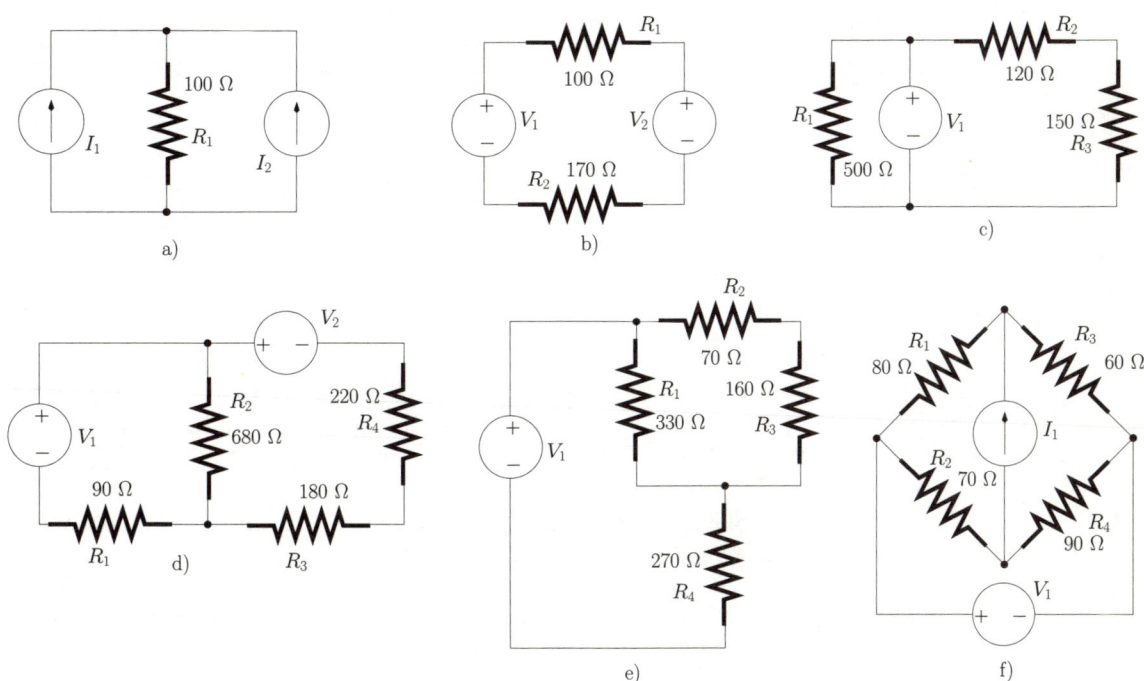

Figure 1.33: *Circuits of problem 1.3.*

1.4 Show that the equivalent resistance for the series combination of N resistances is given by eqn 1.34.

1.5 Show that the equivalent conductance for the parallel combination of N resistances satisfies eqn 1.39.

1.6 For each circuit of figure 1.34 determine the equivalent resistance and conductance between points A and B.

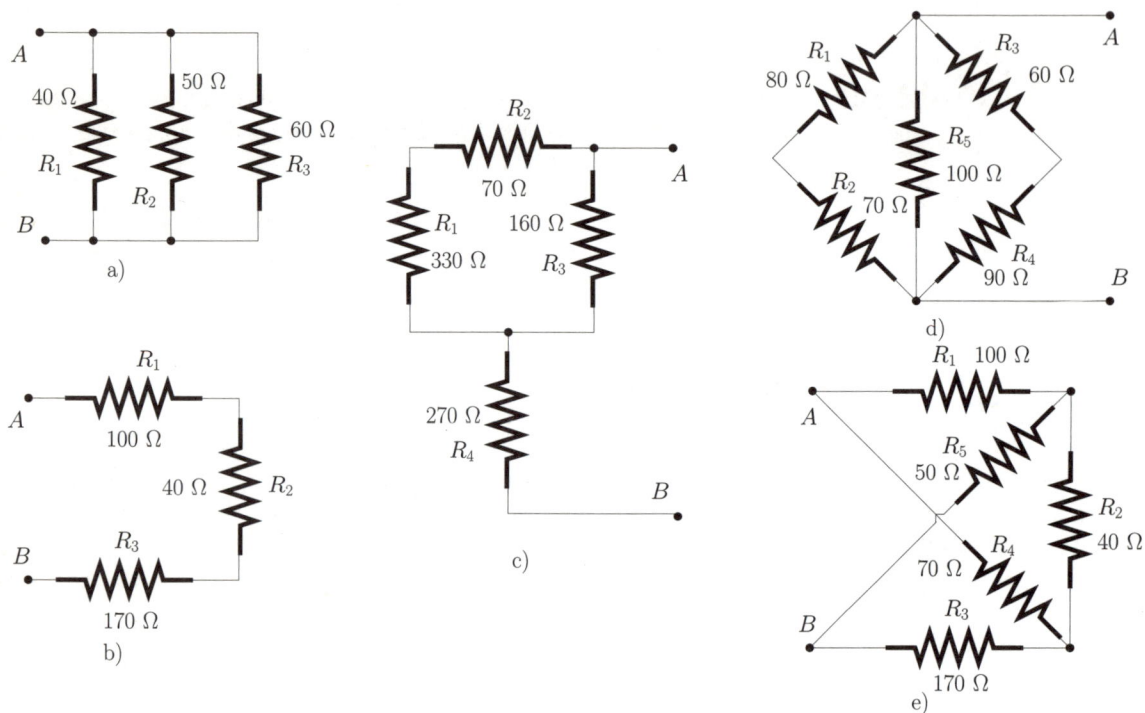

Figure 1.34: *Circuits of problem 1.6.*

1.7 Show that the equivalent capacitance for the series combination of N capacitances satisfies eqn 1.47.

1.8 Show that the equivalent capacitance for the parallel combination of N capacitances satisfies eqn 1.52.

1.9 For each circuit of figure 1.35 determine the equivalent capacitance between points A and B.

1.10 Show that the equivalent inductance for the series combination of N inductors is given by eqn 1.57.

1.11 Show that the equivalent inductance for the parallel combination of N inductors satisfies eqn 1.63.

1. Elementary electrical circuit analysis

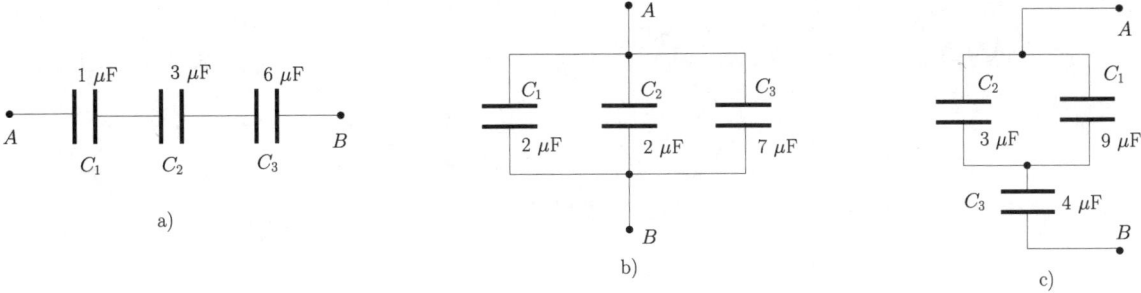

Figure 1.35: *Circuits of problem 1.9.*

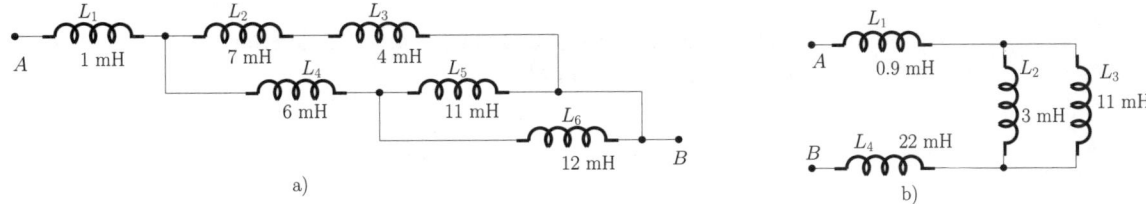

Figure 1.36: *Circuits of problem 1.12.*

1.12 For each circuit of figure 1.36 determine the equivalent inductance between points A and B.

1.13 For each circuit of figure 1.37 determine the voltage across and the current through R_o.

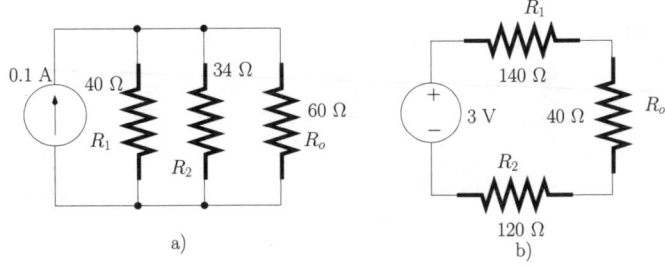

Figure 1.37: *Circuits of problem 1.13.*

1.14 For each circuit of figure 1.38 determine the voltage across and the current through R_1.

1.15 For the circuits b) and c) of figure 1.38 determine the Thévenin equivalent circuits at points A and B.

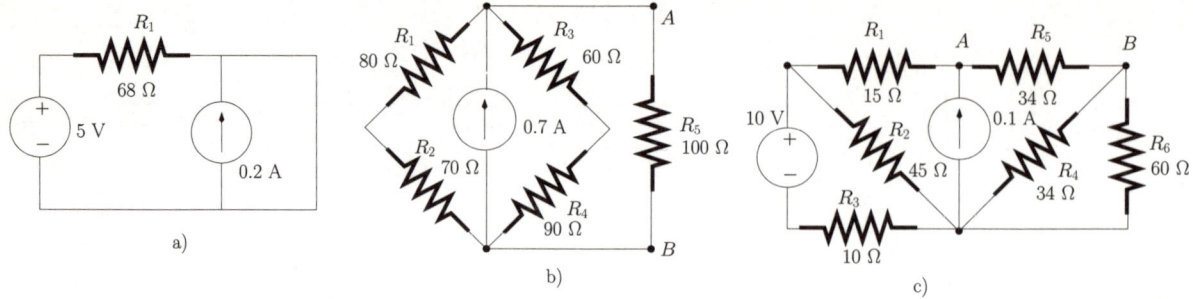

Figure 1.38: *Circuits of problems 1.14, 1.15 and 1.16.*

1.16 For the circuits b) and c) of figure 1.38 determine the Norton equivalent circuits at points A and B.

1.17 For the circuits of figure 1.39 determine the voltage across and the current through R_3. Use values of $A_i = 12$, $G_m = 0.5$ S, $A_v = 10$ and $R_m = 40\ \Omega$.

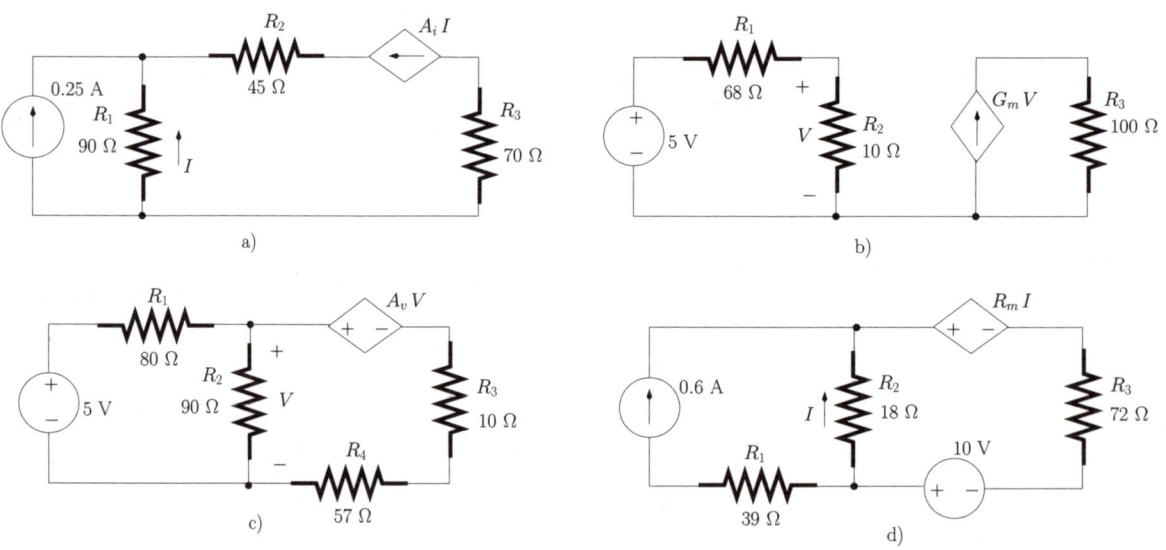

Figure 1.39: *Circuits of problem 1.17.*

1.18 Apply the superposition theorem to the circuits of figure 1.40 to determine the voltage across and the current through R_2. $G_m = 0.9$ S, $A_v = 10$.

1. Elementary electrical circuit analysis

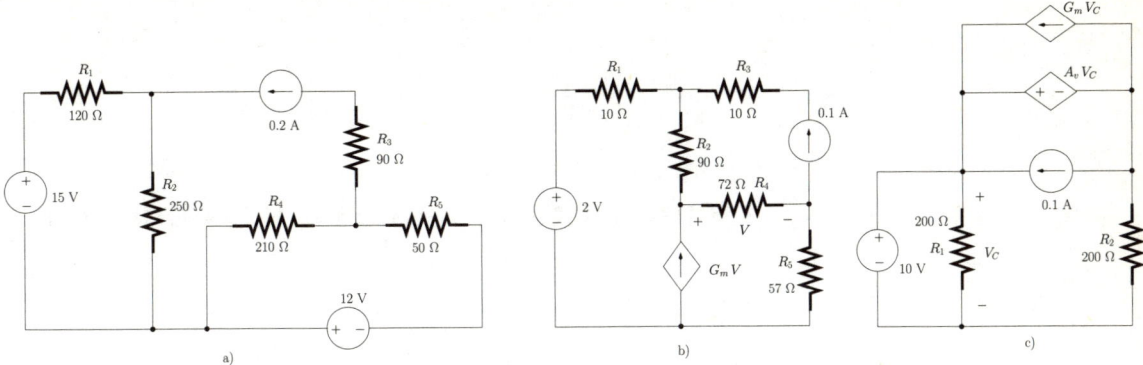

Figure 1.40: *Circuits of problem 1.18.*

2 Complex numbers: An introduction

2.1 Introduction

Complex numbers play a major role in alternating current (AC) circuit analysis through the use of the phasor concept and associated analysis. This simplifies the analysis of circuits by representing voltage and current quantities in terms of magnitude and phase. Phasor analysis is also the foundation of frequency domain signal analysis and is used extensively in the remaining chapters of this book.

Phasors are basically a convenient representation of complex numbers. In this chapter we introduce complex numbers and the different ways of representing them. Following this introduction, we define complex numbers. In section 2.3 we describe the elementary algebraic operations for these types of numbers. Then, in section 2.4 we discuss the polar representation of complex numbers and in section 2.5 we introduce the exponential representation which is basically the phasor representation. Finally, in section 2.6, we present the calculation of powers and roots of complex numbers.

2.2 Definition

Real numbers can be integers (e.g. $-1, 0, +2$), fractional numbers (e.g. $-1/2$, $1/3, 5/6$) and irrational numbers (e.g. $\sqrt{3}, \pi$). We can represent all real numbers on a single axis, the so-called real axis as illustrated in figure 2.1 a).

Complex numbers are quantities which are represented in a *plane* as shown in figure 2.1 b). This plane is called the 'complex plane' and it is defined by two orthogonal axes, X and Y; the real axis and the imaginary axis[1], respectively. The representation of the complex plane using two orthogonal axes is also called the Argand diagram. Every complex number, z, can be defined by a pair of real numbers (or pair of coordinates), x and y, which identify the position of z in the complex plane:

$$z = (x, y) \tag{2.1}$$

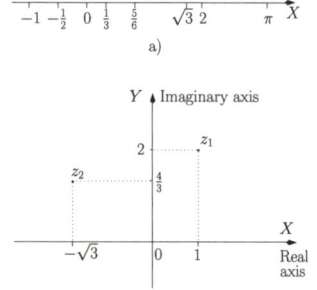

Figure 2.1: *a) The real axis. b) The complex plane.*

In eqn 2.1 x is called the 'real part' of the complex number while y is called the '*imaginary* part' of the complex number. In figure 2.1 b) we illustrate the representation of the complex numbers $z_1 = (1, 2)$ and $z_2 = (-\sqrt{3}, 4/3)$. It should be noted that all real numbers can be represented as complex numbers where the y coordinate is zero and they have the general form $(x, 0)$.

---

[1]The imaginary axis is also represented here by jY.

2. Complex numbers: An introduction

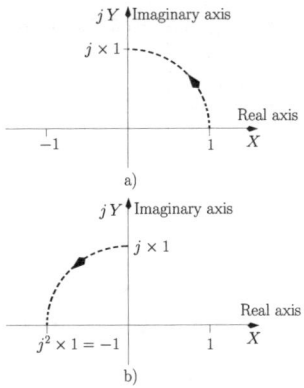

Figure 2.2: *a) The number 1 multiplied by j. b) The number $(j \times 1)$ multiplied by j.*

The complex number j

The complex number j is introduced here as an 'operator' so that when a number is multiplied by j the outcome is that number rotated by 90 degrees ($\pi/2$ radians) counter-clockwise in the complex plane. Let us consider the number 1 multiplied by j. According to the definition of j, the multiplication of 1 by j results in a 90 degrees counter-clockwise rotation of this number in the complex plane, as shown in figure 2.2 a). If we multiply $(j \times 1)$ by j again there is another 90 degrees counter-clockwise rotation, as shown in figure 2.2 b). From this figure we arrive at the central definition of the complex number j:

$$j^2 = -1 \qquad (2.2)$$

which means that $j = \sqrt{-1}$. Note that j is a complex number located on the imaginary axis, that is, its real part is zero. This is expected since real numbers do not encompass square roots of negative numbers. Complex numbers located on the imaginary axis (with zero real part) are usually referred to as 'imaginary numbers'.

The introduction of the j number allows the representation of complex numbers, which were formerly represented as $z = (x, y)$, as shown below:

$$z = x + jy \qquad (2.3)$$

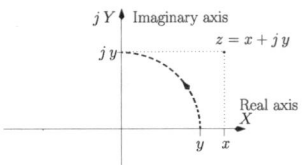

Figure 2.3: *Representation of $z = x + jy$.*

This is called the *Cartesian* (or rectangular) representation of complex numbers. We emphasise that the representation of complex numbers given by eqn 2.3 is equivalent to the representation of complex numbers given by eqn 2.1. In fact, eqn 2.3 indicates that the complex number z is the addition of a real number x and an imaginary number jy, the latter results from the 90 degrees counter-clockwise rotation of the real number y as illustrated by figure 2.3. Also, note that the imaginary axis, jY, can be seen as the counter-clockwise 90 degrees rotation of the real axis, X.

Example 2.2.1 Show that the multiplication of a number by $(-j)$ is equivalent to the rotation of this number by 90 degrees in the clockwise direction.

<u>Solution</u>: We illustrate the operation mentioned above using the number 1. Thus, $1 \times (-j)$ can be written as $j \times (-1)$ which, in turn, can be expressed, according to eqn 2.2, as $(1 \times j) \times (j^2)$. Hence the multiplication by j three times corresponds to the rotation by 270 degrees counter-clockwise in the complex plane. This is equivalent to rotating by 90 degrees clockwise, as shown in figure 2.4.

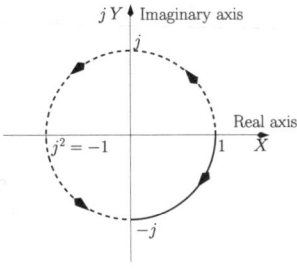

Figure 2.4: *Multiplication by $-j$.*

Equality of two complex numbers

Two complex numbers $z_1 = x_1 + jy_1$ and $z_2 = x_2 + jy_2$ are said to be equal when both their real and imaginary parts are equal, that is:

$$z_1 = z_2 \Leftrightarrow \begin{cases} \text{Real}(z_1) = \text{Real}(z_1) & \text{i.e.} \quad x_1 = x_2 \\ \text{Imag}(z_1) = \text{Imag}(z_1) & \text{i.e.} \quad y_1 = y_2 \end{cases} \qquad (2.4)$$

where Real (z_1) indicates the real part of z_1 and Imag (z_1) indicates the imaginary part of z_1.

2.3 Elementary algebra

We discuss now the addition, subtraction, multiplication and division of complex numbers.

2.3.1 Addition

The addition of two complex numbers $z_1 = x_1 + j\, y_1$ and $z_2 = x_2 + j\, y_2$ results in a third complex number $z_3 = x_3 + j\, y_3$ whose real part, x_3, is the sum of the real parts of z_1 and z_2 and its imaginary part, y_3, is the sum of the imaginary parts of z_1 and z_2. Hence we can write:

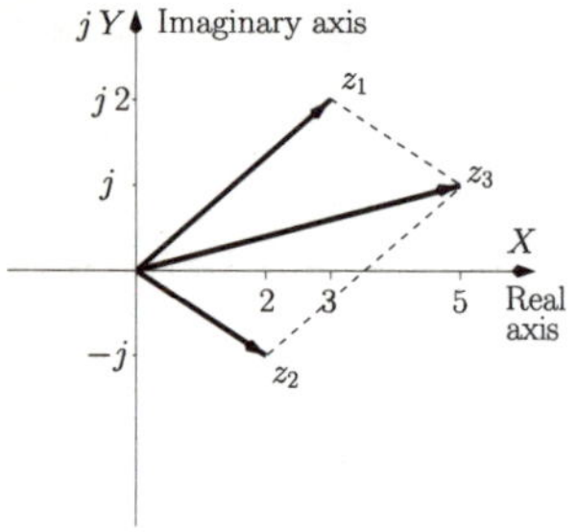

$$\begin{aligned} z_3 &= z_1 + z_2 \\ &= (x_1 + x_2) + j\,(y_1 + y_2) \end{aligned} \quad (2.5)$$

that is:

$$x_3 = x_1 + x_2 \quad (2.6)$$
$$y_3 = y_1 + y_2 \quad (2.7)$$

It is possible to represent the addition of two complex numbers on the Argand diagram. Figure 2.5 illustrates the addition of $z_1 = 3 + j\, 2$ with $z_2 = 2 - j$ which is equal to $z_3 = 5 + j\, 1$. Note that the addition of these two numbers is similar to the addition of two vectors, each is defined by one of the complex numbers, using the parallelogram rule.

Figure 2.5: *Addition of $z_1 = 3 + j\, 2$ and $z_2 = 2 - j$.*

2.3.2 Subtraction

The subtraction of two complex numbers $z_1 = x_1 + j\, y_1$ and $z_2 = x_2 + j\, y_2$ results in a complex number $z_3 = x_3 + j\, y_3$. x_3 is the subtraction of the real parts of z_1 and z_2 and y_3 is the subtraction of the imaginary parts of z_1 and z_2. Thus, we can write:

$$\begin{aligned} z_3 &= z_1 - z_2 \\ &= (x_1 - x_2) + j\,(y_1 - y_2) \end{aligned} \quad (2.8)$$

that is:

$$x_3 = x_1 - x_2 \quad (2.9)$$
$$y_3 = y_1 - y_2 \quad (2.10)$$

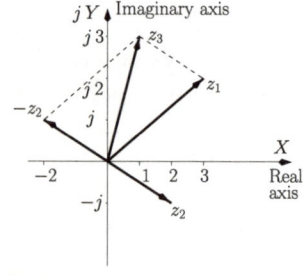

Figure 2.6: *Subtraction of $z_2 = 2 - j$ from $z_1 = 3 + j\, 2$.*

Figure 2.6 illustrates the subtraction of $z_2 = 2 - j$ from $z_1 = 3 + j\, 2$ which is equal to $z_3 = 1 + j\, 3$. In order to be able to apply the parallelogram rule we first need to represent the vector defined by $(-z_2)$ in the Argand plane. Then, we can add z_1 with $(-z_2)$ as described above. Note that $(-z_2) = j^2\, z_2$ can be represented in the Argand diagram by rotating z_2 by 180 degrees.

2.3.3 Multiplication

We discuss now the multiplication of complex numbers where we distinguish between three situations: multiplication of a complex number by a real number, multiplication of a complex number by an imaginary number and multiplication of a complex number by another complex number.

Multiplication by a real number

The multiplication of a complex number $z_1 = x_1 + j\, y_1$ by a real number $z_2 = x_2$ results in a complex number $z_3 = x_3 + j\, y_3$ whose real part, x_3, is the multiplication of the real part of z_1 with x_2 and its imaginary part, y_3, is the multiplication of the imaginary part of z_1 with x_2. Hence we can write:

$$\begin{aligned} z_3 &= z_1 \times z_2 \\ &= (x_1\, x_2) + j\, (y_1\, x_2) \end{aligned} \qquad (2.11)$$

that is:

$$x_3 = x_1\, x_2 \qquad (2.12)$$
$$y_3 = y_1\, x_2 \qquad (2.13)$$

Figure 2.7 illustrates the multiplication of $z_1 = 2 + j$ with $z_2 = 2$ which is equal to $z_3 = 4 + j\, 2$. This multiplication is similar to scaling the magnitude of the vector defined by z_1 by an amount given by x_2.

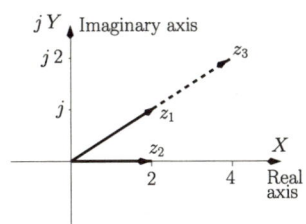

Figure 2.7: *Multiplication of $z_1 = 2 + j$ with $z_2 = 2$.*

Multiplication by an imaginary number

The multiplication of a complex number $z_1 = x_1 + j\, y_1$ by an imaginary number $z_2 = j\, y_2$ results in a complex number $z_3 = x_3 + j\, y_3$:

$$\begin{aligned} z_3 &= z_1 \times z_2 \\ &= x_1\, j\, y_2 + j^2\, y_1\, y_2 \\ &= (-y_1\, y_2) + j\, (x_1\, y_2) \quad \text{(recall that } j^2 = -1\text{)} \end{aligned} \qquad (2.14)$$

that is:

$$x_3 = -y_1\, y_2 \qquad (2.15)$$
$$y_3 = x_1\, y_2 \qquad (2.16)$$

Figure 2.8 illustrates the multiplication of $z_1 = 2 + j$ with $z_2 = j\, 2$ which is equal to $z_3 = -2 + j\, 4$. This multiplication is effectively a 90 degrees rotation of the vector defined by z_1 followed by a scaling of this vector by an amount given by $|y_2| = 2$. The 90 degrees rotation of the vector defined by z_1 is counter clockwise if $y_2 > 0$. If $y_2 < 0$ the 90 degrees rotation of this vector is clockwise.

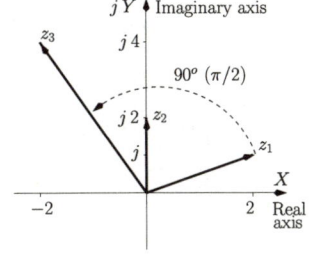

Figure 2.8: *Multiplication of $z_1 = 2 + j$ with $z_2 = j\, 2$.*

Multiplication by a complex number

The multiplication of a complex number $z_1 = x_1 + j\,y_1$ by another complex number $z_2 = x_2 + j\,y_2$ results in a different complex number $z_3 = x_3 + j\,y_3$. The result is clearly the combination of the two multiplication cases considered above. In other words it is addition of the multiplication of a complex number by a real number and the multiplication of a complex number by an imaginary number. z_3 can be calculated as follows:

$$\begin{aligned} z_3 &= z_1 \times z_2 \\ &= (x_1 + j\,y_1) \times (x_2 + j\,y_2) \\ &= x_1\,x_2 + x_1\,j\,y_2 + j\,y_1\,x_2 + j^2\,y_1\,y_2 \\ &= (x_1\,x_2 - y_1\,y_2) + j\,(x_1\,y_2 + y_1\,x_2) \end{aligned} \tag{2.17}$$

that is

$$x_3 = x_1\,x_2 - y_1\,y_2 \tag{2.18}$$
$$y_3 = x_1\,y_2 + x_2\,y_1 \tag{2.19}$$

For example the multiplication of $z_1 = 2 + j$ by $z_2 = 2 + j\,2$ is

$$\begin{aligned} z_3 &= (2+j) \times (2+j\,2) \\ &= 4 + j\,4 + j\,2 + j^2\,2 \\ &= 2 + j\,6 \end{aligned}$$

Complex conjugate

Two complex numbers are said to be the conjugate of each other when they have the same real part but have imaginary parts of opposite sign. For example, the complex conjugate of $z_1 = 1 + j\,2$ is $1 - j\,2$ and the complex conjugate of $z_2 = -3 - j\,\sqrt{2}$ is $-3 + j\,\sqrt{2}$ as illustrated by figure 2.9. It is common to represent the complex conjugate of z by z^*. It is interesting to note the following results which apply to complex conjugates:

- The addition of a complex number, z, to its conjugate is a real number equal to twice the real part of z.

$$\begin{aligned} z + z^* &= (x + j\,y) + (x - j\,y) \\ &= 2\,x \end{aligned} \tag{2.20}$$

- The subtraction of a complex number, z, from its conjugate is an imaginary number equal to twice the imaginary part of z.

$$\begin{aligned} z - z^* &= (x + j\,y) - (x - j\,y) \\ &= 2\,j\,y \end{aligned} \tag{2.21}$$

- The multiplication of a complex number, z, with its conjugate is a real number equal to the addition of the squares of its real part and imaginary

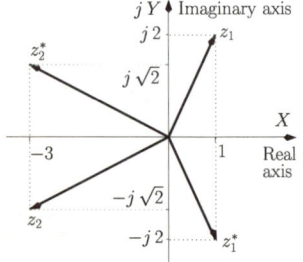

Figure 2.9: *Complex numbers and their conjugates.*

2. Complex numbers: An introduction

part.

$$\begin{aligned} z \times z^* &= (x+jy) \times (x-jy) \\ &= x^2 - xjy + jyx + (jy)(-jy) \\ &= x^2 - j^2 y^2 \quad \text{(recall that } j^2 = -1\text{)} \\ &= x^2 - (-y^2) \\ &= x^2 + y^2 \end{aligned} \qquad (2.22)$$

2.3.4 Division

The division of a complex number $z_1 = x_1 + jy_1$ by a complex number $z_2 = x_2 + jy_2$ results in another complex number $z_3 = x_3 + jy_3$. In order to calculate z_3 we use the fact that a complex number multiplied by its conjugate is a real number (see also eqn 2.22). $z_3 = z_1/z_2$ can then be calculated as follows:

$$\begin{aligned} z_3 &= \frac{z_1}{z_2} \\ &= \frac{z_1}{z_2} \times \frac{z_2^*}{z_2^*} \\ &= \frac{(x_1 + jy_1)(x_2 - jy_2)}{(x_2 + jy_2)(x_2 - jy_2)} \end{aligned} \qquad (2.23)$$

Using eqn 2.22 we can write eqn 2.23 as follows:

$$z_3 = \frac{(x_1 + jy_1)(x_2 - jy_2)}{x_2^2 + y_2^2} \qquad (2.24)$$

Expanding the numerator we obtain:

$$\begin{aligned} z_3 &= \frac{x_1 x_2 - j y_2 x_1 + j y_1 x_2 - j^2 y_1 y_2}{x_2^2 + y_2^2} \\ &= \frac{x_1 x_2 + y_1 y_2}{x_2^2 + y_2^2} + j \frac{y_1 x_2 - y_2 x_1}{x_2^2 + y_2^2} \end{aligned} \qquad (2.25)$$

that is

$$x_3 = \frac{x_1 x_2 + y_1 y_2}{x_2^2 + y_2^2} \qquad (2.26)$$

$$y_3 = \frac{y_1 x_2 - y_2 x_1}{x_2^2 + y_2^2} \qquad (2.27)$$

For example, the division of $3 + j4$ by $1 - j2$ can be calculated as follows:

$$\begin{aligned} z_3 &= \frac{3 + j4}{1 - j2} \\ &= \frac{3 + j4}{1 - j2} \times \frac{1 + j2}{1 + j2} \\ &= \frac{3 + j6 + j4 + j^2 8}{1 + 4} \\ &= -1 + j2 \end{aligned}$$

2.3.5 Complex equations

The solution of complex equations involves the calculation of two unknown quantities; the real part and the imaginary part. For example, let us solve the following complex equation:

$$3z = 4z^* + 9 + 5j \tag{2.28}$$

If we expand z into its components x and jy, the last eqn can be written as follows:

$$3x + j3y = 4x - 4jy + 9 + 5j \tag{2.29}$$

Grouping the real parts and the imaginary parts we have:

$$(3x - 4x - 9) + j(3y + 4y - 5) = 0 \tag{2.30}$$

Both real and imaginary parts must each be equal to zero:

$$\begin{cases} 3x - 4x - 9 = 0 \\ 3y + 4y - 5 = 0 \end{cases} \tag{2.31}$$

that is

$$\begin{cases} x = -9 \\ y = \frac{5}{7} \end{cases} \tag{2.32}$$

2.3.6 Quadratic equations

Quadratic equations have the general form:

$$ax^2 + bx + c = 0 \tag{2.33}$$

where a, b, and c are real numbers and x is the unknown variable which is to be determined. This eqn has two solutions which can be expressed as follows:

$$x = \frac{-b \pm \sqrt{b^2 - 4ac}}{2a} \tag{2.34}$$

The solutions are real numbers if $b^2 - 4ac \geq 0$. However, when $b^2 - 4ac < 0$ the square root of a negative number is required. Using complex numbers the solutions can be written as follows:

$$\begin{aligned} x &= \frac{-b \pm \sqrt{-(-b^2 + 4ac)}}{2a} \\ &= \frac{-b \pm \sqrt{j^2(4ac - b^2)}}{2a} \end{aligned} \tag{2.35}$$

where now we have $4ac - b^2 > 0$. Therefore, the last eqn can be written as

$$\begin{aligned} x &= \frac{-b \pm \sqrt{j^2}\sqrt{4ac - b^2}}{2a} \\ &= \frac{-b \pm j\sqrt{4ac - b^2}}{2a} \end{aligned} \tag{2.36}$$

Equation 2.36 indicates that when $b^2 - 4ac < 0$ the two solutions for this quadratic eqn are complex numbers. It should also be noted that these are complex conjugates.

Example 2.3.1 Solve the following quadratic eqns:

1. $3z^2 + z + 5 = 0$
2. $3z^2 - 3z - 6 = 0$

Solution:

1. We have $a = 3$, $b = 1$, $c = 5$, and $b^2 - 4ac = -59$. Therefore the two solutions are complex and given by eqn 2.36:

$$z_1 = -\frac{1}{6} + j\frac{\sqrt{59}}{6}$$
$$z_2 = -\frac{1}{6} - j\frac{\sqrt{59}}{6}$$

2. Now we have $a = 3$, $b = -3$, $c = -6$, and $b^2 - 4ac = 81$. Therefore the two solutions are real and given by eqn 2.34

$$z_1 = -1$$
$$z_2 = 2$$

2.4 Polar representation

So far the representation of a complex number z, has used a Cartesian (or rectangular) representation where a complex number is identified by its coordinates on the real and the imaginary axes using one of the following notations: $z = (x, y) = x + jy$. This representation is illustrated again in figure 2.10. A complex number also defines a vector represented by its length, r, and by the angle, θ, of the vector with the real axis. When z is represented by r and θ (usually written as $z = r\angle\theta$) it is said to be represented by its 'polar coordinates'. The length of the vector, r, is often called the 'modulus', $|z|$, or magnitude of the complex number z. The angle θ is usually called the 'argument' of z. When z is represented in its Cartesian form ($z = x + jy$) the modulus of z can be determined from Pythagoras's theorem:

$$|z| = r = \sqrt{x^2 + y^2} \qquad (2.37)$$

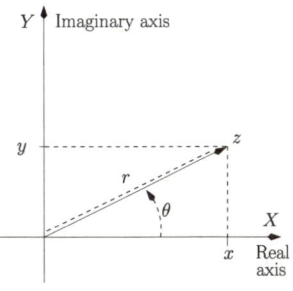

Figure 2.10: *Cartesian and polar representations for complex numbers.*

Note that $|z|$ can also be obtained as follows (see also eqn 2.22);

$$|z| = \sqrt{z \times z^*} \qquad (2.38)$$

The argument of z, θ, can be determined using the following trigonometric relationship:

$$\tan(\theta) = \frac{y}{x} \qquad (2.39)$$

that is:

$$\theta = \tan^{-1}\left(\frac{y}{x}\right) \qquad (2.40)$$

Example 2.4.1 Represent $z = 5 + j\sqrt{3}$ in polar coordinates

Solution: From eqn 2.37, r is the magnitude of z;

$$r = \sqrt{25 + 3} = \sqrt{28}$$

and from 2.40

$$\theta = \tan^{-1}\left(\frac{\sqrt{3}}{5}\right) = 0.3 \text{ rad}$$

Thus $z = \sqrt{28} \angle 0.3$ rad.

Conversion between polar and Cartesian

Using basic trigonometry, a complex number in polar form, $z = r\angle\theta$, can be converted to Cartesian form, $z = (x, y) = x + jy$, as follows:

$$x = r\cos(\theta) \tag{2.41}$$
$$y = r\sin(\theta) \tag{2.42}$$

Example 2.4.2 Represent $z = 3 \angle \pi/4$ rad in rectangular coordinates

Solution: According to eqns 2.41 and 2.42 z can be written as:

$$z = 3\cos\left(\frac{\pi}{4}\right) + j3\sin\left(\frac{\pi}{4}\right) = 2.1 + j\,2.1$$

2.4.1 Multiplication and division

The polar representation is very attractive since it considerably simplifies the multiplication and division of complex numbers. To multiply two complex numbers in the polar form we multiply the moduli and we add the arguments, that is, if we want to multiply $z_1 = r_1\angle\theta_1$ with $z_2 = r_2\angle\theta_2$ we obtain a complex number $z_3 = r_3\angle\theta_3$ where:

$$z_3 = r_1 \times r_2 \angle(\theta_1 + \theta_2) \tag{2.43}$$

that is

$$r_3 = r_1 \times r_2 \tag{2.44}$$
$$\theta_3 = \theta_1 + \theta_2 \tag{2.45}$$

For example, the multiplication of $z_1 = 2.3 \angle 2.3$ rad with $z_2 = 4.0 \angle 0.4$ rad is equal to $z_3 = 9.2 \angle 2.7$ rad.

2. Complex numbers: An introduction

The division of two complex numbers represented in the polar form is done by dividing the moduli and subtracting the arguments, that is, dividing $z_1 = r_1 \angle \theta_1$ by $z_2 = r_2 \angle \theta_2$ gives a complex number $z_3 = r_3 \angle \theta_3$ where:

$$z_3 = \frac{r_1}{r_2} \angle (\theta_1 - \theta_2) \tag{2.46}$$

that is

$$r_3 = \frac{r_1}{r_2} \tag{2.47}$$
$$\theta_3 = \theta_1 - \theta_2 \tag{2.48}$$

For example, the division of $z_1 = 2.3 \angle 2.3$ rad by $z_2 = 4.0 \angle 0.4$ rad is equal to $z_3 = 0.6 \angle 1.9$ rad.

It is worth mentioning that to add or subtract two complex numbers expressed in polar form it is necessary to convert them first to a Cartesian (rectangular) form. The addition or the subtraction can then be effected as described in section 2.3. The result can, of course, be converted back to a polar representation.

Example 2.4.3 Determine the result of the addition of $z_1 = 4.2 \angle \pi/9$ with $z_2 = 1.5 \angle -3\pi/4$.

Solution: According to eqns 2.41 and 2.42 z_1 and z_2 can be written as follows:

$$z_1 = 3.9 + j\,1.4$$
$$z_2 = -1.1 - j\,1.1$$

Hence $z_1 + z_2 = 2.8 + j\,0.3 = 2.82 \angle 0.11$ rad.

2.5 The exponential form

The exponential form of a complex number is similar to the polar representation discussed above. In order to obtain this exponential form we start by expanding $\cos(\theta)$ and $\sin(\theta)$ in Maclaurin series (see appendix A) we have

$$\cos(\theta) = 1 - \frac{\theta^2}{2!} + \frac{\theta^4}{4!} + \ldots + (-1)^n \frac{\theta^{2n}}{2n!} + \ldots \tag{2.49}$$

$$\sin(\theta) = \theta - \frac{\theta^3}{3!} + \frac{\theta^5}{5!} + \ldots + (-1)^n \frac{\theta^{2n+1}}{(2n+1)!} + \ldots \tag{2.50}$$

Now, $\cos(\theta) + j\sin(\theta)$ can be written as

$$\cos(\theta) + j\sin(\theta) = \left(1 - \frac{\theta^2}{2!} + \frac{\theta^4}{4!} + \ldots \right) + j\left(\theta - \frac{\theta^3}{3!} + \frac{\theta^5}{5!} + \ldots \right)$$

$$= 1 + j\theta - \frac{\theta^2}{2!} - j\frac{\theta^3}{3!} + \frac{\theta^4}{4!} + j\frac{\theta^5}{5!} + \ldots \tag{2.51}$$

The series can be written as

$$\cos(\theta) + j\sin(\theta) = 1 + j\theta + \frac{(j\theta)^2}{2!} + \frac{(j\theta)^3}{3!} + \frac{(j\theta)^4}{4!} + j\frac{(j\theta)^5}{5!} + \ldots$$

$$= \sum_{k=0}^{\infty} \frac{(j\theta)^k}{k!} \tag{2.52}$$

Recognising eqn 2.52 as the Maclaurin series of $\exp(j\theta)$ we arrive at Euler's formula:

$$e^{j\theta} = \cos(\theta) + j\sin(\theta) \tag{2.53}$$

Using eqns 2.41, 2.42 and 2.53 we can write a complex number $z = r \angle \theta$ as

$$\begin{aligned} z &= r\cos(\theta) + jr\sin(\theta) \\ &= r[\cos(\theta) + j\sin(\theta)] \\ &= re^{j\theta} \end{aligned} \tag{2.54}$$

One advantage of the exponential representation of complex numbers is its simplicity. This exponential form is also called the 'phasor representation' of complex numbers and it plays a major role in the representation of signals and systems in the frequency domain as discussed in the next chapter.

It is interesting to note that setting $\theta = \pi/2$ and $\theta = -\pi/2$ in eqn 2.53 we get the useful relationships:

$$\begin{aligned} j &= e^{j\frac{\pi}{2}} \end{aligned} \tag{2.55}$$

$$\begin{aligned} -j &= e^{-j\frac{\pi}{2}} \end{aligned} \tag{2.56}$$

Example 2.5.1 Determine the exponential form of $z = (3 + j5)^{-1}$.

Solution: According to eqns 2.37, 2.40 and 2.54, z can be written as follows:

$$\begin{aligned} z &= \frac{1}{3 + j5} \\ &= \frac{1}{\sqrt{3^2 + 5^2}\, e^{j\,\tan^{-1}(5/3)}} \\ &= \frac{1}{\sqrt{34}\, e^{j\,1.03}} \\ &= \frac{1}{\sqrt{34}}\, e^{-j\,1.03} \end{aligned}$$

2.5.1 Trigonometric functions and the exponential form

It is possible to express the trigonometric functions using the exponential representation of complex numbers. Let us consider a complex number with $r = 1$ and its complex conjugate in the exponential form:

$$e^{j\theta} = \cos(\theta) + j\sin(\theta) \tag{2.57}$$

$$\begin{aligned} e^{-j\theta} &= \cos(-\theta) + j\sin(-\theta) \\ &= \cos(\theta) - j\sin(\theta) \end{aligned} \tag{2.58}$$

Adding eqn 2.57 to eqn 2.58 we obtain:
$$e^{j\theta} + e^{-j\theta} = 2\cos(\theta)$$

that is:
$$\cos(\theta) = \frac{e^{j\theta} + e^{-j\theta}}{2} \qquad (2.59)$$

On the other hand if we subtract eqn 2.58 from eqn 2.57 we obtain:
$$e^{j\theta} - e^{-j\theta} = 2j\sin(\theta)$$

that is:
$$\sin(\theta) = \frac{e^{j\theta} - e^{-j\theta}}{2j} \qquad (2.60)$$

It is a trivial matter to show that:
$$\tan(\theta) = \frac{1}{j}\frac{e^{j\theta} - e^{-j\theta}}{e^{j\theta} + e^{-j\theta}} \qquad (2.61)$$

2.6 Powers and roots

A very useful theorem for the calculation of the powers and roots of complex numbers is De Moivre's theorem which states that:
$$[\cos(\theta) + j\sin(\theta)]^n = \cos(n\theta) + j\sin(n\theta) \qquad (2.62)$$

Therefore, a complex number z^n can be written as:
$$z^n = \left(r\,e^{j\theta}\right)^n \qquad (2.63)$$
$$= r^n\,e^{j\theta n} \qquad (2.64)$$
$$= r^n\cos(n\theta) + j\,r^n\sin(n\theta) \qquad (2.65)$$

These eqns are valid for all real values of n. This means that De Moivre's formula allows us to calculate the powers and roots of complex numbers. However, we must bear in mind that there is usually more than one solution when finding the roots of a complex number.

Powers of a complex number

The calculation of the powers of a complex number results from the straightforward application of eqn 2.64 or 2.65. For example, the calculation of the cube of $z = 3\,e^{j\pi/4}$ is:
$$z^3 = \left(3\,e^{j\pi/4}\right)^3$$
$$= 27\,e^{j\,3\pi/4}$$
$$= 27\cos(3\pi/4) + j\,27\sin(3\pi/4)$$

The powers of complex numbers have many uses in obtaining trigonometric identities, as the following example illustrates.

Example 2.6.1 Express $\sin(2\phi)$ in terms of $\cos(\phi)$ and of $\sin(\phi)$

Solution: From eqn 2.57 we can write

$$e^{j\,2\,\phi} = \cos(2\,\phi) + j\,\sin(2\,\phi) \tag{2.66}$$

therefore, $\sin(2\phi)$ can be expressed as:

$$\begin{aligned}\sin(2\phi) &= \text{Imag}\left[e^{j\,2\,\phi}\right]\\ &= \text{Imag}\left[\left(e^{j\,\phi}\right)^2\right]\\ &= \text{Imag}\left[(\cos(\phi) + j\,\sin(\phi))^2\right]\\ &= \text{Imag}\left[\cos(\phi)^2 + 2\,j\,\cos(\phi)\,\sin(\phi) - \sin(\phi)^2\right]\\ &= 2\,\cos(\phi)\,\sin(\phi)\end{aligned} \tag{2.67}$$

n roots of unity

We consider now the solution of $z^n = 1$ which is equivalent to determining the n-roots of the number 1 in the complex plane. Note that 1 can be seen as a complex number with modulus one and argument zero. Other arguments with multiples of 2π ($\pm 2\pi$, $\pm 4\pi$, $\pm 6\pi$, etc.) are also valid since the addition (or the subtraction) of 2π to the argument of a complex number does not change its position in the complex plane or change its value. Therefore, $z^n = 1$ can be expressed as follows:

$$z^n = e^{j\,2\,\pi\,N}, \quad N \in \{\ldots, -2, -1, 0, +1, +2, \ldots\} \tag{2.68}$$

To solve this eqn we take both sides to the power of $1/n$, that is:

$$(z^n)^{\frac{1}{n}} = e^{j\,2\,\pi\,N/n}, \quad N \in \{\ldots, -2, -1, 0, +1, +2, \ldots\} \tag{2.69}$$

We can find the n different roots by setting N to 0, 1, 2, ..., $(n-1)$ in the last eqn. Note that other values of N result in repeated roots. To illustrate this concept consider the example below.

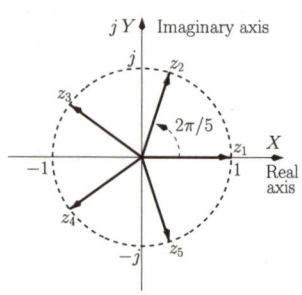

Figure 2.11: *Solutions of $z^5 = 1$.*

Example 2.6.2 Find the solutions of $z^5 = 1$.

Solution: $z^5 = 1$ can be expressed as follows:

$$z^5 = e^{j\,2\,\pi\,N} \tag{2.70}$$

with $N \in \{\ldots, -2, -1, 0, +1, +2, \ldots\}$. Taking the 5th root we obtain

$$z = e^{j\,2\,\pi\,N/5} \tag{2.71}$$

2. Complex numbers: An introduction

Substituting $N \in \{0,1,2,3,4\}$ in the last eqn, we obtain (see also figure 2.11):

$$
\begin{aligned}
N=0 &: & z_1 &= e^{j\,2\pi\,0/5} = 1 \\
N=1 &: & z_2 &= e^{j\,2\pi\,1/5} = e^{j\,2\pi/5} \\
N=2 &: & z_3 &= e^{j\,2\pi\,2/5} = e^{j\,4\pi/5} \\
N=3 &: & z_4 &= e^{j\,2\pi\,3/5} = e^{j\,6\pi/5} \\
N=4 &: & z_5 &= e^{j\,2\pi\,4/5} = e^{j\,8\pi/5}
\end{aligned}
$$

Note that, for this example there is a basic set of five roots. For values of N greater than four we start to obtain repetitions of the roots. For example, by setting $N=5$ in eqn 2.71 we obtain $z = e^{j\,2\pi} = 1$. Note that this is the same root as for $N=0$.

The n roots of a general complex number

The calculation of the n roots of a general complex number, w, can be seen as the calculation of the solutions of the following eqn:

$$z^n = w \qquad (2.72)$$

Expressing w in an exponential form gives:

$$w = r_w\, e^{j\,\theta_w} \qquad (2.73)$$

We can write eqn 2.72 as follows:

$$z^n = r_w\, e^{j\,\theta_w + j\,2\pi N} \qquad (2.74)$$

where we use the fact that the addition of a multiple of 2π to the argument of a complex number does not change its value. Taking the n roots of both sides of eqn 2.74 we obtain

$$z = (r_w)^{\frac{1}{n}}\, e^{j\,\theta_w/n + j\,2\pi N/n} \qquad (2.75)$$

The n different roots are determined by setting N to 0, 1, 2, ..., $(n-1)$.

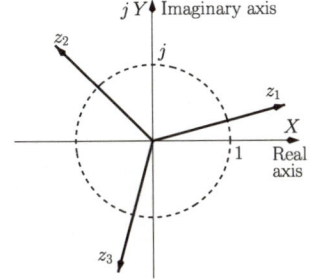

Figure 2.12: *Solutions of* $z^3 = 4 + j\,4$.

Example 2.6.3 Find the solutions of $z^3 = 4 + j\,4$.

Solution: First, we represent $4 + j\,4$ in an exponential form; $4\sqrt{2}\,e^{j\,\pi/4}$. Then, $z^3 = 4 + j\,4$ can be written as follows:

$$z^3 = 4\sqrt{2}\,e^{j\,\pi/4 + j\,2\pi N} \qquad (2.76)$$

taking the cubic root we obtain

$$z = (4\sqrt{2})^{\frac{1}{3}}\, e^{j\,\pi/12 + j\,2\pi N/3} \qquad (2.77)$$

substituting $N \in \{0, 1, 2\}$ in the last eqn we obtain:

$$N = 0 \quad : \quad z_1 = (4\sqrt{2})^{\frac{1}{3}} e^{j\pi/12}$$
$$N = 1 \quad : \quad z_2 = (4\sqrt{2})^{\frac{1}{3}} e^{j3\pi/4}$$
$$N = 2 \quad : \quad z_3 = (4\sqrt{2})^{\frac{1}{3}} e^{j\pi 17/12}$$

Figure 2.12 shows these roots represented in the Argand diagram.

2.7 Bibliography

1. M. Attenborough, *Mathematics for Electrical Engineering and Computing*, 2003 (Newnes).

2. C.R. Wylie and L.C. Barrett, *Advanced Engineering Mathematics*, 1995 (McGraw-Hill International Editions), 6th edition.

2.8 Problems

2.1 Represent the following complex numbers in the Argand diagram:

1. $z_1 = 1 + j\,4$
2. $z_2 = 1 - j\,4$
3. $z_3 = -2 + j\,2.5$
4. $z_4 = -\pi - j\sqrt{3}$

2.2 Perform the following algebraic operations:

1. $(1 + j\,4) + (1 - j\,4)$
2. $(-2 + j\,1) - (-2 - j\,1)$
3. $(j\,2.5) \times (1 - j\,4.5)$
4. $(2 - j\,4)/(-3 - j\,8)$
5. $(-0.45, 4) - (0.8, 3.1)$
6. $(1.4, 2) + (0.8, 3.1)$
7. $(-5, 0) \times (0.8, 3.1)$
8. $(-0.45, 4)/(0.8, 3.1)$

2.3 Solve the following quadratic equations:

1. $z^2 + 3z + 34 = 0$
2. $4z^2 - 2z - 5 = 0$
3. $-2z^2 + z = 5$
4. $z^2 + 6z + 9 = 0$
5. $6z^2 + 3\sqrt{7}z + 27 = 0$

2. Complex numbers: An introduction

2.4 Represent the following numbers in the polar representation:

1. $z = 1 + j\,1$
2. $z = -1 + j\sqrt{2}$
3. $z = 2 - j\,0.3$
4. $z = -\sqrt{7} - j\sqrt{3}$

2.5 Represent the following numbers in the Cartesian representation:

1. $z = 0.5 \angle 2\pi$
2. $z = 1.5 \angle -\pi/3$
3. $z = 0.5 \angle 6\pi/4$
4. $z = 0.5 \angle -3\pi/2$

2.6 Calculate the following:

1. $4\,e^{j\pi/2} \times 0.5\,e^{j\,3\pi/2}$
2. $4\,e^{j\pi/2} / (0.5\,e^{j\,3\pi/2})$
3. $3.4\,e^{-j\pi/5} \times 5\,e^{-j\pi}$
4. $2.1\,e^{-j\pi/5} / (9\,e^{j\,7\pi/5})$
5. $4.8\,e^{j\pi/9} + 6.5\,e^{j\,5\pi/2}$
6. $0.9\,e^{-j\pi/3} - 0.5\,e^{j\,3\pi/2}$

2.7 Solve the following equations:

1. $z^2 = 1$
2. $z^2 = j$
3. $z^3 = 1 + j\,1$
4. $2\,e^{j\pi}\,z^5 = 5\,e^{j\pi/4}$

3 Frequency domain electrical signal and circuit analysis

3.1 Introduction

In this chapter we present the main electrical analysis techniques for time varying signals. We start by discussing sinusoidal alternating current (AC) signals[1] and circuits. Phasor analysis is presented and it is shown that this greatly simplifies this analysis since it allows the introduction of the 'generalised impedance'. The generalised impedance allows us to analyse AC circuits using all the circuit techniques and methods for DC circuits discussed in Chapter 1. In section 3.3 we extend the phasor analysis technique to analyse circuits driven by non-sinusoidal signals. This is done by first discussing the Fourier series which presents periodic signals as a sum of phasors. The Fourier series is a very important tool since it forms the basis of fundamental concepts in signal processing such as spectra and bandwidth. Finally, we present the Fourier transform which allows the analysis of virtually any time-varying signal (periodic and non periodic) in the frequency domain.

3.2 Sinusoidal AC electrical analysis

AC sinusoidal electrical sources are time-varying voltages and currents described by functions of the form:

$$v_s(t) = V_s \sin(\omega t) \quad (3.1)$$
$$i_s(t) = I_s \sin(\omega t) \quad (3.2)$$

where V_s and I_s are the peak-amplitudes of the voltage and of the current waveforms, respectively, as illustrated in figure 3.1. ω represents the *angular* frequency, in radians/second, equal to $2\pi/T$ where T is the period of the waveform in seconds. The repetition rate of the waveform, that is the *linear* frequency, is equal to $1/T$ in hertz. The quantity (ωt) is an angle, in radians, usually called the instantaneous phase. Note that ωT corresponds to 2π rad.

Here we interchangeably use the terms voltage/current sinusoidal signal or waveform, to designate the AC sinusoidal quantities.

By definition, all transient phenomena (such as those resulting, for example, from switching-on the circuit) have vanished in an AC circuit in its steady-state condition. Thus, the time origin in eqns 3.1 and 3.2 can be 'moved' so

---

[1] Any signal varying with time is effectively an AC signal. We limit our definition of an AC signal here to a sinusoidal signal at specific frequency. This is particularly helpful to calculate impedances at specific frequencies as will be seen later in this chapter.

3. Frequency domain electrical signal and circuit analysis

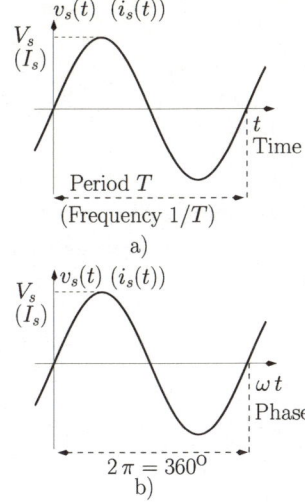

Figure 3.1: *a) AC voltage (current) waveform versus time. b) AC voltage (current) waveform versus phase.*

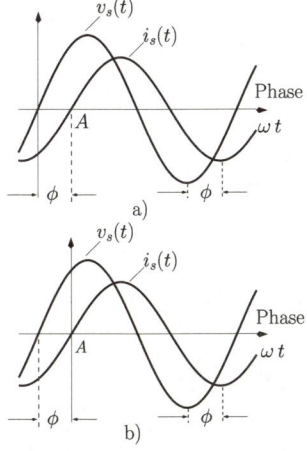

Figure 3.2: *Phase difference ($\phi = \pi/3$) between an AC voltage and an AC current. a) The current lags the voltage. b) The voltage leads the current.*

$v_s(t)$ and $i_s(t)$ are equally well described by cosine functions, that is:

$$v_s(t) = V_s \cos(\omega t) \qquad (3.3)$$
$$i_s(t) = I_s \cos(\omega t) \qquad (3.4)$$

While the choice of the absolute time origin is of no relevance in AC analysis, the relative time difference between waveforms, which can also be quantified in terms of phase difference, is of vital importance. Figure 3.2 a) illustrates the constant phase difference between a voltage waveform and a current waveform at the same angular frequency ω. If any two AC electrical waveforms have different angular frequencies, ω_1 and ω_2, then the phase difference between these two waveforms is a linear function of time; $(\omega_1 - \omega_2)\,t$. Assuming a time origin for the voltage waveform we can write the waveforms of figure 3.2 a) as:

$$v_s(t) = V_s \sin(\omega t) \qquad (3.5)$$
$$i_s(t) = I_s \sin(\omega t - \phi) \qquad (3.6)$$

where $\phi = \pi/3$. In this situation it is said that the current waveform **lags** the voltage waveform by ϕ. In fact, the current waveform crosses the phase axis (point A) later than the voltage waveform. On the other hand, if we choose the time origin for the current waveform, as illustrated in figure 3.2 b), we can write these waveforms as follows:

$$v_s(t) = V_s \sin(\omega t + \phi) \qquad (3.7)$$
$$i_s(t) = I_s \sin(\omega t) \qquad (3.8)$$

and it is said that the voltage waveform **leads** the current waveform.

3.2.1 Effective electrical values

By definition, the effective value of any voltage waveform is the DC voltage that, when applied to a resistance, would produce as much power dissipation (heat) as that caused by that voltage waveform. Hence, if we represent the AC voltage waveform by $V_s \sin(\omega t)$ and the effective voltage by V_{eff}, then, according to eqn 1.20, we can write:

$$\frac{1}{T}\int_0^T \frac{V_{eff}^2}{R}\,dt = \frac{1}{T}\int_0^T \frac{v_s^2(t)}{R}\,dt$$
$$\Leftrightarrow \frac{1}{T}\int_0^T \frac{V_{eff}^2}{R}\,dt = \frac{1}{T}\int_0^T \frac{V_s^2 \sin^2(\omega t)}{R}\,dt \qquad (3.9)$$

The last eqn can be written as follows:

$$\frac{1}{T}\frac{V_{eff}^2}{R}T = \frac{1}{T}\frac{V_s^2}{R}\int_0^T \frac{1-\cos(2\omega t)}{2}\,dt$$
$$\Leftrightarrow V_{eff}^2 = \frac{V_s^2}{2T}\left[t - \frac{1}{2\omega}\sin(2\omega t)\right]_0^T \qquad (3.10)$$

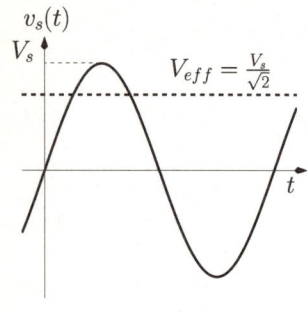

Figure 3.3: *Voltage AC waveforms and its corresponding effective voltage.*

Since $\omega = 2\pi/T$ the last eqn can be written as:

$$V_{eff}^2 = \frac{V_s^2}{2} \qquad (3.11)$$

or $V_{eff} = V_s/\sqrt{2} \simeq 0.707\, V_s$. Figure 3.3 illustrates the effective voltage of an AC voltage waveform.

In a similar way it can be shown that the effective value of a sinusoidal current with peak-amplitude I_s is $I_{eff} = I_s/\sqrt{2}$. The effective value of a sinusoidal voltage and/or current is also called the root-mean-square (RMS) value.

Example 3.2.1 Show that the effective value of a triangular voltage waveform, like that shown in figure 3.4, with peak amplitude V_s is $V_{eff} = V_s/\sqrt{3}$.

Solution: Following the procedure described above we can write:

$$\frac{1}{T}\int_0^T \frac{V_{eff}^2}{R}\, dt = \frac{1}{T}\int_0^T \frac{v_s^2(t)}{R}\, dt$$

Looking at figure 3.3 we see that the triangular waveform is symmetrical. Therefore, it is sufficient to consider the period of integration from $t = 0$ to $t = T/4$, giving

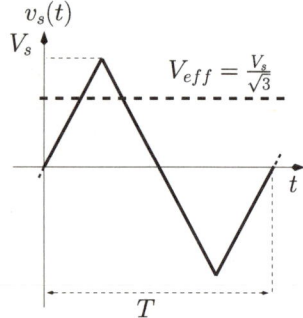

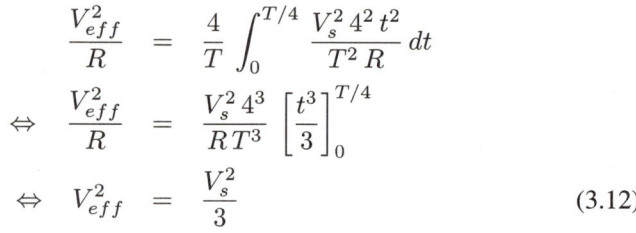

$$V_{eff}^2 = \frac{V_s^2}{3} \qquad (3.12)$$

that is, $V_{eff} = V_s/\sqrt{3} \simeq 0.577\, V_s$.

Figure 3.4: *Triangular voltage waveform and its corresponding effective voltage.*

3.2.2 I–V characteristics for passive elements

We now study the AC current–voltage (I–V) relationships for the main passive elements, presented in Chapter 1. We use cosine functions to represent AC currents and voltages waveforms. However, the same results would be obtained if sine functions were used instead.

Resistance

Assuming a current, $i(t) = I_x \cos(\omega\, t)$ passing through a resistance R, the voltage developed across its terminals is, according to Ohm's law:

$$\begin{aligned} v_R(t) &= R\, i(t) \\ &= R\, I_x\, \cos(\omega\, t) \\ &= V_r\, \cos(\omega\, t) \end{aligned} \qquad (3.13)$$

3. Frequency domain electrical signal and circuit analysis

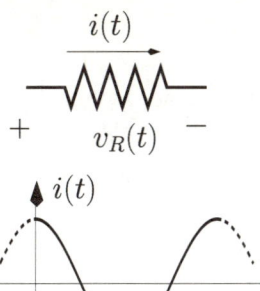

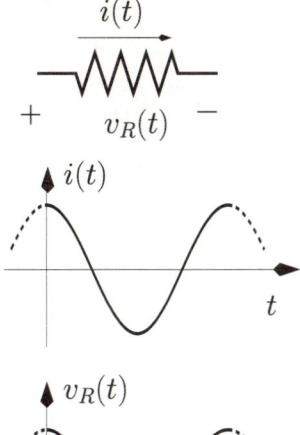

Figure 3.5: *Voltage and current in a resistance.*

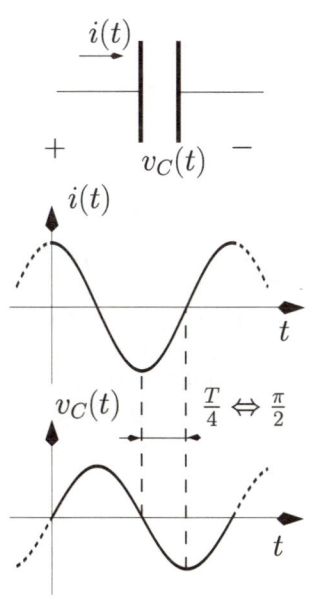

Figure 3.6: *Voltage and current in a capacitor.*

with
$$V_r = R I_x \tag{3.14}$$

Dividing both sides by $\sqrt{2}$ we obtain the RMS (or effective) value for the AC voltage as

$$\begin{aligned} V_{r_{eff}} &= R \frac{I_x}{\sqrt{2}} \\ &= R I_{x_{eff}} \end{aligned} \tag{3.15}$$

where $I_{x_{eff}}$ is the RMS (or effective) value for the AC current. From eqn 3.13 and figure 3.5 we observe that the voltage and the current are in phase, that is, the phase difference between the voltage and the current is zero.

Capacitance

If a current, $i(t) = I_x \cos(\omega t)$ passes through a capacitance C, the voltage developed across its terminals is (see also eqn 1.24)

$$v_C(t) = \frac{1}{C} \int_0^t i(t)\, dt + V_{co} \tag{3.16}$$

Note that since we are assuming steady-state conditions in the AC analysis we may set the initial condition $V_{co} = 0$, that is

$$v_C(t) = \frac{1}{C} \int_0^t I_x \cos(\omega t)\, dt \tag{3.17}$$

Performing the integration we obtain:

$$\begin{aligned} v_C(t) &= \frac{I_x}{\omega C} \sin(\omega t) \\ &= \frac{I_x}{\omega C} \cos\left(\omega t - \frac{\pi}{2}\right) \\ &= V_c \cos\left(\omega t - \frac{\pi}{2}\right) \end{aligned} \tag{3.18}$$

where
$$V_c = \frac{I_x}{\omega C} \tag{3.19}$$

In terms of RMS magnitudes we have:

$$\begin{aligned} V_{c_{eff}} &= \frac{I_{x_{eff}}}{\omega C} \\ &= X_C I_{x_{eff}} \end{aligned} \tag{3.20}$$

where $I_{x_{eff}} = I_x / \sqrt{2}$. The quantity $X_C = (\omega C)^{-1}$ is called the **capacitive reactance** and is measured in ohms. It is important to note that the amplitude of $v_C(t)$ is inversely proportional to the capacitance and the angular frequency of the AC current. From eqn 3.18 and figure 3.6 we observe that the voltage waveform lags the current waveform by $\pi/2$ radians or 90 degrees.

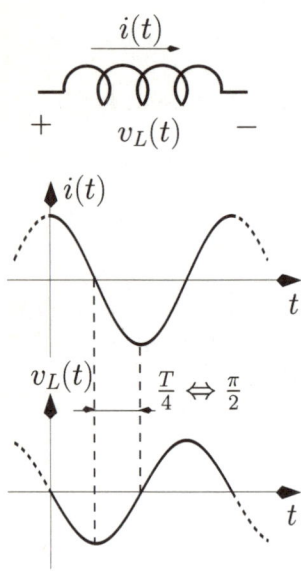

Figure 3.7: *Voltage and current in an inductor.*

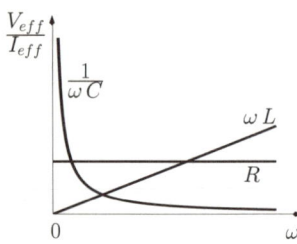

Figure 3.8: V_{eff}/I_{eff} versus ω for passive elements.

Inductance

When a current, $i(t) = I_x \cos(\omega t)$ passes through an inductance L, the voltage developed across its terminals is, according to eqn 1.26, given by:

$$
\begin{aligned}
v_L(t) &= L \frac{i(t)}{dt} \\
&= -L\omega I_x \sin(\omega t)\, dt \\
&= L\omega I_x \cos\left(\omega t + \frac{\pi}{2}\right) \\
&= V_l \cos\left(\omega t + \frac{\pi}{2}\right)
\end{aligned} \quad (3.21)
$$

with $V_l = L\omega I_x$. In terms of RMS values we have:

$$
\begin{aligned}
V_{l_{eff}} &= I_{x_{eff}} \omega L \\
&= X_L I_{x_{eff}}
\end{aligned} \quad (3.22)
$$

where $I_{x_{eff}} = I_x/\sqrt{2}$. The quantity $X_L = \omega L$ is called the **inductive reactance** which is also measured in Ohms. Note that now the amplitude of the voltage $v_L(t)$ is proportional to the inductance and the angular frequency of the AC current. From eqn 3.21 and figure 3.7 we observe that the voltage waveform leads the current waveform by $\pi/2$ radians or 90 degrees.

Figure 3.8 illustrates the ratio V_{eff}/I_{eff} versus the frequency, ω, for the three passive elements discussed above. It is interesting to note that at DC ($\omega = 0$) the capacitor behaves as an open-circuit and the inductor behaves as a short-circuit. On the other hand, for very high frequencies ($\omega \to \infty$) the capacitor behaves as a short-circuit and the inductor behaves as an-open circuit.

A note about voltage polarity and current direction in AC circuits

Although voltages and currents in AC circuits continuously change polarity and direction it is important to set references for these two quantities. The convention we follow in this book is illustrated above. When the current flows from the positive to the negative terminal of a circuit element it is implied that the current and voltage are in phase for a resistor as in figure 3.5; the current leads the voltage by 90 degrees for a capacitor as in figure 3.6 and lags by the same amount for an inductor as in figure 3.7.

Kirchhoff's laws

Kirchhoff's laws presented in Chapter 1 (see section 1.4) can be applied to determine the voltage across or the current through any circuit element. However, we must bear in mind that the voltages and the currents in AC circuits will, in general, exhibit phase differences when capacitors or inductors are present.

3. Frequency domain electrical signal and circuit analysis

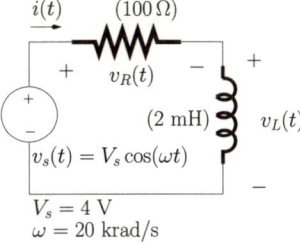

Figure 3.9: *RL circuit*.

Example 3.2.2 Determine the amplitude of the current $i(t)$ in the RL circuit of figure 3.9. Also, determine the phase difference between this current and the voltage source.

Solution: Since the circuit contains an inductor we expect that the current will exhibit a phase difference, ϕ, with respect to the source voltage. Hence, the current $i(t)$ can be expressed as follows:

$$i(t) = I_s \cos(\omega t + \phi) \tag{3.23}$$

This current flows through the resistance inducing a voltage difference at its terminals which is in phase with $i(t)$:

$$\begin{aligned} v_R(t) &= R\, i(t) \\ &= R\, I_s \cos(\omega t + \phi) \end{aligned} \tag{3.24}$$

On the other hand, the flow of $i(t)$ through the inductor causes a voltage difference across its terminals which is in quadrature with $i(t)$, as expressed by eqn 3.21:

$$v_L(t) = X_L I_s \cos\left(\omega t + \phi + \frac{\pi}{2}\right) \tag{3.25}$$

with $X_L = \omega L$. According to Kirchhoff's voltage law we can write:

$$\begin{aligned} v_s(t) &= v_R(t) + v_L(t) \\ &= R I_s \cos(\omega t + \phi) + X_L I_s \cos\left(\omega t + \phi + \frac{\pi}{2}\right) \\ &= R I_s \cos(\omega t + \phi) + X_L I_s \cos(\omega t + \phi) \cos\left(\frac{\pi}{2}\right) \\ &\quad - X_L I_s \sin(\omega t + \phi) \sin\left(\frac{\pi}{2}\right) \\ &= R I_s \cos(\omega t + \phi) - X_L I_s \sin(\omega t + \phi) \end{aligned} \tag{3.26}$$

The last eqn can be written as follows (see also appendix A):

$$V_s \cos(\omega t) = \sqrt{R^2 + X_L^2}\, I_s \cos(\omega t + \phi + \psi) \tag{3.27}$$

where

$$\psi = \tan^{-1}\left(\frac{X_L}{R}\right) \tag{3.28}$$

In order for eqn 3.27 to be an equality the amplitude and the phase of the cosine functions on both sides of this eqn must be equal. That is:

$$\begin{cases} V_s = \sqrt{R^2 + X_L^2}\, I_s \\ \omega t = \omega t + \phi + \psi \end{cases} \tag{3.29}$$

Solving the last set of eqns in order to obtain I_s and ϕ we have:

$$I_s = \frac{V_s}{\sqrt{R^2 + \omega^2 L^2}} \tag{3.30}$$
$$= 37 \text{ mA}$$
$$\phi = -\psi \tag{3.31}$$
$$= -0.38 \text{ rad } (-21.8°)$$

3.2.3 Phasor analysis

In principle any AC circuit can be analysed by applying Kirchhoff's laws with the trigonometric rules, as in the example 3.2.2 above. However, the application of these trigonometric rules to analyse complex AC circuits can be a cumbersome task. Fortunately, the use of the complex exponential (the phasor) and complex algebra, discussed in the previous chapter, provides a considerable simplification of AC circuit analysis.

From Euler's formula (see also section 2.5) a cosine alternating voltage waveform can be represented using the complex exponential function as follows:

$$V_s \cos(\omega t + \phi) = V_s \frac{e^{j(\omega t + \phi)} + e^{-j(\omega t + \phi)}}{2} \tag{3.32}$$

where we can see that the voltage expressed by eqn 3.32 is the addition of two complex conjugated exponential functions (phasors). Note that either of these two complex exponential functions carries all the phase information, ωt and ϕ, of the voltage waveform. In fact, the simplicity of analysis using phasors arises from each AC voltage and current being mathematically represented and manipulated as a **single** complex exponential function. However, in order to obtain the corresponding time domain waveform we must take the real part of the complex exponential waveform. Thus, the voltage waveform of eqn 3.32 can be expressed as:

$$V_s \cos(\omega t + \phi) = \text{Real}\left[V_s e^{j(\omega t + \phi)}\right] \tag{3.33}$$

In order to illustrate that phasor analysis is similar to AC analysis using trigonometric rules we reconsider the current–voltage relationships for the passive elements using the complex exponential representation. We determine the voltage developed across each element when an AC current, $i(t)$, flows through them, $i(t)$ being expressed by its complex exponential representation, $I(j\omega, t)$, as follows:

$$i(t) = \text{Real}\left[I(j\omega, t)\right] \tag{3.34}$$
$$I(j\omega, t) = I_x e^{j\omega t} \tag{3.35}$$

Resistance

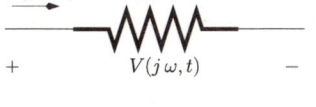

Figure 3.10: *Complex V–I relationship for a resistance.*

The complex voltage (see also figure 3.10) across the resistance terminals is determined by applying Ohm's law to the phasors representing the voltage across and the current flowing through the resistance, that is:

$$V_R(j\omega, t) = RI(j\omega, t)$$
$$= RI_x e^{j\omega t} \quad (3.36)$$

Taking the real part of $V_R(j\omega, t)$ we obtain the corresponding voltage waveform;

$$v_R(t) = RI_x \cos(\omega t) \quad (3.37)$$

This eqn is the same as eqn 3.13.

Capacitance

Assuming a complex representation for the current flowing through a capacitor, $I(j\omega, t)$, the complex voltage across the capacitance is given by:

$$V_C(j\omega, t) = \frac{1}{C} \int_0^t I(j\omega, t)\, dt \quad (3.38)$$
$$= \frac{1}{j\omega C} I_x e^{j\omega t} \quad (3.39)$$
$$= \frac{1}{j\omega C} I(j\omega, t) \quad (3.40)$$

The quantity $(j\omega C)^{-1}$ is called the capacitive (complex) *impedance*. This impedance can be seen as[2] $(-j)$ times the capacitive reactance $X_C = (\omega C)^{-1}$ discussed in section 3.2.2. Note that $(-j)$ accounts for the $-90°$ phase difference between the voltage and the current.

Taking the real part of $V_C(j\omega, t)$ we obtain the corresponding voltage waveform at the capacitor terminals;

$$v_C(t) = \text{Real}\left[\frac{1}{j\omega C} I_x e^{j\omega t}\, dt\right]$$
$$= \text{Real}\left[\frac{1}{\omega C} I_x e^{j(\omega t - \pi/2)}\, dt\right]$$

where we used the following equalities (see also section 2.5):

$$-j = e^{-j\pi/2} \quad (3.41)$$

Figure 3.11: *Complex V–I relationship for a capacitance.*

Now $v_C(t)$ can be written as

$$v_C(t) = \frac{I_x}{\omega C} \cos\left(\omega t - \frac{\pi}{2}\right) \quad (3.42)$$

Note that eqn 3.42 is the same as eqn 3.18.

---
[2] Recall that $j^{-1} = -j$.

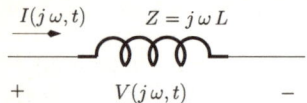

Figure 3.12: *Complex V–I relationship for an inductance.*

Inductance

Assuming a complex representation for the current flowing through the inductor, $I(j\omega, t)$, the complex voltage across the inductance is given by:

$$\begin{aligned}
V_L(j\omega, t) &= L\frac{dI(j\omega, t)}{dt} \\
&= j\omega L\, I_x e^{j\omega t} \quad (3.43)\\
&= j\omega L\, I(j\omega, t) \quad (3.44)
\end{aligned}$$

The quantity $Z = j\omega L$ is called the inductive (complex) *impedance*. This impedance can be seen as j times the inductive reactance $X_L = \omega L$ discussed in section 3.2.2. Note that now j accounts for the 90° phase difference between the voltage and the current. Taking the real part of $V_L(j\omega, t)$ we obtain

$$\begin{aligned}
v_C(t) &= \text{Real}\left[j\omega L\, I_x e^{j\omega t}\right] \\
&= \text{Real}\left[\omega L\, I_x e^{j(\omega t + \pi/2)}\right] \\
&= I_x\,\omega L\,\cos\left(\omega t + \frac{\pi}{2}\right) \quad (3.45)
\end{aligned}$$

We note again that eqn 3.45 is the same as eqn 3.21.

3.2.4 The generalised impedance

The greatest advantage of using phasors in AC circuit analysis is that they allow for an Ohm's law type of relationship between the phasors describing the voltage and the current for each passive element:

$$\frac{V(j\omega, t)}{I(j\omega, t)} = Z \quad (3.46)$$

Where Z is called the *generalised impedance*:

- $Z = R$ for a resistance
- $Z = (j\omega C)^{-1}$ for a capacitance
- $Z = j\omega L$ for an inductance

The generalised impedance concept is of great importance since it permits an extrapolation of the DC circuit analysis techniques discussed in Chapter 1 to the analysis of AC circuits. This means, for example, that we can apply the Nodal analysis technique to analyse AC circuits as illustrated by the next example. Figure 3.13 shows the symbol used to represent a general impedance.

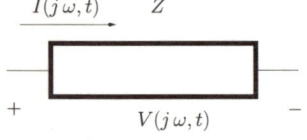

Figure 3.13: *Symbol of the general impedance.*

Example 3.2.3 Using the phasor analysis described above, determine the amplitude and phase of the current in the circuit of figure 3.9 and show that the results are the same as those obtained in example 3.2.2.

3. Frequency domain electrical signal and circuit analysis

Solution: The phasor describing the current can be written as follows:

$$I(j\omega, t) = I_s\, e^{j(\omega t + \phi)} \tag{3.47}$$

Applying Kirchhoff's voltage law we can write:

$$V_s e^{j\omega t} = R I_s\, e^{j(\omega t + \phi)} + j\omega L I_s\, e^{j(\omega t + \phi)} \tag{3.48}$$

or

$$V_s e^{j\omega t} = (R + j\omega L)\, I_s\, e^{j(\omega t + \phi)} \tag{3.49}$$

The impedance $R + j\omega L$ can be expressed in the exponential form (see also section 2.5) as follows

$$R + j\omega L = \sqrt{R^2 + \omega^2 L^2}\, e^{j\tan^{-1}\left(\frac{\omega L}{R}\right)} \tag{3.50}$$

Hence eqn 3.49 can be written as:

$$V_s e^{j\omega t} = \sqrt{R^2 + \omega^2 L^2}\, I_s\, e^{j(\omega t + \phi + \tan^{-1}\left(\frac{\omega L}{R}\right))} \tag{3.51}$$

In order for eqn 3.51 to be an equality the amplitude and the phase of the complex voltages on both sides of this eqn must be equal. That is:

$$\begin{cases} V_s = \sqrt{R^2 + \omega^2 L^2}\, I_s \\ \omega t = \omega t + \phi + \psi \end{cases} \tag{3.52}$$

Solving, we have:

$$I_s = \frac{V_s}{\sqrt{R^2 + \omega^2 L^2}} \tag{3.53}$$
$$= 37 \text{ mA}$$

$$\phi = -\psi \tag{3.54}$$
$$= -0.38 \text{ rad } (-21.8°)$$

Note that these values are equal to those obtained in example 3.2.2.

The rotating and the stationary phasor

The concept of the rotating phasor arises from the time dependence of the complex exponential which characterises AC voltages and currents. Let us consider the phasor representation for an AC voltage as shown below

$$V(j\omega, t) = V_s\, e^{j(\omega t + \phi)} \tag{3.55}$$

This rotating phasor can be represented in the Argand diagram, as illustrated in figure 3.14 a). Note that each instantaneous value for the rotating phasor, (that is, its position in the Argand diagram) is located on a circle whose radius is given by the voltage amplitude, V_s, with an angle $\omega t + \phi$ at each instant of time. Each position in this circle is reached by the phasor every $2\pi/\omega$ seconds.

The rotating phasor described by eqn 3.55 can be decomposed into the product of a stationary (or static) phasor with a rotating phasor as expressed by the eqn below:

$$V(j\omega, t) = \underbrace{V_s\, e^{j\phi}}_{\text{Static phasor}} \times \underbrace{e^{j\omega t}}_{\text{Rotating phasor}} \qquad (3.56)$$

$$= V_S \times e^{j\omega t} \qquad (3.57)$$

where V_S represents the static phasor. In the rest of this chapter, and unless stated otherwise, static phasors are represented by capital letters with capital sub-scripts.

In AC circuits where currents and voltages feature the same single tone or angular frequency, ω, both sides of the eqns describing the voltage and current relationships contain the complex exponential describing the rotating phasor, $\exp(j\omega t)$, as illustrated by eqns 3.48, 3.49, and 3.51 of example 3.2.3. Thus, the phasor analysis of an AC circuit can be further simplified if we apply Ohm's law and the concept of the generalised impedance to only the static phasor to represent AC voltages and currents. Note that this mathematical manipulation is reasonable since, in AC circuits, what is important is to determine the *amplitude* and the *relative phase difference* between the AC quantities, both described by the *static phasor*. In the rest of this chapter a phasor will mean a static phasor.

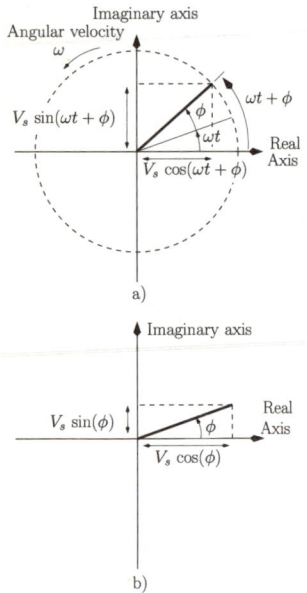

Figure 3.14: *The complex phasor represented in the Argand diagram.*
a) Instantaneous value of the rotating phasor.
b) The stationary phasor.

Example 3.2.4 Determine the amplitude and phase of the current in the circuit of figure 3.9 using the *static* phasor concept described above and show that the results are the same as those obtained in example 3.2.2.

Solution: The static phasor describing the current can be written as follows:

$$I_S = I_s\, e^{j\phi} \qquad (3.58)$$

while the static phasor describing the source voltage can be written as:

$$V_S = V_s\, e^{j0}$$
$$= V_s \qquad (3.59)$$

Applying Kirchhoff's voltage law we can write:

$$V_S = (R + j\omega L)\, I_S \qquad (3.60)$$
$$= \sqrt{R^2 + \omega^2 L^2}\, e^{j\tan^{-1}\left(\frac{\omega L}{R}\right)}\, I_S \qquad (3.61)$$

that is

$$I_S = \frac{V_S}{\sqrt{R^2 + \omega^2 L^2}}\, e^{-j\tan^{-1}\left(\frac{\omega L}{R}\right)}$$
$$= \frac{V_s}{\sqrt{R^2 + \omega^2 L^2}}\, e^{-j\tan^{-1}\left(\frac{\omega L}{R}\right)}$$
$$= 37.0 \times 10^{-3}\, e^{-j\,0.38} \text{ A}$$

3. Frequency domain electrical signal and circuit analysis 59

Note that this result is equivalent to those obtained in examples 3.2.2 and 3.2.3.

Series and parallel connection of complex impedances

As mentioned previously the concept of the generalised impedance greatly simplifies the analysis of AC circuits. It is also important to note that the series of various impedances Z_k, $k = 1, 2, \ldots N$, can be characterised by an equivalent impedance, Z_{eq}, which is the sum of these impedances:

$$Z_{eq} = \sum_{k=1}^{N} Z_k \qquad (3.62)$$

For example, in the circuit of figure 3.9 we observe that the impedance of the resistance is in a series connection with the impedance of the inductor. Hence, an equivalent impedance for this connection can be obtained adding them:

$$Z_{eq} = R + j\omega L \qquad (3.63)$$

The real part of an impedance is called the resistance while the imaginary part of the impedance is called the reactance.

For a parallel connection of various electrical elements it is sometimes easier to work with the inverse of the complex impedance, the 'admittance', Y;

$$Y = \frac{1}{Z} \qquad (3.64)$$

The parallel connection of admittances Y_k, $k = 1, 2, \ldots N$, can be characterised by an equivalent admittance, Y_{eq}, which is equal to their sum:

$$Y_{eq} = \sum_{k=1}^{N} Y_k \qquad (3.65)$$

It follows that the parallel connection of two impedances Z_1 and Z_2 can be represented by an equivalent impedance Z_{eq} given by

$$Z_{eq} = \frac{Z_1 Z_2}{Z_1 + Z_2} \qquad (3.66)$$

Example 3.2.5 Consider the AC circuit represented in figure 3.15 a). Determine the amplitude and the phase of the voltage across the resistance R_2. Then, determine the average power dissipated in R_2.

Solution: $v_{s1}(t)$ and $i_{s2}(t)$ can be expressed in their phasor representations as follows:

$$v_{s1}(t) = \text{Real}\left[V_{S1}\, e^{j\omega t}\right]$$
$$V_{S1} = V_{s1}\, e^{j\frac{\pi}{4}}$$
$$i_{s2}(t) = \text{Real}\left[I_{S2}\, e^{j\omega t}\right]$$
$$I_{S2} = I_{s2}\, e^{-j\frac{\pi}{2}}$$

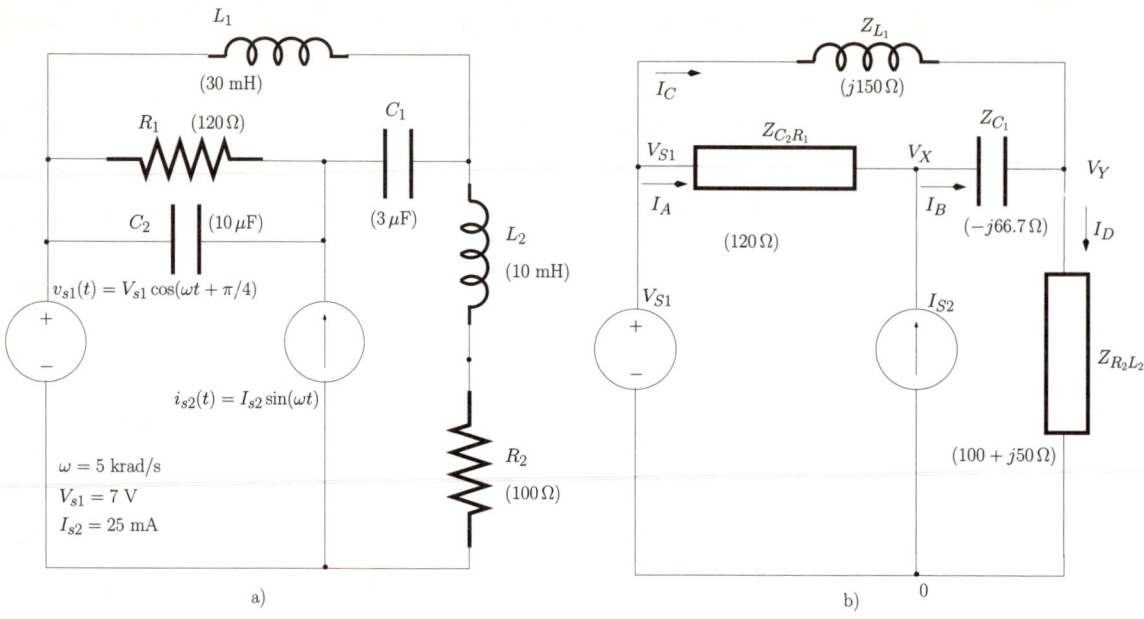

Figure 3.15: a) AC circuit. b) Equivalent circuit represented as complex impedances.

where we have used the following equality: $\sin(\omega t) = \cos(\omega t - \pi/2)$. The impedances associated with the two inductances and two capacitances are calculated as follows:

$$\begin{aligned}
Z_{L_1} &= j\omega L_1 \Big|_{\omega=5\times 10^3 \text{ rad/s}} \\
&= j\,150\ \Omega \\
Z_{L_2} &= j\omega L_2 \Big|_{\omega=5\times 10^3 \text{ rad/s}} \\
&= j\,50\ \Omega \\
Z_{C_1} &= \frac{1}{j\omega C_1}\Big|_{\omega=5\times 10^3 \text{ rad/s}} \\
&= -j\,66.7\ \Omega \\
Z_{C_2} &= \frac{1}{j\omega C_2}\Big|_{\omega=5\times 10^3 \text{ rad/s}} \\
&= -j\,20\ \Omega
\end{aligned}$$

From figure 3.15 a) we observe that the impedance associated with the capacitance C_2 is in a parallel connection with the resistance R_1. We can determine an equivalent impedance for this parallel connection as follows (see eqn 3.66):

$$\begin{aligned}
Z_{C_2 R_1} &= \frac{Z_{C_2}\,R_1}{Z_{C_2} + R_1} \\
&= 3.2 - j\,19.5\ \Omega
\end{aligned}$$

Also, we can see that R_2 is in a series connection with the inductance L_2. The equivalent impedance for this connection can be calculated as shown below:

$$\begin{aligned} Z_{R_2L_2} &= R_2 + Z_{L_2} \\ &= 100 + j\,50\ \Omega \end{aligned}$$

Figure 3.15 b) shows the reduced AC circuit with the various impedances associated with the inductances and capacitances as well as the phasor currents and phasor voltages at each node referenced to node 0. Applying Kirchhoff's current law we can write:

$$\begin{aligned} I_A + I_{S2} &= I_B \\ I_C + I_B &= I_D \end{aligned}$$

These can be rewritten after applying Ohm's law to the various impedances as shown below:

$$\begin{aligned} \frac{V_{S1} - V_X}{Z_{C_2R_1}} + I_{S2} &= \frac{V_X - V_Y}{Z_{C_1}} \\ \frac{V_{S1} - V_Y}{Z_{L_1}} + \frac{V_X - V_Y}{Z_{C_1}} &= \frac{V_Y}{Z_{R_2L_2}} \end{aligned}$$

Solving in order to obtain V_Y, we have:

$$V_Y = Z_{R_2L_2} \frac{V_{S1}(Z_{L_1} + Z_{C_1} + Z_{C_2R_1}) + Z_{L_1}I_{S2}Z_{C_2R_1}}{Z_{R_2L_2}(Z_{L_1} + Z_{C_1} + Z_{C_2R_1}) + Z_{L_1}(Z_{C_2R_1} + Z_{C_1})}$$

Substituting complex values in the last eqn we obtain:

$$V_Y = 3.5\,e^{j\,2.3}\ \text{V}$$

The current that flows through R_2 is I_D given by:

$$\begin{aligned} I_D &= \frac{V_Y}{Z_{R_2L_2}} \\ &= 32 \times 10^{-3}\,e^{j\,1.80}\ \text{A} \end{aligned}$$

and the voltage across the resistance R_2 is given by:

$$\begin{aligned} V_{R_2} &= R_2 I_D \\ &= 3.2\,e^{j\,1.80}\ \text{V} \end{aligned}$$

that is, the AC voltage across the resistance R_2 has a peak amplitude of 3.2 V. The phase of this voltage is 1.80 rad (103°).

The average power dissipated by R_2 can be calculated according to eqn 1.20 (see also section 1.3):

$$P_{AV_{R_2}} = \frac{1}{R_2 T} \int_{t_o}^{t_o+T} v_{R_2}^2(t)\,dt \qquad (3.67)$$

with $T = 2\pi/\omega = 1.3 \times 10^{-3} = 1.3$ ms. t_o is chosen to be zero. $v_{R_2}(t)$ can be obtained from its phasor value as follows:

$$\begin{aligned} v_{R_2}(t) &= \text{Real}\left[V_{R_2}\, e^{j\omega t}\right] \\ &= \text{Real}\left[3.2\, e^{j1.80}\, e^{j\omega t}\right] \\ &= 3.2\cos(\omega t + 1.80)\text{ V} \end{aligned}$$

The average power dissipated by R_2 can be calculated as shown below:

$$P_{AV_{R2}} = \frac{3.2^2}{T\, R_2} \int_0^T \cos^2(\omega t + 1.80)\, dt$$

It is left to the reader to show that the $P_{AV_{R2}}$ is equal to:

$$\begin{aligned} P_{AV_{R2}} &= \frac{3.2^2}{2}\frac{1}{R_2} \\ &= 0.05\text{ W} \end{aligned}$$

It is important to note that the average power dissipated in the resistance can also be calculated directly from the phasor representation of the current flowing through and the voltage across R_2 as follows (see problem 3.2):

$$\begin{aligned} P_{AV_{R2}} &= \frac{1}{2}\text{Real}\left[V_{R_2}\, I_{R_2}^*\right] = \frac{1}{2}\text{Real}\left[V_{R_2}^*\, I_{R_2}\right] & (3.68) \\ &= \frac{1}{2}\frac{|V_{R_2}|^2}{R_2} & (3.69) \\ &= \frac{1}{2}|I_{R_2}|^2\, R_2 & (3.70) \\ &= 0.05\text{ W} \end{aligned}$$

where the current flowing through R_2 is $I_{R_2} = I_D$.

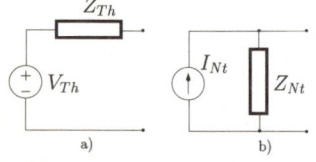

Figure 3.16: *a) Thévenin equivalent AC circuit. b) Norton equivalent AC circuit.*

Thévenin and Norton theorems

Thévenin and Norton equivalent AC circuits can be obtained in a way similar to that described for DC resistive circuits. The main difference is that now the Thévenin equivalent AC circuit comprises an ideal AC voltage source in series with a complex impedance as shown in figure 3.16 a). The Norton equivalent AC circuit is constituted by an ideal AC current source in parallel with a complex impedance as illustrated in figure 3.16 b).

Example 3.2.6 Consider the AC circuit represented in figure 3.17 a). Determine the Thévenin equivalent AC circuit at the terminals X and Y.

<u>Solution</u>: Figure 3.17 b) shows the equivalent circuit for the calculation of the open-circuit voltage between terminals X and Y. Firstly, the impedances for

3. Frequency domain electrical signal and circuit analysis

the capacitance and inductance are calculated for $\omega = 10^4$ rad/s, as shown below:

$$Z_L = j\omega L \Big|_{\omega=10^4 \text{ rad/s}}$$
$$= j\,400\ \Omega$$
$$Z_C = \frac{1}{j\omega C}\Big|_{\omega=10^4 \text{ rad/s}}$$
$$= -j\,1000\ \Omega$$

The phasor associated with the voltage $v_s(t)$ is $V_S = 3\,e^{-j\pi/5}$ V.

Note that the impedance associated with the capacitance is in a parallel connection with the resistance. Hence, we can replace these two impedances by an equivalent impedance given by:

$$Z_{RC} = \frac{Z_C\,R}{Z_C + R} \qquad (3.71)$$
$$= 400 - j\,800\ \Omega$$

The voltage between terminals X and Y can be obtained from the voltage

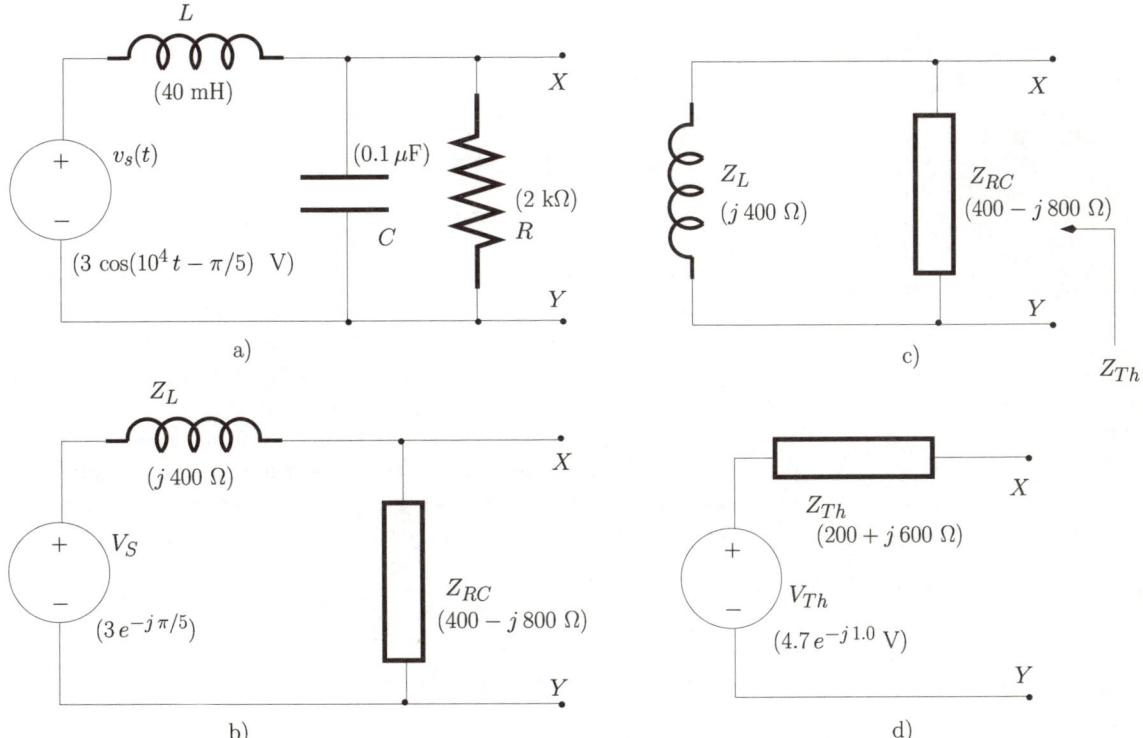

Figure 3.17: *a) AC circuit. b) Calculation of the Thévenin voltage. c) Calculation of the Thévenin Impedance. d) Equivalent Thévenin circuit.*

impedance divider (see also section 1.4.4) formed by the impedances Z_{RC} and Z_L as follows:

$$\begin{aligned} V_{Th} &= V_S \frac{Z_{RC}}{Z_{RC}+Z_L} \\ &= 4.7\, e^{-j\,1.0} \text{ V} \end{aligned}$$

Figure 3.17 c) shows the equivalent circuit for the calculation of the Thévenin impedance, where the AC voltage source has been replaced by a short-circuit. From this figure it is clear that the impedance Z_L is in a parallel connection with Z_{RC}. Hence Z_{Th} can be calculated as follows:

$$\begin{aligned} Z_{Th} &= \frac{Z_{RC}\,Z_L}{Z_{RC}+Z_L} \\ &= 200+j\,600 \; \Omega \end{aligned}$$

Figure 3.17 d) shows the Thévenin equivalent circuit for the circuit of 3.17 a). The Thévenin voltage $v_{Th}(t)$ can be determined from its phasor, V_{Th}, as follows:

$$\begin{aligned} v_{Th}(t) &= \text{Real}\left[V_S\, e^{j\,\omega\, t}\right]_{\omega=10^4 \text{ rad/s}} \\ &= 4.7\,\cos(10^4\,t-1.0) \text{ V} \end{aligned}$$

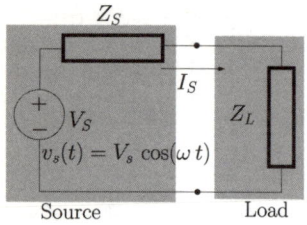

Figure 3.18: *Circuit model to derive maximum power transfer.*

3.2.5 Maximum power transfer

Whenever an AC signal is processed by an electrical network containing at least one resistance there is loss of power in the resistances. Since it is often important to ensure that this loss is minimal we consider the conditions which ensure maximum power transfer from two adjacent parts of a circuit. For this purpose we consider the circuit shown in figure 3.18 where the section of the circuit providing the power is modelled as an AC voltage source with an output impedance Z_S and the section where the power is transmitted is modelled as an impedance Z_L. We assume that the source impedance Z_S has a resistive part given by R_S and a reactive part described by $j\,X_S$. Similarly, the load impedance has a resistive component, R_L and a reactive component given by $j\,X_L$. The current I_S supplied by the source is given by

$$I_S = \frac{V_L}{Z_L+Z_S} \tag{3.72}$$

and the average power dissipated in the load, P_L, is given by (see eqn 3.70):

$$\begin{aligned} P_L &= \frac{|I_S|^2}{2}\,R_L \\ &= \frac{V_s^2}{2}\,\frac{R_L}{(R_S+R_L)^2+(X_S+X_L)^2} \end{aligned} \tag{3.73}$$

From the last eqn we observe that the value of X_L which maximises the average power in the load is such that it minimises the denominator, that is:

$$X_L = -X_S \tag{3.74}$$

Under this condition the average power in the load is given by

$$P_L = \frac{V_s^2}{2} \frac{R_L}{(R_S + R_L)^2} \tag{3.75}$$

In order to find the value of R_L which maximises the power in the load we calculate dP_L/dR_L and then we determine the value of R_L for which dP_L/dR_L is zero;

$$\frac{dP_L}{dR_L} = \frac{V_s^2}{2} \frac{(R_S + R_L) - 2R_L}{(R_S + R_L)^3} \tag{3.76}$$

Clearly, the value for R_L which sets $dP_L/dR_L = 0$ is

$$R_L = R_S \tag{3.77}$$

Hence, the maximum average power delivered to the load is;

$$P_{L\,max} = \frac{V_s^2}{8 R_L} \tag{3.78}$$

It is clear that maximum power transfer occurs when $Z_L = Z_S^*$.

3.3 Generalised frequency domain analysis

The analysis presented in the previous sections can be considered as a particular case of frequency domain analysis of single frequency signals. As discussed previously, those single frequency signals can be expressed in terms of phasors which, in turn, give rise to phasor analysis. It was seen that phasor analysis allows the application of Ohm's law to the generalised impedance associated with any passive element considerably simplifying electrical circuit analysis.

The analysis of circuits where the signal sources can assume other time-varying (that is non-sinusoidal) waveforms can be a cumbersome task since this gives rise to differential-integral equations. Therefore, it would be most convenient to be able to apply phasor analysis to such circuits. This analysis can indeed be employed using the 'Fourier transform' which allows us to express almost any time varying voltage and current waveform as a 'sum' of phasors.

For reasons of simplicity, before we discuss the Fourier transform we present the Fourier series which can be seen as a special case of the Fourier transform.

The term 'signal' will be used to express either a voltage or a current waveform and we use the terms signal, waveform or function interchangeably to designate voltage or current quantities, which vary with time.

3.3.1 The Fourier series

The Fourier series is used to express periodic signals in terms of sums of sine and cosine waveforms or in terms of sums of phasors. A periodic signal, with

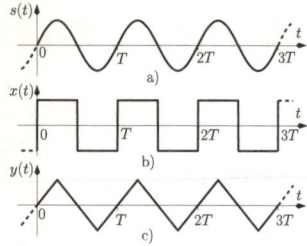

Figure 3.19: *Periodic waveforms. a) Sine. b) Rectangular. c) Triangular.*

period T, is by definition a signal which repeats its shape and amplitude every T seconds, that is:

$$x(t \pm kT) = x(t), \quad k = 1, 2, \ldots \quad (3.79)$$

Examples of periodic waveforms are presented in figure 3.19 where we have drawn a sine wave, a periodic rectangular waveform, and a periodic triangular waveform. From this figure it is clear that the waveforms repeat their shape and amplitude every T seconds.

In order to show how the Fourier series provides representations of periodic waves as sums of sine or cosine waves we present, in figure 3.20 a), the first two non-zero terms (sine waves) of the Fourier series for the periodic rectangular waveform of figure 3.19 b). Figure 3.20 b) shows that the sum of these two sine waves starts to resemble the rectangular waveform. It will be shown that the addition of all the terms (harmonics) of a particular series converges to the periodic rectangular waveform. In a similar way, figure 3.20 c) represents the first two non-zero terms of the Fourier series of the triangular waveform. Figure 3.20 d) shows that the sum of just these two sine waves produces a good approximation to the triangular waveform.

Since sine and cosine functions can be expressed as a sum of complex exponential functions (phasors), the Fourier series of a periodic waveform $x(t)$ with period T can be expressed as a weighted sum, as shown below:

$$x(t) = \sum_{n=-\infty}^{\infty} C_n \, e^{j \, 2\pi \frac{n}{T} t} \quad (3.80)$$

where the weights or Fourier coefficients, C_n, of the series can be determined as follows:

$$C_n = \frac{1}{T} \int_{t_o}^{t_o + T} x(t) \, e^{-j \, 2\pi \frac{n}{T} t} \, dt \quad (3.81)$$

Here t_o is a time instant which can be chosen to facilitate the calculation of these coefficients.

The existence of a convergent Fourier series of a periodic signal $x(t)$ requires only that the area of $x(t)$ per period to be finite and that $x(t)$ has a finite number of discontinuities and a finite number of maxima and minima per period. All periodic signals studied here and are to be found in any electrical system satisfy these requirements and, therefore, have a convergent Fourier series.

From eqn 3.80 we observe that the phasors which compose the periodic signal $x(t)$ have an angular frequency $2\pi n/T$ which, for $|n| > 1$ is a multiple, or harmonic, of the fundamental angular frequency $\omega = 2\pi/T$. Note that, for $n = 0$ the coefficient C_0 is given by:

$$C_0 = \frac{1}{T} \int_{t_o}^{t_o + T} x(t) \, dt \quad (3.82)$$

This eqn indicates that C_0 represents the average value of the waveform over its period T and represents the DC component of $x(t)$.

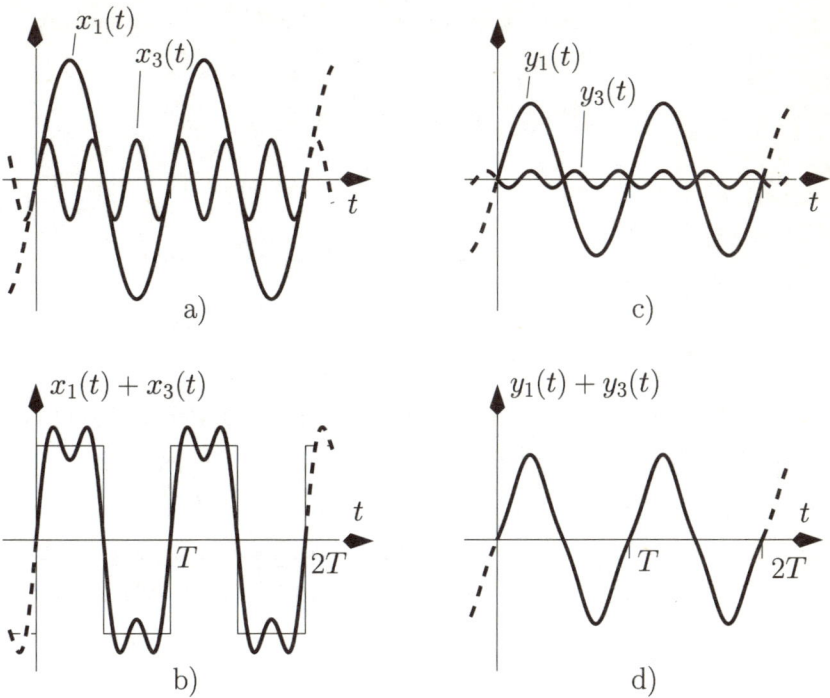

Figure 3.20: a) The first two non-zero terms of the Fourier series for the periodic rectangular waveform. b) The sum of first two non-zero terms of the Fourier series as an approximation to the periodic rectangular waveform. c) The first two non-zero terms of the Fourier series for the periodic triangular waveform. d) The sum of first two non-zero terms of the Fourier series as an approximation to the periodic triangular waveform.

As an example we determine the Fourier series of the periodic rectangular waveform shown in figure 3.19 b). Using eqn 3.81 with $t_o = 0$ we can write:

$$
\begin{aligned}
C_n &= \frac{1}{T}\int_0^T x(t)\,e^{-j\,2\,\pi\,\frac{n}{T}\,t}\,dt \\
&= \frac{1}{T}\left(\int_0^{T/2} A\,e^{-j\,2\,\pi\,\frac{n}{T}\,t}\,dt + \int_{T/2}^T (-A)\,e^{-j\,2\,\pi\,\frac{n}{T}\,t}\,dt\right) \quad (3.83)
\end{aligned}
$$

where A is the peak amplitude. The last eqn can be written as follows:

$$
\begin{aligned}
C_n &= \frac{1}{T}\left(\frac{A\,T}{-j\,2\,\pi\,n}\,e^{-j\,2\,\pi\,\frac{n}{T}\,t}\bigg|_0^{T/2} + \frac{-A\,T}{-j\,2\,\pi\,n}\,e^{-j\,2\,\pi\,\frac{n}{T}\,t}\bigg|_{T/2}^T\right) \\
&= \frac{A}{-j\,2\,\pi\,n}\left(e^{-j\,2\,\pi\,\frac{n}{T}\,\frac{T}{2}} - 1\right) + \frac{-A}{-j\,2\,\pi\,n}\left(e^{-j\,2\,\pi\,\frac{n}{T}\,T} - e^{-j\,2\,\pi\,\frac{n}{T}\,\frac{T}{2}}\right) \\
&= \frac{2\,A}{j\,2\,\pi\,n}\left(1 - e^{-j\,\pi\,n}\right) \quad (3.84)
\end{aligned}
$$

where we have used the following equality:

$$e^{-j\,2\,\pi\,n} = 1, \quad n = 0, \pm 1, \pm 2, \pm 3, \ldots \tag{3.85}$$

However, we note that:

$$e^{-j\,\pi\,n} = \begin{cases} -1 & \text{if } n = \pm 1, \pm 3, \pm 5, \ldots \\ 1 & \text{if } n = 0, \pm 2, \pm 4, \pm 6, \ldots \end{cases} \tag{3.86}$$

and, therefore, the coefficients given by eqn 3.84 can be written as follows:

$$C_n = \frac{A}{j\,\pi\,n} \times \begin{cases} 2 & \text{if } n = \pm 1, \pm 3, \pm 5, \ldots \\ 0 & \text{if } n = 0, \pm 2, \pm 4, \pm 6, \ldots \end{cases} \tag{3.87}$$

Note that for $n = 0$ the last eqn cannot be determined as the result would be a non-defined number; $0/0$. Hence, C_0 must be determined from eqn 3.82:

$$C_0 = \frac{1}{T}\int_0^{T/2} A\,dt - \frac{1}{T}\int_{T/2}^{T} A\,dt = 0 \tag{3.88}$$

confirming that the average value of $x(t)$ is zero as is clear from figure 3.19 b). From the above, eqn 3.87, can be written as follows:

$$C_n = \begin{cases} \frac{2\,A}{j\,\pi\,n} & \text{if } |n| \text{ is odd} \\ 0 & \text{if } |n| \text{ is even} \end{cases} \tag{3.89}$$

It is clear that all even harmonics of the Fourier series are zero. Also, we observe that $C_n = C^*_{-n}$, a fact that applies to any real (non-complex) periodic signal. The coefficients C_n can be written, in a general form, using the complex exponential form as follows:

$$C_n = |C_n|\,e^{j\,\angle(C_n)} \tag{3.90}$$

and eqn 3.80 can be written as follows:

$$x(t) = \sum_{n=-\infty}^{\infty} |C_n|\,e^{j\,2\,\pi\,\frac{n}{T}\,t + j\,\angle(C_n)} \tag{3.91}$$

$$= C_0 + \sum_{n=1}^{\infty} 2\,|C_n|\,\frac{e^{j\,2\,\pi\,\frac{n}{T}\,t + j\,\angle(C_n)} + e^{-j\,2\,\pi\,\frac{n}{T}\,t - j\,\angle(C_n)}}{2}$$

$$= C_0 + \sum_{n=1}^{\infty} 2\,|C_n|\,\cos\left(2\,\pi\,\frac{n}{T}\,t + \angle(C_n)\right) \tag{3.92}$$

Expressing the coefficients C_n of eqn 3.89 in a complex exponential form (see also eqn 3.90) we have:

$$C_n = \begin{cases} \frac{2\,A}{\pi\,n}\,e^{-j\,\frac{\pi}{2}} & \text{if } |n| \text{ is odd} \\ 0 & \text{if } |n| \text{ is even} \end{cases} \tag{3.93}$$

Hence, $x(t)$ can be written, using eqn 3.92, as shown below:

$$x(t) = \sum_{\substack{n=1 \\ (n\text{ odd})}}^{\infty} \frac{4\,A}{\pi n}\,\cos\left(2\,\pi\,\frac{n}{T}\,t - \frac{\pi}{2}\right) \tag{3.94}$$

3. Frequency domain electrical signal and circuit analysis

Figure 3.20 a) shows the first and the third harmonics of the rectangular signal as:

$$x_1(t) = \frac{4A}{\pi} \cos\left(2\pi \frac{1}{T} t - \frac{\pi}{2}\right) \tag{3.95}$$

$$x_3(t) = \frac{4A}{3\pi} \cos\left(2\pi \frac{3}{T} t - \frac{\pi}{2}\right) \tag{3.96}$$

from which figure 3.20 b) was derived.

Example 3.3.1 Determine the Fourier series of the periodic triangular waveform, $y(t)$, shown in figure 3.19 c).

Solution: From figure 3.19 b) we observe that the average value of this waveform is zero. Hence, $C_0 = 0$. Using eqn 3.81 with $t_o = 0$ we can write:

$$\begin{aligned} C_n &= \frac{1}{T} \int_0^T y(t) \, e^{-j 2\pi \frac{n}{T} t} \, dt \\ &= \frac{1}{T} \left(\int_0^{T/4} \frac{4At}{T} e^{-j 2\pi \frac{n}{T} t} \, dt \right. \\ &\quad + \int_{T/4}^{3T/4} \left(2A - \frac{4At}{T}\right) e^{-j 2\pi \frac{n}{T} t} \, dt \\ &\quad \left. + \int_{3T/4}^{T} \left(\frac{4At}{T} - 4A\right) e^{-j 2\pi \frac{n}{T} t} \, dt \right) \end{aligned} \tag{3.97}$$

with A representing the peak amplitude of the triangular waveform. Solving the integrals the coefficients can be written as follows:

$$C_n = \frac{A}{\pi^2 n^2} \left(2 e^{-j\pi \frac{n}{2}} - 1 - 2 e^{-j\pi \frac{3n}{2}} + e^{-j 2\pi n} \right) \tag{3.98}$$

Using the result of eqn 3.85 we express the coefficients C_n as follows:

$$C_n = \frac{2A}{\pi^2 n^2} e^{-j\pi \frac{n}{2}} \left(1 - e^{-j\pi n}\right) \tag{3.99}$$

and using the result of eqn 3.86 we can write these coefficients as:

$$C_n = \begin{cases} \frac{4A}{\pi^2 n^2} e^{-j\pi \frac{n}{2}} & \text{if } |n| \text{ is odd} \\ 0 & \text{if } |n| \text{ is even} \end{cases} \tag{3.100}$$

From eqn 3.92 the Fourier series for the triangular periodic waveform can be written as:

$$y(t) = \sum_{\substack{n=1 \\ (n \text{ odd})}}^{\infty} \frac{8A}{\pi^2 n^2} \cos\left(2\pi \frac{n}{T} t - \frac{\pi n}{2}\right) \tag{3.101}$$

Figure 3.20 c) shows the first and the third harmonics given by

$$y_1(t) = \frac{8A}{\pi^2} \cos\left(2\pi\frac{1}{T}t - \frac{\pi}{2}\right) \tag{3.102}$$

$$y_3(t) = \frac{8A}{3\pi^2} \cos\left(2\pi\frac{3}{T}t - \frac{3\pi}{2}\right) \tag{3.103}$$

Figure 3.20 d) clearly shows that the sum of these two harmonics, $y_1(t)+y_3(t)$, approximates the triangular periodic signal.

Normalised power

As discussed in section 1.3.1 the instantaneous power dissipated in a resistance R with a voltage $v(t)$ applied to its terminals is $v^2(t)/R$ while the instantaneous power dissipated caused by a current $i(t)$ is $i^2(t)R$. Since signals can be voltages or currents it is appropriate to define a normalised power by setting $R = 1\,\Omega$. Then, the instantaneous power associated with a signal $x(t)$ is equal to:

$$p(t) = x^2(t) \tag{3.104}$$

Thus, if $x(t)$ represents a voltage, the instantaneous power dissipated in a resistance R is obtained by dividing $p(t)$ by R while if $x(t)$ represents a current the instantaneous power dissipated in that resistance R is obtained by multiplying $p(t)$ by R. It is also relevant to define a normalised average power (once again, $R = 1\,\Omega$) by integrating eqn 3.104 as follows:

$$P_{AV} = \frac{1}{T}\int_{t_o}^{t_o+T} x^2(t)\,dt \tag{3.105}$$

Example 3.3.2 Determine an expression for the average power associated with the periodic rectangular waveform shown in figure 3.19 c).

Solution: The average power associated with the periodic rectangular waveform is the normalised average power ($R = 1\,\Omega$) which can be determined according to eqn 3.105, that is:

$$\begin{aligned} P_{AV} &= \frac{1}{T}\left(\int_0^{T/2} A^2\,dt + \int_{T/2}^T (-A)^2\,dt\right) \\ &= A^2 \text{ (Watts)} \end{aligned} \tag{3.106}$$

where A is the amplitude of the waveform.

3. Frequency domain electrical signal and circuit analysis

Parseval's power theorem

Parseval's theorem relates the average power associated with a periodic signal, $x(t)$, with its Fourier coefficients, C_n:

$$\frac{1}{T}\int_{t_o}^{t_o+T} x^2(t)\, dt = C_0^2 + \sum_{n=1}^{\infty} 2|C_n|^2 \qquad (3.107)$$

The proof of this theorem can be obtained as follows: The Fourier series indicates that $x(t)$ can be seen as a sum of a DC component with sinusoidal components as indicated by eqn 3.92. Hence, the average power associated with $x(t)$ can be seen as the addition of the average power associated with the DC component with the average power associated with each of these components. It is known that the average power associated with a DC signal is the square of the amplitude of that DC signal. Also, it is known that the average power associated with a sinusoidal component is equal to half the square of its peak amplitude. Since the amplitude of each Fourier component of $x(t)$ is equal to $2|C_n|$ then the average power associated with each of these AC components is equal to:

$$\begin{aligned} P_{C_n} &= \frac{(2|C_n|)^2}{2}, & n \geq 1 \\ &= 2|C_n|^2, & n \geq 1 \end{aligned} \qquad (3.108)$$

and the total average power of $x(t)$ is:

$$P_{AV_x} = C_0^2 + \sum_{n=1}^{\infty} 2|C_n|^2 \qquad (3.109)$$

Example 3.3.3 Show that the fundamental and the third harmonic of the Fourier series of the periodic rectangular waveform, shown in figure 3.19 c), contain approximately 90% of the power associated with this waveform.

Solution: According to eqn 3.106 the power associated with the rectangular periodic waveform with amplitude $\pm A$ is A^2 W. From eqn 3.108 the power associated with the fundamental component and the third harmonic of the Fourier series of the periodic rectangular waveform can be calculated as follows (see also eqn 3.89):

$$\begin{aligned} P &= 2\left(\frac{2A}{\pi}\right)^2 + 2\left(\frac{2A}{3\pi}\right)^2 \\ &= 0.9\, A^2 \text{ (W)} \end{aligned}$$

Time delay

If a periodic signal $x(t)$ has a Fourier series with coefficients C_n we can obtain the Fourier series coefficients, C_n', of a replica of $x(t)$ delayed by τ seconds, i.e. $x(t-\tau)$ with $|\tau| < T/2$, as follows:

$$C'_n = \int_{t_o}^{t_o+T} x(t-\tau) e^{-j 2\pi \frac{n}{T} t} dt \qquad (3.110)$$

using the change of variable $t' = t - \tau$, we can write:

$$dt = dt'$$
$$t = t_o \quad ; \quad t' = t_o - \tau$$
$$t = t_o + T \quad ; \quad t' = t_o - \tau + T$$

and eqn 3.110 can be written as

$$\begin{aligned}C'_n &= \int_{t'_o}^{t'_o+T} x(t') e^{-j 2\pi \frac{n}{T} t'} dt' \, e^{-j 2\pi \frac{n}{T} \tau} \\ &= C_n e^{-j 2\pi \frac{n}{T} \tau}\end{aligned} \qquad (3.111)$$

where $t'_o = t_o - \tau$. Note that the delay τ adds an extra linear phase to the Fourier series coefficients C_n.

3.3.2 Fourier coefficients, phasors and line spectra

Each phasor which composes the Fourier series of a periodic signal can be seen as the product of a static phasor with a rotating phasor as indicated below:

$$|C_n| e^{j 2\pi \frac{n}{T} t + j \angle(C_n)} = \underbrace{|C_n| e^{j \angle(C_n)}}_{\text{Static phasor}} \times \underbrace{e^{j 2\pi \frac{n}{T} t}}_{\text{Rotating phasor}} \qquad (3.112)$$

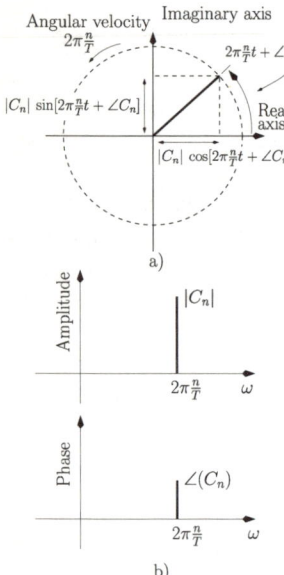

Figure 3.21: *a) Phasor. b) Line spectrum of a phasor.*

Comparing this eqn with eqn 3.56 we can identify each complex coefficient, C_n, as the static phasor corresponding to a rotating phasor with angular frequency $\omega = 2\pi n/T$. The phasor (static and rotating components), which is shown in figure 3.21 a) can be represented in the *frequency domain* by associating its amplitude, $|C_n|$, and its phase, $\angle(C_n)$, with its angular frequency $\omega = 2\pi n/T$ (or with its linear frequency $f = n/T$). This gives rise to the so called line-spectrum, as illustrated in figure 3.21 b). This frequency representation consists of two plots; amplitude versus frequency and phase versus frequency.

Since the Fourier series expresses periodic signals as a sum of phasors we are now in a position to represent the line spectrum of any periodic signal. As an example, the line spectrum of the periodic square wave with period T can be represented with C_n given by eqn 3.89. Figure 3.22 shows the line spectrum representing the fundamental component, the third and the fifth harmonics for this waveform. As mentioned previously, all the frequencies represented are integer multiples of the fundamental frequency $\omega = 2\pi/T$. Hence, the spectral lines have a uniform spacing of $2\pi/T$. It is also important to note that the line spectrum of figure 3.22 has positive and negative frequencies. Negative frequencies have no physical meaning and their appearance is a consequence of the mathematical representation of sine and of cosine functions by complex

3. Frequency domain electrical signal and circuit analysis

exponentials because these trigonometric functions (sine and cosine) are represented by the sum of a pair of complex conjugated phasors (see also eqns 2.59, 2.60 and 3.92). We also note that the line spectrum has been plotted as a function of the angular frequency $\omega = 2\pi f$. However, we frequently plot line spectra versus the linear frequency $f = \omega/(2\pi)$.

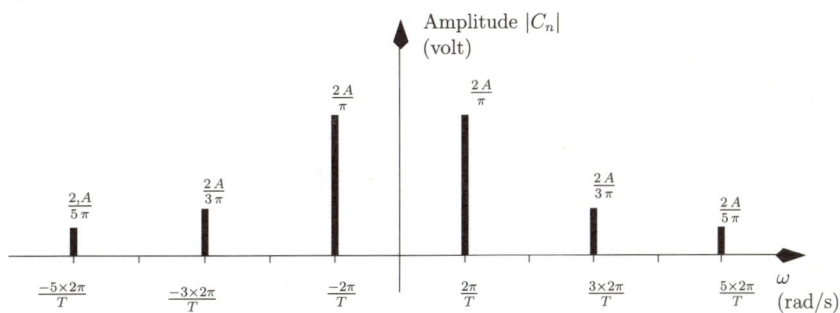

Figure 3.22: *Line spectrum of the rectangular waveform.*

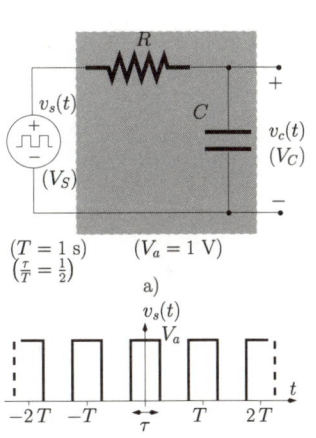

Figure 3.23: *Periodic voltage applied to an RC circuit. b) The periodic voltage $v(t)$.*

3.3.3 Electrical signal and circuit bandwidths

We discuss now the concepts of signal and electrical system bandwidths. In order to do so we consider the RC circuit of figure 3.23 which is driven by a square-wave voltage $v_s(t)$ as shown in figure 3.23 b). This voltage waveform can be expressed as:

$$v_s(t) = \sum_{k=-\infty}^{\infty} V_a \, \text{rect}\left(\frac{t - kT}{\tau}\right) \tag{3.113}$$

where V_a (V) is the amplitude and T is the period. τ/T is called the 'duty-cycle' of the waveform and is equal to $1/2$ in this case. The function $\text{rect}(t/\tau)$ is defined as follows:

$$\text{rect}\left(\frac{t}{\tau}\right) = \begin{cases} 1, & -\frac{1}{2} < \frac{t}{\tau} < \frac{1}{2} \\ 0, & \text{elsewhere} \end{cases} \tag{3.114}$$

The Fourier coefficients for $v_s(t)$, V_{S_n}, can be obtained from eqn 3.81 where t_o is chosen to be $-T/2$, that is:

$$\begin{aligned} V_{S_n} &= \frac{1}{T} \int_{-T/2}^{T/2} V_a \operatorname{rect}\left(\frac{t}{\tau}\right) e^{-j 2 \pi \frac{n}{T} t} \, dt \\ &= \frac{1}{T} \int_{-\tau/2}^{\tau/2} V_a \, e^{-j 2 \pi \frac{n}{T} t} \, dt \\ &= \frac{T V_a}{-j T 2 \pi n} \left[e^{-j 2 \pi \frac{n}{T} t} \right]_{-\tau/2}^{\tau/2} \\ &= \frac{V_a}{\pi n} \frac{e^{j \pi \frac{n}{T} \tau} - e^{-j \pi \frac{n}{T} \tau}}{2 j} \\ &= \frac{V_a}{\pi n} \sin\left(\pi \frac{n}{T} \tau\right) \end{aligned} \qquad (3.115)$$

The last eqn can be written as follows:

$$\begin{aligned} V_{S_n} &= \frac{V_a \tau}{T} \frac{\sin\left(\pi \frac{n}{T} \tau\right)}{\frac{\pi n \tau}{T}} & (3.116) \\ &= \frac{V_a \tau}{T} \operatorname{sinc}\left(\frac{n \tau}{T}\right) & (3.117) \end{aligned}$$

where the function $\operatorname{sinc}(x)$ is defined as follows:

$$\operatorname{sinc}(x) \triangleq \frac{\sin(\pi x)}{\pi x} \qquad (3.118)$$

Since $\tau/T = 1/2$, eqn 3.117 can be further simplified to:

$$V_{S_n} = \frac{V_a}{2} \operatorname{sinc}\left(\frac{n}{2}\right) \qquad (3.119)$$

It is left to the reader to show that the DC component of $v_s(t)$, V_{S_0}, is equal to $V_a \tau/T = V_a/2$.

The voltage signal $v_s(t)$ can be written as follows:

$$v_s(t) = \sum_{n=-\infty}^{\infty} \frac{V_a}{2} \operatorname{sinc}\left(\frac{n}{2}\right) e^{j 2 \pi \frac{n}{T} t} \qquad (3.120)$$

Once again, it is left to the reader to show that the periodic square waveform of figure 3.19 b) can be seen as a particular case of the rectangular waveform of figure 3.23 b) when $\tau/T = 1/2$. Hint, assume that the average (or DC) component is zero and use a delay of $T/4$.

The **signal bandwidth** is a very important characteristic of any time varying waveform since it indicates the spectral content and, of course, its minimum and maximum frequency components. From eqn 3.120 we observe that the spectrum and therefore the bandwidth of the periodic square wave is infinite. However, it is clear that very high order harmonics have very small amplitudes and its impact on the series can be neglected. So a question arises;

3. Frequency domain electrical signal and circuit analysis

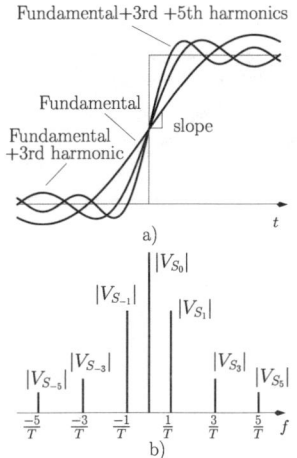

Figure 3.24: *Rectangular periodic waveform. a) Approximation by the various components. b) Line spectrum of the approximation.*

where do we truncate the Fourier series in order to determine the significant bandwidth of the signal? The criteria to perform such a truncation can vary depending on the application. One of these can be stated as the range of frequencies which contain a large percentage of the average power associated with this signal. For example, if this criterion defines this percentage as 95% of the total, then the bandwidth for the signal of figure 3.23 b) is $3/T$. In fact, $|V_{S_0}|^2 + 2|V_{S_1}|^2 + 2|V_{S_3}|^2 = 0.95 \times V_a^2/2$ where $V_a^2/2$ is the total average power associated with this signal. It is also important to realise that the signal bandwidth is a measure of how fast a signal varies in time. In order to illustrate this idea we consider figure 3.24 a) where we see that the addition of higher order harmonics increases the 'slope' of the reconstructed signal and that it varies more rapidly with time.

Now that we have determined the Fourier components of the input voltage signal, $v_s(t)$, of the circuit of figure 3.23 a) we are in a position to determine the output voltage $v_c(t)$. This voltage can be determined using the AC phasor analysis, discussed in section 3.2.3, and then applying the superposition theorem to all the voltage components (phasors) of the input signal $v_s(t)$.

The voltage phasor at the terminals of the capacitor, V_C, is determined using phasor analysis. This voltage can be obtained noting that the impedance associated with the capacitor and the resistor form an impedance voltage divider. Thus V_C can be expressed as follows:

$$V_C = \frac{Z_c}{Z_c + R} V_S \qquad (3.121)$$

where $Z_c = (j\omega C)^{-1}$ is the impedance associated with the capacitor. Hence, we can write:

$$V_C = \frac{1}{1 + j\omega RC} V_S \qquad (3.122)$$

If we divide the phasor which represents the circuit output quantity, V_C, by the phasor which represents the circuit input quantity, V_S, we obtain the circuit **transfer function** which, for the circuit of figure 3.23 a), can be written as follows:

$$H(\omega) = \frac{1}{1 + j\omega RC} \qquad (3.123)$$

or

$$H(f) = \frac{1}{1 + j2\pi f RC} \qquad (3.124)$$

The transfer function of a circuit is of particular relevance to electrical and electronic circuit analysis since it relates the output with the input by indicating how the amplitude and phase of the input phasors are modified. Figure 3.25 shows the magnitude (on a logarithmic scale) and phase of $H(f)$, given by eqn 3.124, versus the frequency f, also on a logarithmic scale, for various values of the product RC. RC is called the 'time constant' of the circuit. Close inspection of the transfer function $H(f)$ allows us to identify two distinct

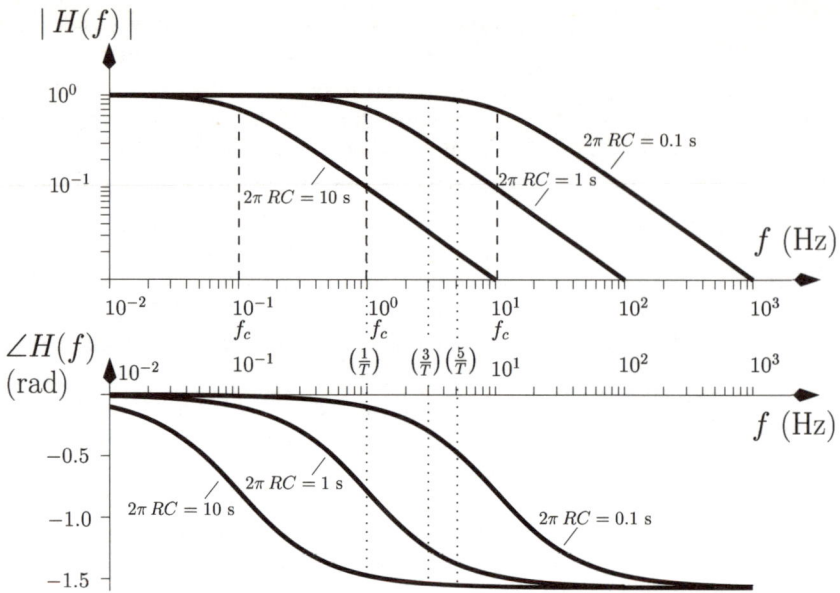

Figure 3.25: *Magnitude and phase of the transfer function of the RC circuit of figure 3.23.*

frequency ranges. The first is for $2\pi f RC \ll 1$, that is for $f \ll (2\pi RC)^{-1}$. Over this frequency range we can write:

$$H(f) \simeq 1 \quad \text{for } f \ll (2\pi RC)^{-1} \quad (3.125)$$

indicating that the circuit does not significantly change the amplitudes or phases of those components of the input signal with frequencies smaller than $(2\pi RC)^{-1}$.

The second frequency range is identified as $2\pi f RC \gg 1$. Now we can write:

$$H(f) \simeq \frac{1}{j 2\pi f RC} \quad \text{for } f \gg (2\pi RC)^{-1} \quad (3.126)$$

indicating that the circuit significantly attenuates the amplitudes of those components of the input signal with frequencies larger than $(2\pi RC)^{-1}$. The attenuation of these high frequency components means that the circuit preferentially allows the passage of low-frequency components. Hence, this circuit is also called a low-pass filter. The frequency $f_c = (2\pi RC)^{-1}$ is called the cut-off frequency of the filter and it establishes its bandwidth. A more detailed discussion of the definition of circuit bandwidth is presented in section 3.3.5. Note that for frequencies $f \gg f_c$ this circuit introduces a phase shift of $-\pi/2$.

We are now in a position to apply the superposition theorem in order to obtain the output voltage. This can be effected by substituting the phasor V_S in eqn 3.124 by the sum of phasors (Fourier series) which represents the square wave and by evaluating the circuit transfer function at each frequency $f =$

3. Frequency domain electrical signal and circuit analysis

n/T. That is:

$$V_{C_n} = [H(f)]_{f=\frac{n}{T}} \times V_{S_n} \qquad (3.127)$$

$$= \left[\frac{1}{1+j2\pi f RC}\right]_{f=\frac{n}{T}} \times V_{S_n}$$

$$= \frac{1}{1+j2\pi \frac{n}{T} RC} V_{S_n}$$

where the phasors V_{C_n} are the coefficients of the Fourier series representing the voltage $v_c(t)$ and the phasors V_{S_n} are the coefficients representing the periodic square voltage $v_s(t)$. The phasors V_{C_n} can be written as:

$$V_{C_n} = \frac{1}{1+j2\pi \frac{n}{T} RC} \frac{V_a}{2} \operatorname{sinc}\left(\frac{n}{2}\right) \quad (V) \qquad (3.128)$$

Which can also be written in the complex exponential form as:

$$V_{C_n} = \begin{cases} \dfrac{V_a\, e^{j\left(\frac{n\pi}{2}-\frac{\pi}{2}\right)-j\tan^{-1}\left(2\pi \frac{n}{T} RC\right)}}{\pi n \sqrt{1+\left(\frac{2\pi n}{T} RC\right)^2}} & \text{for } |n| \text{ odd} \\ \dfrac{V_a}{2} & \text{for } n=0 \\ 0 & \text{for } |n| \text{ even and } |n|>1 \end{cases} \qquad (3.129)$$

Figure 3.25 shows that if the low-pass filter features a time constant such that $2\pi RC = 10$ s, corresponding to $f_c = 0.1$ Hz, all frequency components of the input signal, with the exception of the DC component, are severely attenuated. Although for $2\pi RC = 1$ s ($f_c = 1$ Hz) the fundamental frequency component is slightly attenuated, all higher order harmonics are considerably attenuated. This implies that for both situations described above the output voltage will be significantly different from the input voltage. On the other hand, for $2\pi RC = 0.1$ s ($f_c = 10$ Hz) the fundamental, the third and the fifth order frequency components are hardly attenuated although higher-order harmonics suffer great attenuation. Note that, for this last situation ($f_c = 10$ Hz), the significant bandwidth of the input voltage signal does not suffer significant attenuation. This means that the output voltage is very similar to the input voltage.

Since the Fourier coefficients of $v_c(t)$ are known, this voltage can be written using eqn 3.92, that is:

$$v_c(t) = \frac{V_a}{2} + \sum_{\substack{n=1 \\ (n \text{ odd})}}^{\infty} \frac{2V_a}{\pi n \sqrt{1+\left(\frac{2\pi n}{T} RC\right)^2}} \cos\left(2\pi \frac{n}{T} t + \phi_n\right) \qquad (3.130)$$

with

$$\phi_n = \begin{cases} \frac{n\pi}{2} - \frac{\pi}{2} - \tan^{-1}\left(2\pi \frac{n}{T} RC\right) & \text{for } n \text{ odd} \\ 0 & \text{for } n \text{ even} \end{cases} \qquad (3.131)$$

Figure 3.26 illustrates the output voltage $v_c(t)$ for the three time constants discussed above. As expected, for the two situations where $2\pi RC = 10$ s and

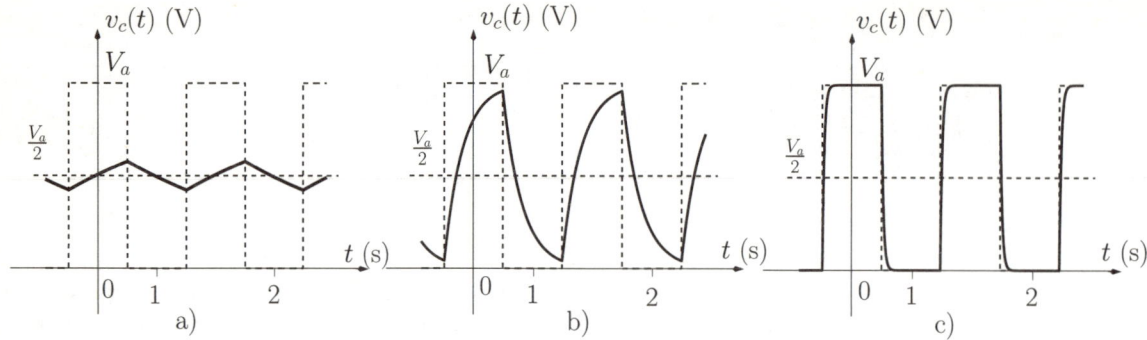

Figure 3.26: *Waveforms for $v_c(t)$. a) $2\pi RC = 0.1$ s. b) $2\pi RC = 1$ s. c) $2\pi RC = 10$ s.*

$2\pi RC = 1$ s the output voltage $v_c(t)$ is very different from the input voltage due to the filtering effect of the input signal frequency components. However, for $2\pi RC = 0.1$ s the output voltage is very similar to the input signal since the main frequency components are not significantly attenuated.

It is also interesting to note that the effect of filtering all frequency components ($2\pi RC = 10$ s) of the square voltage waveform results in a near-triangular periodic waveform, such as that of figure 3.19 c), with an average value (DC component) equal to the DC value of the input square wave input voltage (see next example).

The waveforms of $v_c(t)$ illustrated in figure 3.26 can be interpreted as the repetitive charging (towards V_a) and discharging (towards 0) of the capacitor. At the higher cut-off frequency ($2\pi RC = 0.1$ s) the capacitor can charge and discharge in a rapid manner almost following the input signal. However, as the cut-off frequency (or bandwidth) of the filter is decreased the charging and discharging of the capacitor takes more time. It is as if the output voltage is suffering from an 'electrical inertia' which opposes to the time-variations of that signal. In fact, the **bandwidth of a circuit** can actually be viewed as a qualitative measure of this 'electrical inertia'.

Example 3.3.4 Consider the circuit of figure 3.23 a). Show that if the cut-off frequency is such that $f_c \ll T^{-1}$ then the resulting output voltage is a near-triangular waveform as shown in figure 3.26 a).

Solution: If $(2\pi RC)^{-1} \ll T^{-1}$ this means that;

$$\frac{2\pi n RC}{T} \gg 1, \quad n \geq 1 \tag{3.132}$$

and we can write the Fourier coefficients of the output voltage, expressed by eqn 3.128, as follows:

$$V_{C_n} \simeq \begin{cases} \frac{V_a}{j2\frac{2\pi n}{T}RC} \operatorname{sinc}\left(\frac{n}{2}\right) & \text{if } n \neq 0 \\ \frac{V_a}{2} & \text{if } n = 0 \end{cases} \tag{3.133}$$

3. Frequency domain electrical signal and circuit analysis

This eqn can be written in exponential form as follows:

$$V_{c_n} = \begin{cases} \frac{V_a T}{2RC} \frac{1}{\pi^2 n^2} e^{-j n \pi /2} & \text{if } |n| \text{ is odd} \\ \frac{V_a}{2} & \text{if } n = 0 \\ 0 & \text{if } |n| \text{ is even and } |n| > 0 \end{cases} \quad (3.134)$$

Comparing the last eqn for $|n|$ odd with eqn 3.100 for $|n|$ odd we observe that they are similar in the sense that they exhibit the same behaviour as $|n|$ increases (note the existence of the term $1/n^2$ in both equations). The difference lies in the amplitude and in the average value for the output triangular waveform which now is $V_a/2$.

3.3.4 Linear distortion

Linear distortion is usually associated with the unwanted filtering of a signal while non-linear distortion is associated with non-linear effects in circuits. To illustrate linear distortion let us consider the transmission of a periodic signal $y(t)$ through an electrical channel with a transfer function $H(f)$. The output signal, $z(t)$, is said undistorted if it is a replica of $y(t)$, that is if $z(t)$ differs from $y(t)$ by a multiplying constant A, representing an amplification ($A > 1$) or attenuation ($A < 1$), and a time delay, t_d. Hence, $z(t)$ can be written as

$$z(t) = A\, y(t - t_d) \quad (3.135)$$

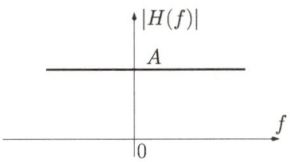

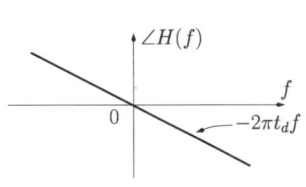

Figure 3.27: *Magnitude and phase of a transfer function of a distortionless system.*

The relevant question is: what must $H(f)$ be in order to have such a distortionless transmission? To answer this we assume that $y(t)$ has a Fourier series given by:

$$y(t) = \sum_{n=-\infty}^{\infty} C_{Y_n}\, e^{j 2\pi \frac{n}{T} t}\, dt \quad (3.136)$$

From eqn 3.135 and from the time delay property of Fourier series (see eqn 3.111) we can write the Fourier coefficients of $z(t)$ as follows:

$$C_{Z_n} = A\, C_{Y_n}\, e^{j 2\pi \frac{n}{T} t_d} \quad (3.137)$$

From eqn 3.127 we can determine $H(f)$ as follows:

$$[H(f)]_{f=\frac{n}{T}} = \frac{C_{Z_n}}{C_{Y_n}}$$
$$= A\, e^{j 2\pi \frac{n}{T} t_d} \quad (3.138)$$

that is

$$H(f) = A\, e^{j 2\pi f t_d} \quad (3.139)$$

Figure 3.27 shows the magnitude and the phase of this transfer function. From this figure we conclude that a distortionless system must provide the same amplification (or attenuation) to all frequency components of the input signal *and* must provide a linear phase shift to all these components.

The application of a sequence of rectangular pulses to an RC circuit illustrates what can be considered as linear distortion. Now, let us consider the transmission of those same pulses through an electrical channel which is modelled as the RC circuit of figure 3.23 a). From the discussion above we saw that if the cut-off frequency of the RC circuit is smaller than the third harmonic frequency of the input signal, then the output signal is significantly different from the input signal. Severe linear distortion occurs since the various frequency components of the input signal are attenuated by different amounts *and* suffer different phase shifts. However, if the cut-off frequency of the RC circuit is larger than the third harmonic frequency then the output signal is approximately equal to the input signal, as illustrated by figure 3.26 c). This is because the most significant frequency components of the input signal are affected by the same (unity) gain. Note that, in this situation, the phase shift is zero indicating that there is no delay between the input and output signals.

3.3.5 Bode plots

In the previous section we saw that the complex[3] nature of a transfer function, $H(f)$ (or $H(\omega)$), implies that the graphical representation of $H(f)$ requires two plots; the magnitude of $H(f)$, $|H(f)|$, and the phase of $H(f)$, $\angle H(f)$, versus frequency, as illustrated in figure 3.25.

Often, it is advantageous to represent the transfer function, $|H(f)|$, on a logarithmic scale, given by

$$|H_{\text{dB}}(f)| = 20 \log_{10} |H(f)| \quad \text{(dB)} \tag{3.140}$$

Here, $|H_{\text{dB}}(f)|$ and frequency are represented on logarithmic scales. The unit of the transfer function expressed in such a logarithmic scale is the decibel (dB).

The main advantage of this representation is that we can determine the asymptotes of the transfer function which, in turn facilitate its graphical representation. Note that the logarithmic operation also emphasises small differences in the transfer function which, if plotted in the linear scale, would not be so clearly visible. In order to illustrate this we again consider the transfer function of the RC circuit of figure 3.23, given by

$$H(f) = \frac{1}{1 + j2\pi f RC} \tag{3.141}$$

We can express this as

$$|H_{\text{dB}}(f)| = 20 \log_{10} \left| \frac{1}{1 + j2\pi f RC} \right|$$

---
[3] As in complex numbers described in Chapter 2.

3. Frequency domain electrical signal and circuit analysis

$$\begin{aligned}
&= 20 \log_{10} \frac{1}{\sqrt{1 + (2\pi f RC)^2}} \\
&= 20 \log_{10}(1) - 20 \log_{10}\left(1 + (2\pi f RC)^2\right)^{\frac{1}{2}} \\
&= -10 \log_{10}\left(1 + (2\pi f RC)^2\right) \quad (3.142)
\end{aligned}$$

We can now identify the two *asymptotes* of $|H_{\text{dB}}(f)|$, noting that

$$1 + (2\pi f RC)^2 \simeq 1 \quad \text{if } 2\pi f RC \ll 1 \quad (3.143)$$
$$1 + (2\pi f RC)^2 \simeq (2\pi f RC)^2 \quad \text{if } 2\pi f RC \gg 1 \quad (3.144)$$

Hence, we can write

$$\begin{aligned}
|H_{\text{dB}}(f)| &\simeq -10 \log_{10}(1) \\
&\simeq 0 \text{ dB} \quad &&\text{if } f \ll \frac{1}{2\pi RC} \quad (3.145)\\
|H_{\text{dB}}(f)| &\simeq -10 \log_{10}(2\pi f RC)^2 \\
&\simeq -20 \log_{10}(2\pi f RC) \quad &&\text{if } f \gg \frac{1}{2\pi RC} \quad (3.146)
\end{aligned}$$

The phase of $H(f)$ is given by

$$\angle H(f) = e^{-j \tan^{-1}(2\pi f RC)} \quad (3.147)$$

and it can also be approximated by asymptotes:

$$\angle H(f) \simeq \begin{cases} 0 & \text{if } f < \frac{1}{10} \times \frac{1}{2\pi RC} \\ -\frac{\pi}{4} \log_{10}(2\pi f RC) - \frac{\pi}{4} & \text{if } \frac{1}{10 \times 2\pi RC} < f < \frac{10}{2\pi RC} \\ -\frac{\pi}{2} & \text{if } f > \frac{10}{2\pi RC} \end{cases} \quad (3.148)$$

Figure 3.28 a) shows $|H_{\text{dB}}(f)|$ versus the frequency. In this figure we also show the corresponding values of $|H(f)|$. A gain of -20 dB (corresponding to an attenuation of 20 dB) is equivalent to a *linear* gain of 0.1 (or an attenuation of 10 times).

The two asymptotes given by eqns 3.145 and 3.146 are represented in figure 3.28 a), by dashed lines. Since the X-axis is also logarithmic the asymptote given by eqn 3.146 is represented as a line whose slope is -20 dB/decade. A decade is a frequency range over which the ratio between the maximum and minimum frequency is 10. Note that this slope can be inferred by inspection of figure 3.28 a) where we observe that for $f = (2\pi RC)^{-1}$ the asymptote given by eqn 3.146 indicates 0 dB. From this figure we observe that these two asymptotes approximately describe the entire transfer function. The maximum error, Δ, between $H(f)$ and the asymptotes occurs at the frequency $f = (2\pi RC)^{-1}$. It is given by

$$\begin{aligned}
\Delta &= -20 \log_{10}(2\pi f RC)_{f=(2\pi RC)^{-1}} - |H_{\text{dB}}(f)|_{f=(2\pi RC)^{-1}} \\
&= 0 + 10 \log_{10}(2) \\
&\simeq 3 \text{ dB}
\end{aligned}$$

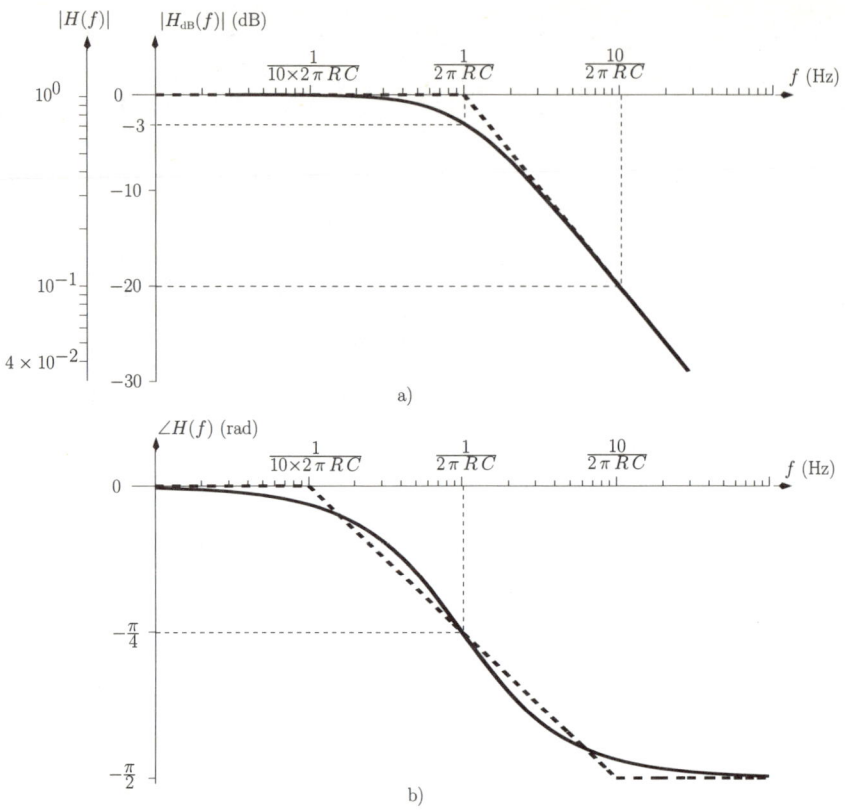

Figure 3.28: *Magnitude and phase of the transfer function of the RC circuit of figure 3.23 (solid lines) and asymptotes (dashed lines).*

The **circuit or system bandwidth** is very often defined as the range of positive frequencies for which the magnitude of its transfer function is above the 3 dB attenuation value. This 3 dB value is equivalent to voltage or current output to input ratio of $1/\sqrt{2} \simeq 71\%$ (see figure 3.28 a)) or, alternatively, output to input power ratio of 50%. Hence, the bandwidth for the RC circuit is from DC to $f = (2\pi RC)^{-1}$, the *cut-off* frequency.

Figure 3.28 b) shows the angle of the transfer function, $\angle H(f)$, and also its asymptotes given by eqn 3.148. From this figure we observe that for frequencies smaller than one tenth of the cut-off frequency the phase of the transfer function is close to zero. At the cut-off frequency $f = (2\pi RC)^{-1}$ the phase of the transfer function is $-\pi/4$ and for frequencies significantly greater than this, the phase of the transfer function tends to $-\pi/2$.

Poles and zeros of a transfer function

In general, a circuit transfer function can be written as follows:

$$H(f) = A \frac{(1+j2\pi f/z_1)(1+j2\pi f/z_2)\ldots(1+j2\pi f/z_n)}{(1+j2\pi f/p_1)(1+j2\pi f/p_2)\ldots(1+j2\pi f/p_m)} \quad (3.149)$$

Each z_i, $i = 1, \ldots, n$, is called a zero of the transfer function, and, for $j2\pi f = -z_i$ the transfer function is zero. Each p_i, $i = 1, \ldots, m$, is called a pole of the transfer function. At $j2\pi f = -p_i$ the transfer function is not defined since $H(j\, p_i/(2\pi)) \to \pm\infty$ depending on the sign of the DC gain, A. For a practical circuit $m \geq n$ and m, the number of poles, is called the order of the transfer function.

This representation of a transfer function is quite advantageous when all the poles and zeros are *real numbers* since, in this situation, it greatly simplifies the calculation of $|H_{\text{dB}}(f)|$. In fact, if all the poles and zeros of $H(f)$ are real numbers we can write:

$$|H_{\text{dB}}(f)| = \sum_{i=1}^{n} 10 \log_{10}\left[1 + \left(\frac{2\pi f}{z_i}\right)^2\right] - \sum_{k=1}^{m} 10 \log_{10}\left[1 + \left(\frac{2\pi f}{p_k}\right)^2\right] \quad (3.150)$$

Let us consider the CR circuit of figure 3.29. Note the new positions of the resistor and capacitor. It can be shown (see problem 3.6) that the transfer function of this circuit, $H_{CR}(f) = V_R/V_S$, can be written as:

$$H_{CR}(f) = \frac{j2\pi f RC}{1 + j2\pi f RC} \quad (3.151)$$

Relating this transfer function with eqn 3.149 we observe that $H_{CR}(f)$ has one pole, equal to $(RC)^{-1}$, and a zero located at the origin. Since the pole and the zero are real numbers we can use eqn 3.150 to determine $|H_{CR_{\text{dB}}}(f)|$ as follows:

$$|H_{CR_{\text{dB}}}(f)| = 20 \log_{10}(2\pi f RC) - 10 \log_{10}\left(1 + (2\pi f RC)^2\right) \quad (3.152)$$

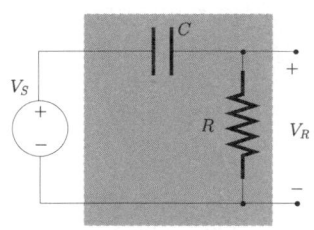

Figure 3.29: *CR circuit.*

We can identify the two asymptotes of $|H_{CR_{\text{dB}}}(f)|$ (see also eqns 3.143 and 3.144) which are given by:

$$|H_{CR_{\text{dB}}}(f)| \simeq 20 \log_{10}(2\pi f RC) \text{ dB} \qquad \text{if } f \ll \frac{1}{2\pi RC} \quad (3.153)$$

$$|H_{CR_{\text{dB}}}(f)| \simeq 0 \text{ dB}, \qquad \text{if } f \gg \frac{1}{2\pi RC} \quad (3.154)$$

The phase of $H_{CR}(f)$ is given by

$$\angle H_{CR}(f) = e^{j\frac{\pi}{2} - j\tan^{-1}(2\pi f RC)} \quad (3.155)$$

and it can also be approximated by asymptotes:

$$\angle H_{CR}(f) \simeq \begin{cases} \frac{\pi}{2} & \text{if } f < \frac{1}{10} \times \frac{1}{2\pi RC} \\ \frac{\pi}{4} - \frac{\pi}{4}\log_{10}(2\pi f RC) & \frac{1}{10 \times 2\pi RC} < f < \frac{10}{2\pi RC} \\ 0 & f > \frac{10}{2\pi RC} \end{cases} \quad (3.156)$$

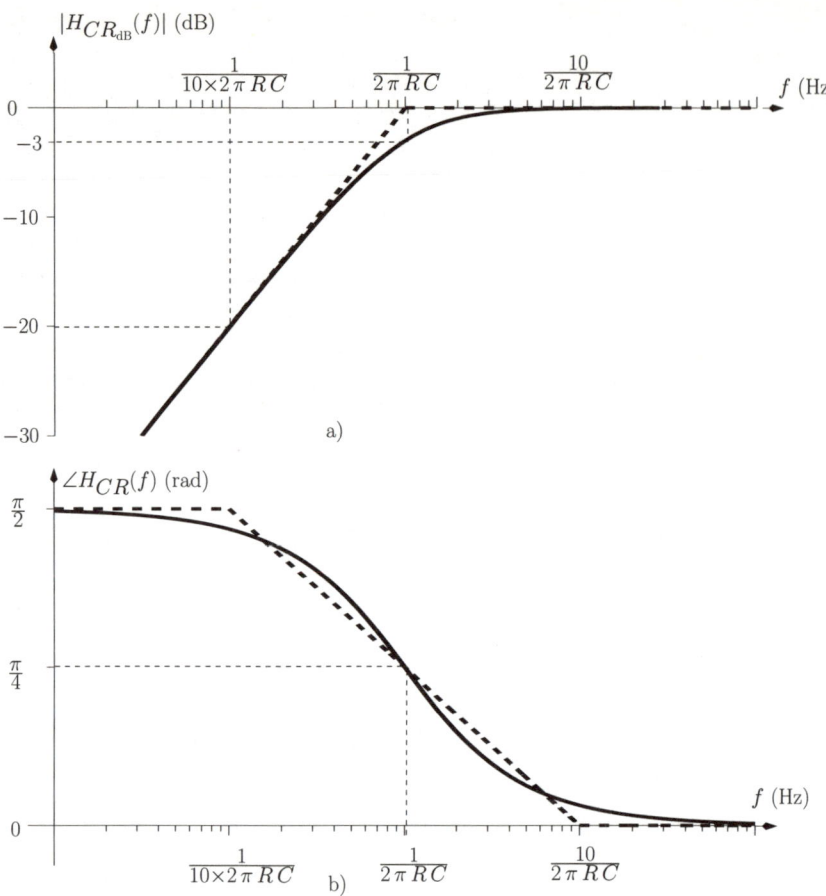

Figure 3.30: *Magnitude and phase of the transfer function of the CR circuit of figure 3.29 (solid lines) and asymptotes (dashed lines).*

Figure 3.30 a) shows the magnitude, in dB, of this transfer function given by eqn 3.152 and the asymptotes given by eqns 3.153 and 3.154. We observe that this circuit attenuates frequencies smaller than the cut-off frequency, $f_c = (2\pi RC)^{-1}$, while it passes the frequency components higher than f_c. Hence, this circuit is called a *high-pass filter*. Note that, in theory, the bandwidth of this filter is infinity, although in practice unwanted circuit elements set a maximum operating frequency to this circuit.

Figure 3.30 b) shows the phase of the transfer function. The three asymptotes for this phase given by eqn 3.156 are also shown. At frequencies smaller than $f = (2\pi RC\, 10)^{-1}$ the circuit imposes a phase of $\pi/2$ while at frequencies higher than $f = 10\,(2\pi RC)^{-1}$ the circuit does not change the phase.

3. Frequency domain electrical signal and circuit analysis

Signal filtering as signal shaping

Signal filtering can act as signal shaping as illustrated in example 3.3.4 where a triangular waveform was obtained from the low-pass filtering of a square-wave. This shaping is accomplished using at least one energy storage element in an electronic network, that is by using capacitors or inductors. Capacitive and inductive impedances are frequency dependent and different frequency components of a periodic signal suffer different amounts of attenuation (or amplification) and different amounts of phase shift giving rise to modified signals.

To further illustrate this idea let us consider the circuit of figure 3.31 where a square-wave voltage is applied (see figure 3.19 b)). The purpose of this circuit is to reshape the input signal in order to obtain a sine wave voltage.

The output voltage, $v_o(t)$, is the voltage across the capacitor and inductor. Since the input voltage $v_s(t)$ can be decomposed as a sum of phasors the voltage $v_o(t)$ can be determined using AC phasor analysis together with the superposition theorem. We start by calculating the voltage at the output, V_O, using phasor analysis. Since the capacitor is in a parallel connection with the inductor we can determine an equivalent impedance,

$$Z_{LC} = \frac{Z_L Z_C}{Z_L + Z_C} \tag{3.157}$$

with

$$Z_C = \frac{1}{j\omega C} \tag{3.158}$$

$$Z_L = j\omega L \tag{3.159}$$

that is:

$$Z_{LC} = \frac{j\omega L}{1 - \omega^2 LC} \tag{3.160}$$

From figure 3.31 b) we observe that Z_{LC} and the resistor form an impedance voltage divider. Thus the voltage V_O can be expressed as follows:

$$\begin{aligned} V_O &= \frac{Z_{LC}}{Z_{LC} + R} V_S \\ &= \frac{\frac{j\omega L}{1-\omega^2 LC}}{\frac{j\omega L}{1-\omega^2 LC} + R} \\ &= \frac{j\omega L}{R(1 - \omega^2 LC) + j\omega L} V_S \end{aligned} \tag{3.161}$$

The transfer function is, therefore,

$$H_{RLC}(\omega) = \frac{j\omega L}{R(1 - \omega^2 LC) + j\omega L} \tag{3.162}$$

Clearly, this can also be written as

$$H_{RLC}(f) = \frac{j2\pi f L}{R(1 + (j2\pi f)^2 LC) + j2\pi f L} \tag{3.163}$$

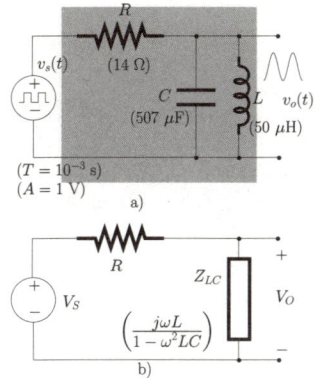

Figure 3.31: *a) RLC circuit. b) AC equivalent circuit.*

The two poles of $H_{RLC}(f)$ can be determined by setting the denominator of eqn 3.163 to zero and solving this eqn in order to obtain $j\,2\,\pi\,f$, that is

$$R(1+(j\,2\,\pi\,f)^2\,L\,C) + j\,2\,\pi\,f\,L = 0 \tag{3.164}$$

and since $L^2 - 4\,L\,C\,R^2 < 0$ we obtain

$$j\,2\,\pi\,f_i = \frac{-L \pm j\,\sqrt{4\,L\,C\,R^2 - L^2}}{2\,R\,L\,C}, \quad i = 1, 2 \tag{3.165}$$

The two poles of the transfer function are obtained from the last eqn (see also eqn 3.149) as

$$\begin{aligned} p_i &= -j\,2\,\pi\,f_i \\ &= \frac{+L \mp j\,\sqrt{4\,L\,C\,R^2 - L^2}}{2\,R\,L\,C}, \quad i = 1, 2 \end{aligned} \tag{3.166}$$

The two poles given by the last eqn are complex conjugated. This means that we cannot apply eqn 3.150 and we must determine $|H_{RLC_{\text{dB}}}(f)|$ using the standard procedure, that is:

$$\begin{aligned} |H_{RLC_{\text{dB}}}(f)| &= 20\,\log_{10}\left|\frac{j\,2\,\pi\,f\,L}{R(1-(2\,\pi\,f)^2\,L\,C) + j\,2\,\pi\,f\,L}\right| \\ &= 20\,\log_{10}(2\pi\,f\,L) \\ &\quad - 10\,\log_{10}\left[R^2(1-(2\,\pi\,f)^2\,L\,C)^2 + (2\,\pi\,f\,L)^2\right] \end{aligned} \tag{3.167}$$

Figure 3.32 shows a plot of $|H_{RLC_{\text{dB}}}(f)|$. This figure indicates that the RLC circuit does not attenuate the component $f = (2\pi\sqrt{LC})^{-1} = 1$ kHz since $|H_{RLC_{\text{dB}}}((2\pi\sqrt{LC})^{-1})| = 0$ dB. However, it attenuates all frequency components around this frequency. Thus, this circuit is called a *band-pass* filter. The (3 dB) bandwidth of this circuit is 22 Hz centred in 1 kHz. For band-pass filters the Quality Factor, Q, is defined as the ratio of the central frequency, f_o, to its bandwidth, BW, that is

$$Q = \frac{f_o}{BW} \tag{3.168}$$

The quality factor is a measure of the sharpness of the response of the circuit. A high quality factor indicates a high frequency selectivity of the band-pass filter. For this circuit the quality factor is $Q = 45$. Note that the third and the fifth harmonics suffer an attenuation greater than 40 dB resulting from the frequency selectivity of the circuit. This means that these frequency components have an amplitude (at least) 100 times smaller at the output of the circuit compared to its original amplitude at the input of the circuit.

We are now in a position to apply the superposition theorem to obtain $v_o(t)$. This can be effected by substituting the phasor V_S in eqn 3.163 by the Fourier series which represents the periodic square wave and by evaluating the transfer function of the circuit, $H_{RLC}(f)$, at *each* frequency of these phasors, that is:

$$\begin{aligned} V_{O_n} &= [H_{RLC}(f)]_{f=\frac{n}{T}} \times V_{S_n} \\ &= \frac{j\,2\,\pi\,\frac{n}{T}\,L}{R(1 - 4\,\pi^2\,\frac{n^2}{T^2}\,L\,C) + j\,2\,\pi\,\frac{n}{T}\,L}\,V_{S_n} \end{aligned} \tag{3.169}$$

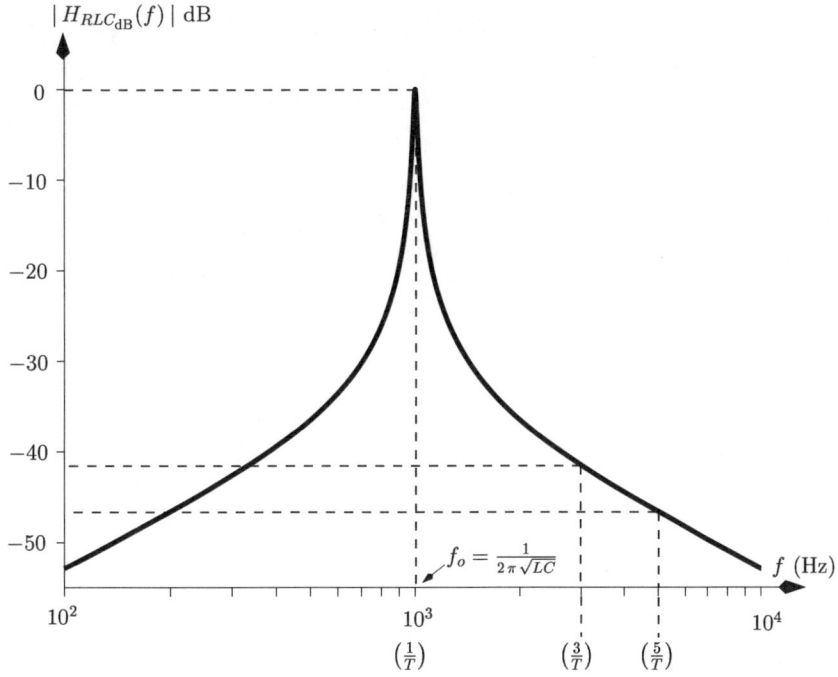

Figure 3.32: *Magnitude of the transfer function of the RLC circuit of figure 3.31.*

where the phasors V_{O_n} are the coefficients of the Fourier series representing $v_o(t)$ and the phasors V_{S_n} are the coefficients of the Fourier series representing $v_s(t)$. Clearly, V_{S_n} coincide with C_n given by eqn 3.87. However, the units for these coefficients are volts. The phasors V_{O_n} can be written as:

$$V_{O_n} = \begin{cases} \dfrac{j2\pi\dfrac{n}{T}L}{R(1-4\pi^2\dfrac{n^2}{T^2}LC)+j2\pi\dfrac{n}{T}L}\dfrac{2A}{j\pi n} \text{ (V)} & \text{for } |n| \text{ odd} \\ 0 & \text{for } |n| \text{ even} \end{cases} \quad (3.170)$$

Figures 3.33 a) and 3.33 b) show the magnitude and the phase of the spectral components of $v_s(t)$, respectively, while figures 3.33 c) and 3.33 d) show the magnitude and the phase of the components of $v_o(t)$, respectively. It is clear that the fundamental component (at $f = 1/T$) is present in the output voltage but that higher order harmonics are severely attenuated[4]. Comparing figures 3.33 b) and 3.33 d) it is also clear that the circuit changes the phase of the higher order harmonics of the input signal.

---

[4]These harmonics appear to be zero given the resolution of figure 3.33 c).

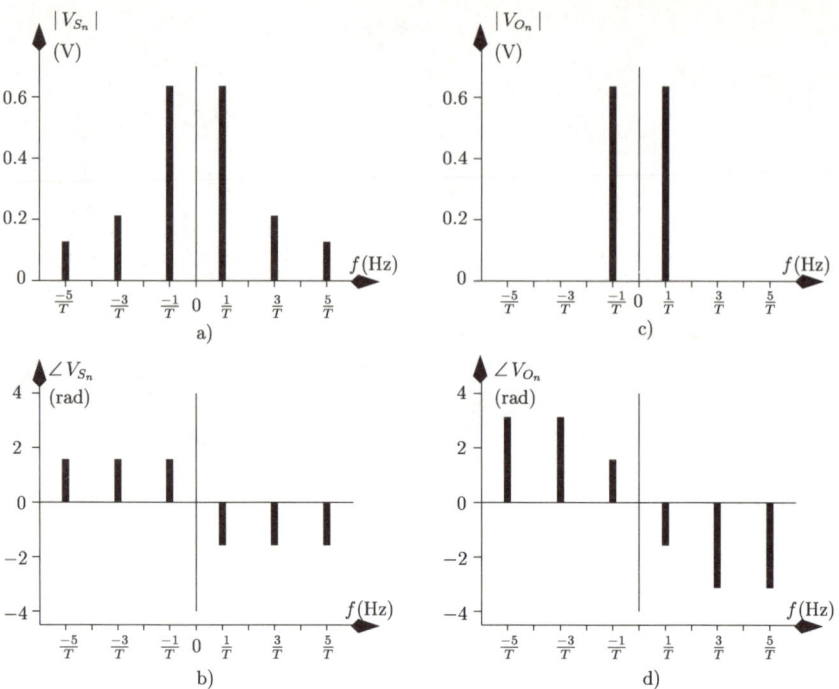

Figure 3.33: *Spectral representations of: a) magnitude of $v_s(t)$; b) phase of $v_s(t)$; c) magnitude of $v_o(t)$; d) phase of $v_o(t)$.*

The voltage $v_o(t)$ can now be written using eqn 3.92 as:

$$v_o(t) = \sum_{\substack{n=1 \\ (n \text{ odd})}}^{\infty} 2|V_{O_n}| \cos\left(2\pi\frac{n}{T}t + \text{angle}(V_{O_n})\right) \quad (3.171)$$

Since the harmonics, at frequencies higher than the fundamental, are strongly attenuated, we can write $v_o(t)$ as

$$\begin{aligned} v_o(t) &\simeq 2|V_{O_1}| \cos\left(2\pi\frac{1}{T}t + \text{angle}(V_{O_1})\right) \\ &= \frac{4}{\pi} \cos\left(2\pi\frac{1}{T}t - \frac{\pi}{2}\right) \end{aligned} \quad (3.172)$$

Finally figure 3.34 shows $v_s(t)$ and $v_o(t)$ given by eqn 3.171. From this figure it is clear that the output voltage is a sine wave corresponding to the fundamental component of the input periodic voltage signal $v_s(t)$.

3.3.6 The Fourier transform

In the previous section we have seen that the Fourier series is a very powerful signal analysis tool since it allows us to decompose periodic signals into a sum

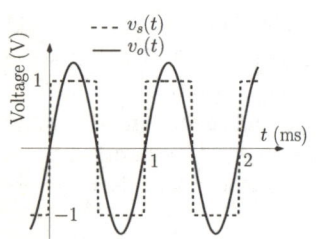

Figure 3.34: $v_s(t)$ *and* $v_o(t)$.

3. Frequency domain electrical signal and circuit analysis

of phasors. Such a decomposition, in turn, allows the analysis of electrical circuits using the AC phasor technique with the superposition theorem. Whilst the Fourier series applies only to periodic waveforms the Fourier transform is a far more powerful tool since, in addition to periodic signals, it can represent non-periodic signals as a 'sum' of phasors. In order to illustrate the difference between the Fourier series and the Fourier transform we recall the Fourier series of a rectangular waveform like that depicted in figure 3.35 with amplitude V_a and duty-cycle τ/T. Figure 3.36 a) shows the waveform and its correspondent line spectrum (magnitude). If we now increase the period T (maintaining τ and the amplitude constant) we observe that the density of phasors increases (figure 3.36 b) and 3.36 c)). Note that the amplitude of these phasors decreases

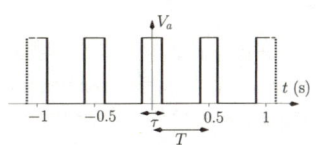

Figure 3.35: *Periodic voltage rectangular waveform.*

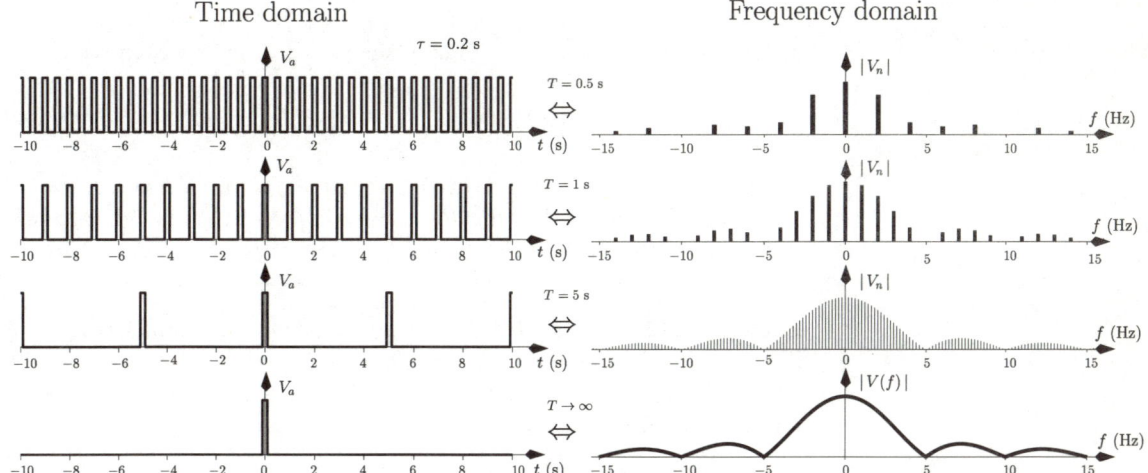

Figure 3.36: *The Fourier transform of a rectangular pulse.*

since the power of the signal decreases. If we let the period tend to infinity this is equivalent to having a non-periodic signal, that is, we have a situation where the signal $v(t)$ is just a single rectangular pulse. In this situation the signal spectrum is no longer discrete and no longer constituted by equally spaced discrete phasors. Instead the spectrum becomes *continuous*. In this situation the spectrum is often referred to as having a continuous spectral density.

The procedure described above, where the period T is increased, can be written, in mathematical terms, as follows:

$$v(t) = \lim_{T\to\infty} \sum_{n=-\infty}^{\infty} V_n\left(\frac{n}{T}\right) e^{j\,2\,\pi\,\frac{n}{T}\,t} \tag{3.173}$$

where we indicate the explicit dependency of the Fourier coefficients V_n on the discrete frequency n/T. The last eqn can be written as shown below:

$$v(t) = \lim_{T\to\infty} \sum_{n=-\infty}^{\infty} T\,V_n\left(\frac{n}{T}\right) e^{j\,2\,\pi\,\frac{n}{T}\,t} \Delta f \tag{3.174}$$

where $\Delta f = 1/T$. Equation 3.174 can be written as follows[5]:

$$v(t) = \int_{-\infty}^{\infty} V(f)\, e^{j\,2\pi f t}\, df \tag{3.175}$$

The discrete frequencies are described by the discrete variable, n/T. This variable tends to a continuous variable, f, describing a continuous frequency when $T \to \infty$. $V(f)$, the (continuous) spectrum or the spectral density of $v(t)$, can be calculated as follows:

$$V(f) = \lim_{T \to \infty} T V_n\left(\frac{n}{T}\right) \tag{3.176}$$

$$= \lim_{T \to \infty} \int_{-T/2}^{T/2} v(t)\, e^{-j\,2\pi \frac{n}{T} t}\, dt \tag{3.177}$$

Where we chose $t_o = -T/2$. Finally, the last eqn can be written as:

$$V(f) = \int_{-\infty}^{\infty} v(t)\, e^{-j\,2\pi f t}\, dt \tag{3.178}$$

A sufficient condition (but not strictly necessary) for the existence of the Fourier transform of a signal $x(t)$ is that the integral expressed by eqn 3.178 has a finite value for every value of f.

Example 3.3.5 Consider the single square voltage pulse shown in figure 3.36. Show that the Fourier transform of this pulse is the same as that obtained from eqn 3.176, which is derived from the Fourier series of a periodic sequence of rectangular pulses (see eqn 3.117), when $T \to \infty$.

Solution: Using eqn 3.178 we can write

$$\begin{aligned}
V(f) &= \int_{-\infty}^{\infty} V_a \operatorname{rect}\left(\frac{t}{\tau}\right) e^{-j\,2\pi f t}\, dt \\
&= \int_{-\tau/2}^{\tau/2} V_a\, e^{-j\,2\pi f t}\, dt \\
&= \frac{V_a}{-j\,2\pi f}\, \frac{e^{j\pi f \tau} - e^{-j\pi f \tau}}{2j} \\
&= V_a\, \tau\, \operatorname{sinc}(f\tau) \tag{3.179}
\end{aligned}$$

From eqn 3.176 we can write :

$$\begin{aligned}
V(f) &= \lim_{T \to \infty} T V_n\left(\frac{n}{T}\right) \\
&= \lim_{T \to \infty} T\, \frac{V_a \tau}{T}\, \operatorname{sinc}\left(\frac{n\tau}{T}\right) \\
&= V_a\, \tau\, \operatorname{sinc}(f\tau) \tag{3.180}
\end{aligned}$$

---

[5]Recall that

$$\lim_{\Delta x \to 0} \sum_{i=a}^{b} f(x_i)\, \Delta x = \int_a^b f(x)\, dx$$

3. Frequency domain electrical signal and circuit analysis 91

where $n/T \to f$ as $T \to \infty$.

From the above it should be clear that the Fourier transform, $V(f)$, represents a density of phasors which completely characterise $v(t)$ in the frequency domain. Such a representation is similar to the Fourier series coefficients in the context of periodic signals. However, it is important to note that whilst the unit of the voltage phasors (Fourier coefficients), V_n is the volt, the unit of the spectral density, $V(f)$, is volt/hertz (or volt×second). $v(t)$ and $V(f)$, as given by eqns 3.175 and 3.178 respectively, form the so-called Fourier transform pair:

$$v(t) \overset{\mathfrak{F}}{\longleftrightarrow} V(f) \qquad (3.181)$$

where $\mathfrak{F}$ denotes the Fourier integral operation.

Linearity

The Fourier transform is a linear operator. Given two distinct signals $x_1(t)$ and $x_2(t)$ with Fourier transforms $X_1(f)$ and $X_2(f)$, respectively, then the Fourier transform of $y(t) = a\, x_1(t) + b\, x_2(t)$ is given by;

$$\begin{aligned} V(f) &= \int_{-\infty}^{\infty} [a\, x_1(t) + b\, x_2(t)]\, e^{-j2\pi ft}\, dt \\ &= a\, X_1(f) + b\, X_2(f) \end{aligned} \qquad (3.182)$$

Duality

Another important property of Fourier transform pairs is the so-called duality. Let us consider a signal $x(t)$ with a Fourier transform represented by $X(f)$. If there is a signal $y(t) = X(t)$ then its Fourier transform is given by

$$\begin{aligned} Y(f) &= \int_{-\infty}^{\infty} X(t)\, e^{-j2\pi ft}\, dt \\ &= \int_{-\infty}^{\infty} X(t)\, e^{j2\pi(-f)t}\, dt \end{aligned} \qquad (3.183)$$

and, according to eqn 3.175 we have that

$$Y(f) = x(-f) \qquad (3.184)$$

that is

$$X(t) \overset{\mathfrak{F}}{\longleftrightarrow} x(-f) \qquad (3.185)$$

Example 3.3.6 Use the duality property of Fourier transform pairs to calculate the Fourier transform of $y(t) = A\,\mathrm{sinc}(t\,\eta)$.

Solution: From eqn 3.179 and from eqn 3.185 we can write:

$$Y(f) = \frac{A}{\eta} \text{rect}\left(\frac{f}{\eta}\right) \qquad (3.186)$$

Note that the rectangular function is an even function, that is $\text{rect}(-f) = \text{rect}(f)$.

Time delay

If a function $x(t)$ has a Fourier transform $X(f)$ then the Fourier transform of a delayed replica of $x(t)$ by a time τ, $x(t-\tau)$, is given by:

$$\mathfrak{F}[x(t-\tau)] = \int_{-\infty}^{\infty} x(t-\tau) e^{-j2\pi ft} dt \qquad (3.187)$$

using the change of variable $t' = t - \tau$ we can write:

$$dt = dt'$$
$$t \to -\infty \quad ; \quad t' \to -\infty$$
$$t \to \infty \quad ; \quad t' \to \infty$$

and eqn 3.187 can be written as

$$\mathfrak{F}[x(t-\tau)] = \int_{-\infty}^{\infty} x(t') e^{-j2\pi ft'} dt' \, e^{-j2\pi f\tau}$$
$$= X(f) e^{-j2\pi f\tau} \qquad (3.188)$$

Note that the delay τ causes an addition of a linear phase to $X(f)$. If τ is negative this means that the signal is advanced in time and the linear phase added to the spectrum has a positive slope. It is worth noting the similarity between the delay property of the Fourier transform with the delay property of the Fourier series (see eqn 3.111).

The Dirac delta function

The Dirac delta function, $\delta(t)$ can be visualised as an extremely narrow pulse located at $t = 0$. However, the area of this pulse is unity which implies that its amplitude tends to infinity. A common way of defining this function is to start with a rectangular waveform with unity area, such as that depicted in figure 3.37 a), which can be expressed as follows:

$$z(t) = \frac{1}{\tau} \text{rect}\left(\frac{t}{\tau}\right) \qquad (3.189)$$

with $\tau = 1$. If we now decrease the value of τ, as shown in figures 3.37 b) and c) we observe that the width of the rectangle decreases whilst its amplitude

3. Frequency domain electrical signal and circuit analysis

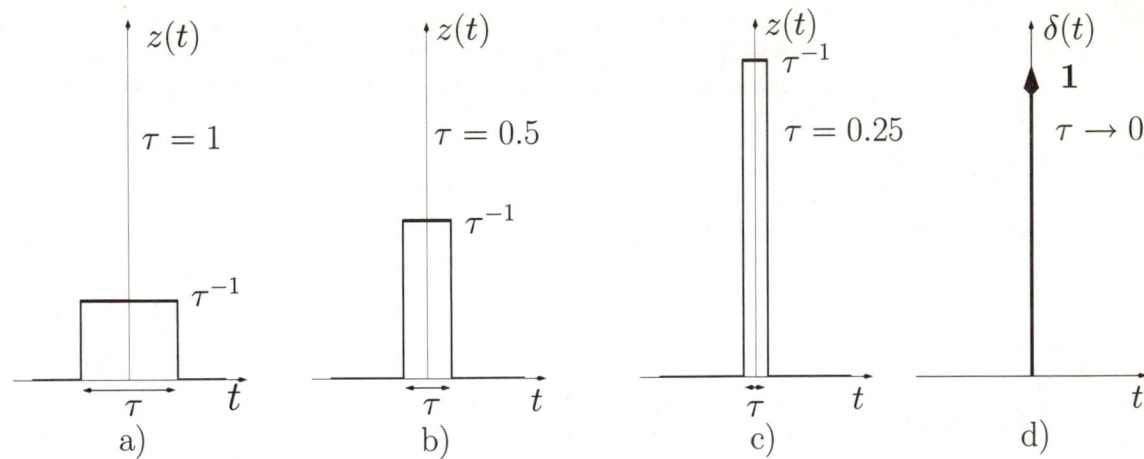

Figure 3.37: *Rectangular function. a) $\tau = 1$. b) $\tau = 0.5$. c) $\tau = 0.25$. d) $\tau \to 0$ (Dirac delta function).*

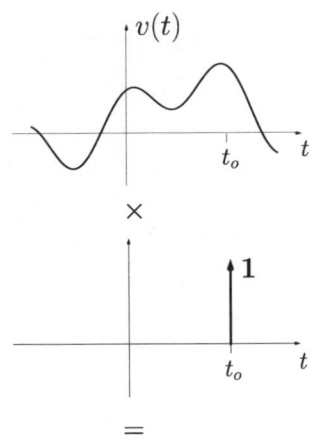

increases in order to preserve unity area. When we let τ tend to zero we obtain the Dirac delta function;

$$\delta(t) = \lim_{\tau \to 0} \frac{1}{\tau} \text{rect}\left(\frac{t}{\tau}\right) \tag{3.190}$$

which is depicted in figure 3.37 d). Note that

$$\int_{-\infty}^{\infty} \delta(t)\, dt = 1 \tag{3.191}$$

The *area* is represented by the bold value next to the arrow representing the delta function. An important property of the Dirac delta function is called the *sampling property* which states that the multiplication of this function, centred at t_o, by a signal $v(t)$ results in a Dirac delta function centred in t_o with an area given by the value of $v(t)$ at $t = t_o$, that is:

$$v(t) \times \delta(t - t_o) = v(t_o) \times \delta(t - t_o) \tag{3.192}$$

We emphasise that the area of $v(t) \times \delta(t - t_o)$ is equal to $v(t_o)$, that is:

$$\int_{-\infty}^{\infty} v(t) \times \delta(t - t_o) = v(t_o) \tag{3.193}$$

Figure 3.38 illustrates this last property expressed by eqns 3.192 and 3.193.

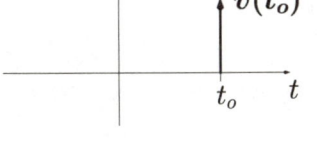

Figure 3.38: *Illustration of the sampling property of the Dirac delta function.*

The Fourier transform of a DC signal

Let us calculate the Fourier transform of a DC signal, $w(t)$, with amplitude A. According to eqn 3.178 this transform would be given by

$$W(f) = \int_{-\infty}^{\infty} A\, e^{-j 2\pi f t}\, dt \tag{3.194}$$

However, the definite integral cannot be determined because it does not converge for any value of f. Hence, the calculation of this Fourier transform requires the following mathematical manipulation. We express the DC value as follows:

$$w(t) = \lim_{\eta \to 0} A \operatorname{sinc}(t\eta) \qquad (3.195)$$

Figure 3.39 a) illustrates eqn 3.195 where we observe that as $\eta \to 0$, $w(t) \to A$. Taking the Fourier transform of $w(t)$, expressed by eqn 3.195, we obtain

$$W(f) = \int_{-\infty}^{\infty} \lim_{\eta \to 0} A \operatorname{sinc}(t\eta) e^{-j 2\pi f t} dt \qquad (3.196)$$

Since the integrand is a continuous function we can change the order of the limit and the integral, that is

$$W(f) = \lim_{\eta \to 0} \int_{-\infty}^{\infty} A \operatorname{sinc}(t\eta) e^{-j 2\pi f t} dt \qquad (3.197)$$

From eqn 3.186 we can write $W(f)$ as follows:

$$W(f) = \lim_{\eta \to 0} \frac{A}{\eta} \operatorname{rect}\left(\frac{f}{\eta}\right) \qquad (3.198)$$

This eqn is, by definition (see eqn 3.190), the Dirac delta function multiplied by A (see also figure 3.39 b)), that is:

$$W(f) = A\delta(f) \qquad (3.199)$$

This type of mathematical manipulation yields what is called the 'generalised Fourier transform' and it allows for the calculation of Fourier transforms of a broad class of functions such as that illustrated in the next example.

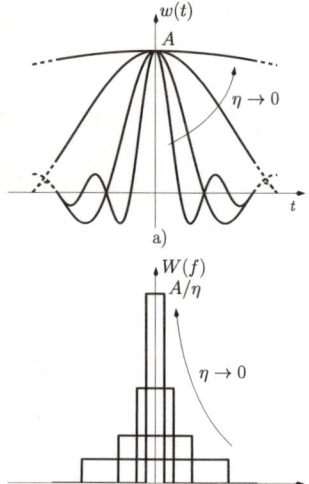

Figure 3.39: *a) Representation of the DC value $w(t) = A$. b) Fourier transform of $w(t)$.*

Example 3.3.7 Determine the Fourier transform of the unit-step function depicted in figure 3.40.

Solution: The unit-step function is defined as follows:

$$u(t) = \begin{cases} 1 & \text{if } t \geq 0 \\ 0 & \text{elsewhere} \end{cases} \qquad (3.200)$$

This function can also be seen as the addition of a DC value of $1/2$ with the signum function multiplied by a factor $1/2$, as illustrated by figure 3.40, and can be written as

$$u(t) = \frac{1}{2} + \frac{1}{2}\operatorname{sign}(t) \qquad (3.201)$$

where the signum function, $\operatorname{sign}(t)$, is defined as:

$$\operatorname{sign}(t) = \begin{cases} 1 & \text{if } t \geq 0 \\ -1 & \text{if } t < 0 \end{cases} \qquad (3.202)$$

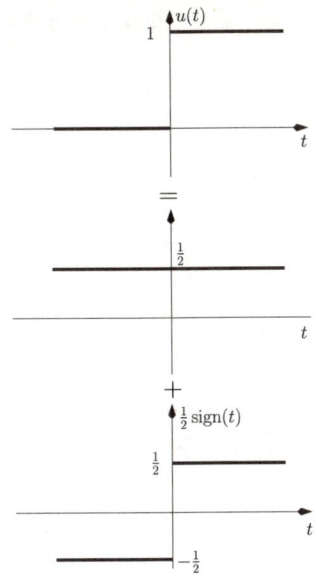

Figure 3.40: *Unit-step function as the addition of a constant value with the signum function.*

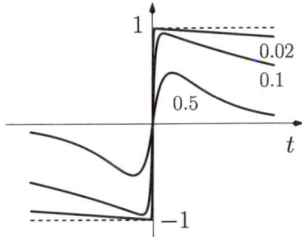

Figure 3.41: *The signum function obtained from eqn 3.203. $\alpha = 0.5$, 0.1 and 0.02.*

The Fourier transform of $u(t)$ is the addition of the Fourier transform of a DC value (discussed above in detail) with the Fourier transform of the signum function. We need a mathematical manipulation so that the calculation of the transform of the signum function converges to its correct value. Figure 3.41 shows that $\text{sign}(t)$ can also be written as follows:

$$\text{sign}(t) = \lim_{\alpha \to 0} \begin{cases} \left(1 - e^{-\frac{t}{\alpha}}\right) e^{-\alpha t} & \text{if } t \geq 0 \\ \left(e^{\frac{t}{\alpha}} - 1\right) e^{\alpha t} & \text{if } t < 0 \end{cases} \quad (3.203)$$

with $\alpha > 0$. Figure 3.41 shows eqn 3.203 for $\alpha = 0.5, 0.1$ and 0.02. From this figure it is clear that as α tends to zero eqn 3.203 tends to eqn 3.202.

The Fourier transform of the signum function can now be calculated as follows:

$$\begin{aligned} \text{Sign}(f) &= \lim_{\alpha \to 0} \int_{-\infty}^{0} \left(e^{\frac{t}{\alpha}} - 1\right) e^{\alpha t} e^{-j 2\pi f t} dt \\ &+ \lim_{\alpha \to 0} \int_{0}^{\infty} \left(1 - e^{-\frac{t}{\alpha}}\right) e^{-\alpha t} e^{-j 2\pi f t} dt \\ &= \lim_{\alpha \to 0} \left[\left(\frac{-1}{\alpha - 2j\pi f} + \frac{\alpha e^{\frac{t}{\alpha}}}{\alpha^2 - 2j\pi f \alpha + 1}\right) e^{\alpha t - j 2\pi f t}\right]_{-\infty}^{0} \\ &+ \lim_{\alpha \to 0} \left[\left(\frac{\alpha e^{-\frac{t}{\alpha}}}{\alpha^2 + 2j\pi f \alpha + 1} - \frac{1}{\alpha + 2j\pi f}\right) e^{-\alpha t - j 2\pi f t}\right]_{0}^{\infty} \\ &= \frac{1}{j\pi f} \quad (3.204) \end{aligned}$$

where we have used the following equalities:

$$\lim_{\alpha \to 0} e^{\frac{t}{\alpha}} = 0 \text{ for } t < 0, \ (\alpha > 0)$$

$$\lim_{\alpha \to 0} e^{\frac{-t}{\alpha}} = 0 \text{ for } t > 0, \ (\alpha > 0)$$

Using eqns 3.201, 3.199 and 3.204, we can write the Fourier transform of the unit step function as follows:

$$\begin{aligned} U(f) &= \frac{1}{2}\delta(f) + \frac{1}{2}\text{Sign}(f) \\ &= \frac{1}{2}\delta(f) + \frac{1}{j 2\pi f} \quad (3.205) \end{aligned}$$

The generalised Fourier transform also allows us to perform the calculation of the Fourier transforms of periodic functions. Let us consider, for example, a periodic voltage signal, $v(t)$ with period T which has a Fourier series such that:

$$v(t) = \sum_{n=-\infty}^{\infty} V_n e^{j 2\pi \frac{n}{T} t} \quad (3.206)$$

The Fourier transform of $v(t)$, $V(f)$, can be related to its Fourier series coefficients, V_n, as follows:

$$\begin{aligned}
V(f) &= \int_{-\infty}^{\infty} v(t)\, e^{-j2\pi f t}\, dt \\
&= \int_{-\infty}^{\infty} \sum_{n=-\infty}^{\infty} V_n\, e^{j2\pi \frac{n}{T} t}\, e^{-j2\pi f t}\, dt \\
&= \sum_{n=-\infty}^{\infty} V_n \int_{-\infty}^{\infty} e^{j2\pi \frac{n}{T} t}\, e^{-j2\pi f t}\, dt \\
&= \sum_{n=-\infty}^{\infty} V_n \int_{-\infty}^{\infty} e^{-j2\pi \left(f - \frac{n}{T}\right) t}\, dt
\end{aligned} \qquad (3.207)$$

This integral can be related to the Fourier transform of a DC quantity. According to eqn 3.199 we have

$$\int_{-\infty}^{\infty} 1 \times e^{-j2\pi f t}\, dt = \delta(f) \qquad (3.208)$$

and, therefore, the integral of eqn 3.207 can be calculated as

$$\int_{-\infty}^{\infty} e^{-j2\pi \left(f - \frac{n}{T}\right) t}\, dt = \delta\left(f - \frac{n}{T}\right) \qquad (3.209)$$

Finally, eqn 3.207 which represents the spectrum of the periodic waveform $v(t)$ can be expressed as

$$V(f) = \sum_{n=-\infty}^{\infty} V_n\, \delta\left(f - \frac{n}{T}\right) \qquad (3.210)$$

which is a discrete series of phasors as expected.

Example 3.3.8 Determine the spectrum $V(f)$ of the periodic voltage waveform, $v(t)$ of figure 3.35 with $\tau = T/3$.

Solution: From eqns 3.117 and 3.210 we can write $V(f)$ as follows:

$$\begin{aligned}
V(f) &= \sum_{n=-\infty}^{\infty} \frac{V_A \tau}{T} \operatorname{sinc}\left(\frac{n\tau}{T}\right) \delta\left(f - \frac{n}{T}\right) \\
&= \sum_{n=-\infty}^{\infty} \frac{V_A}{3} \operatorname{sinc}\left(\frac{n}{3}\right) \delta\left(f - \frac{n}{T}\right)
\end{aligned} \qquad (3.211)$$

Rayleigh's energy theorem

This theorem states that the energy, E_x, of a signal $x(t)$ can be calculated from its spectrum $X(f)$ according to the following eqn:

$$E_x = \int_{-\infty}^{\infty} x(t)^2\, dt = \int_{-\infty}^{\infty} |X(f)|^2\, df \qquad (3.212)$$

3. Frequency domain electrical signal and circuit analysis

Example 3.3.9 Determine the energy of the causal[6] exponential, $w(t)$ shown in figure 3.42, using Rayleigh's energy theorem. Then, show that this result is the same as that obtained from the integration of $w^2(t)$.

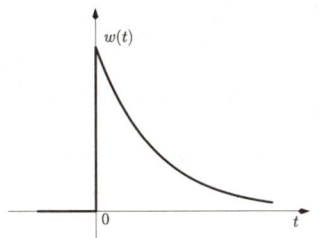

Figure 3.42: *Causal exponential.*

Solution: The causal exponential $w(t)$ of figure 3.42 can be written as:

$$w(t) = \begin{cases} e^{-\sigma t} & \text{for } t \geq 0 \\ 0 & \text{elsewhere} \end{cases} \quad (3.213)$$

where $\sigma > 0$. Hence, the spectrum of $w(t)$ can be calculated as

$$\begin{aligned} W(f) &= \int_{-\infty}^{\infty} w(t) e^{-j2\pi f t} \, dt \\ &= \int_{0}^{\infty} e^{-\sigma t} e^{-j2\pi f t} \, dt \\ &= \left[\frac{1}{-\sigma - j2\pi f} e^{-j2\pi f t} \right]_{0}^{\infty} \\ &= \frac{1}{\sigma + j2\pi f} \end{aligned} \quad (3.214)$$

From eqn 3.212 the energy of $w(t)$ can be calculated as follows,

$$\begin{aligned} E_w &= \int_{-\infty}^{\infty} \frac{1}{\sigma^2 + (2\pi f)^2} \, df \\ &= \left[\frac{1}{\sigma 2\pi} \tan^{-1} \left(\frac{2\pi f}{\sigma} \right) \right]_{-\infty}^{\infty} \\ &= \frac{1}{\sigma 2\pi} \left(\frac{\pi}{2} + \frac{\pi}{2} \right) = \frac{1}{2\sigma} \end{aligned} \quad (3.215)$$

The energy can also be calculated according to:

$$\begin{aligned} E_w &= \int_{-\infty}^{\infty} w^2(t) \, dt \\ &= \int_{0}^{\infty} e^{-2\sigma t} \, dt \\ &= \left[\frac{1}{-2\sigma} e^{-2\sigma t} \right]_{0}^{\infty} \\ &= \frac{1}{2\sigma} \end{aligned} \quad (3.216)$$

This result is the same as that given by Rayleigh's energy theorem.

---

[6]A causal signal $x(t)$ is defined as any signal which is zero for $t < 0$.

3.3.7 Transfer function and impulse response

The transfer function, $H(\omega)$ or $H(f)$, of a circuit has been introduced in section 3.3.3 where we saw that it can be obtained from phasor analysis, more specifically by evaluating the ratio of the phasor of the output signal with that of the input signal for all frequencies, ω or $f = \omega/(2\pi)$. There are four fundamental types of transfer functions:

- **Voltage transfer function**: In this situation both input and output phasors are voltages. The transfer function represents a *voltage* gain (or voltage attenuation if this gain is less than one) versus the frequency. This transfer function is dimensionless.

- **Current transfer function**: Both input and output phasors are currents. Hence, the transfer function represents a *current* gain (or current attenuation if this gain is less than one) versus the frequency. This transfer function is also dimensionless.

- **Impedance transfer function**: In this situation the input phasor is a current whilst the output phasor is a voltage. Note that now the gain versus the frequency has units of ohms. This transfer function is usually called 'transimpedance gain'.

- **Admittance transfer function**: The input phasor is a voltage whilst the output phasor is a current. Now the gain versus the frequency, represented by this transfer function, has units of siemens. This transfer function is usually called 'transconductance gain'.

From the discussion about the Fourier series we have concluded that knowledge of the transfer function of a circuit allows the calculation of the spectrum of the output signal for a given periodic input signal, using eqn 3.127. In similar way, the spectrum of the output signal, $X_o(f)$, for a given input signal with $X_i(f)$ can be calculated as

$$X_o(f) = H(f) \times X_i(f) \quad (3.217)$$

Taking the inverse Fourier transform of $X_o(f)$ and $X_i(f)$ we obtain the time domain representation for the output and input signals respectively. We can also take the inverse Fourier transform of the transfer function, $H(f)$, which is defined as the circuit **impulse response** represented by $h(t)$. The impulse response of a circuit is the circuit response when a Dirac delta function (with unit area) is applied to this circuit.

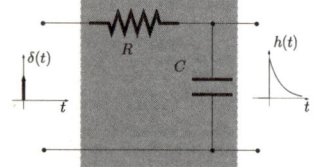

Figure 3.43: *Impulse response of an RC circuit.*

Example 3.3.10 Determine the impulse response of the circuit of figure 3.43.

Solution: The impulse response can be obtained calculating the inverse Fourier transform of the transfer function $H(f)$ which is given by eqn 3.124:

$$H(f) = \frac{1}{1 + j\,2\pi f RC} \quad (3.218)$$

3. Frequency domain electrical signal and circuit analysis

From example 3.3.9 we know that

$$\frac{1}{\sigma + j\,2\pi f} \quad \overset{\mathfrak{F}}{\Longleftrightarrow} \quad \begin{cases} e^{-\sigma t} & \text{for } t \geq 0 \\ 0 & \text{elsewhere} \end{cases} \tag{3.219}$$

Since $H(f)$ in eqn 3.218 can be written as

$$H(f) = \frac{1}{RC} \times \frac{1}{\frac{1}{RC} + j\,2\pi f} \tag{3.220}$$

$h(t)$ is given by

$$h(t) = \begin{cases} \frac{1}{RC} e^{-\frac{t}{RC}} & \text{for } t \geq 0 \\ 0 & \text{elsewhere} \end{cases} \tag{3.221}$$

This eqn can also be written as:

$$h(t) = \frac{1}{RC} e^{-\frac{t}{RC}} u(t) \tag{3.222}$$

where $u(t)$ represents the unit step function defined by eqn 3.200.

From a theoretical point-of-view the impulse response $h(t)$ of a circuit is obtained applying a Dirac delta function, as illustrated in figure 3.43. It should be clear to the reader that, in a practical situation, it is not possible to apply a Dirac delta pulse to a circuit to observe its impulse response; first because extremely narrow pulses with infinite amplitude are physically impossible to create and secondly because if this were possible the circuit would most certainly get damaged with the application of such a pulse! Hence, the application of a Dirac delta pulse should be understood as a mathematical model or abstraction which helps us to identify $h(t)$. However, as we show in example 3.3.11, if we apply a narrow pulse whose bandwidth is much greater than that of the circuit then the output is a very good estimate of its impulse response, $h(t)$.

Example 3.3.11 Show that if we apply a finite narrow pulse, whose bandwidth is much greater than the circuit bandwidth then the output produced by the circuit is a good estimate of its impulse response, $h(t)$.

Solution: Let us consider a circuit with a transfer function $H(f)$ with maximum frequency f_M as illustrated in figure 3.44. If we apply, to the circuit, a narrow rectangular pulse, $x_i(t)$, such that

$$x_i(t) = A \operatorname{rect}\left(\frac{t}{\tau}\right) \tag{3.223}$$

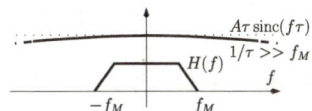

Figure 3.44: *A narrow pulse applied to a circuit. Frequency domain representation.*

with $\tau \ll f_M^{-1}$ then the spectrum of $x_i(t)$, that is $X_i(f) = A\tau \operatorname{sinc}(f\tau)$, is nearly constant in the frequency range $-f_M < f < f_M$, as shown in figure 3.44. Hence, the output spectrum $X_o(f)$ is:

$$\begin{aligned} X_o &= X_i(f) H(f) \\ &\simeq A\tau H(f) \end{aligned} \tag{3.224}$$

Taking the inverse Fourier transform the output signal is $x_o(t) \simeq A\,\tau\,h(t)$.

3.3.8 The convolution operation

The time domain waveform for $x_o(t)$, in eqn 3.217, can be obtained by calculating the following inverse Fourier transform:

$$x_o(t) = \int_{-\infty}^{\infty} H(f)\,X_i(f)\,e^{j\,2\,\pi\,f\,t}\,df \qquad (3.225)$$

Since the input signal, in the time domain, is represented by $x_i(t)$, this can be written as:

$$x_o(t) = \int_{-\infty}^{\infty} H(f) \underbrace{\int_{-\infty}^{\infty} x_i(\lambda)\,e^{j\,2\,\pi\,f\,\lambda}\,d\lambda}_{X_i(f)}\,e^{j\,2\,\pi\,f\,t}\,df \qquad (3.226)$$

Changing the order of integration[7] this eqn can be written as follows:

$$x_o(t) = \int_{-\infty}^{\infty} x_i(\lambda) \underbrace{\int_{-\infty}^{\infty} H(f)\,e^{j\,2\,\pi\,f\,(t-\lambda)}\,df}_{h(t-\lambda)}\,d\lambda \qquad (3.227)$$

Because $H(f)$ has an inverse Fourier transform represented by $h(t)$, then $x_o(t)$ can be calculated as:

$$x_o(t) = \int_{-\infty}^{\infty} x_i(\lambda)\,h(t-\lambda)\,d\lambda \qquad (3.228)$$

This represents the **convolution** operation between $x(t)$ and $h(t)$. This operation is also represented as follows:

$$x_o(t) = x_i(t) * h(t) \qquad (3.229)$$

with $*$ indicating the convolution operation. It can be shown (see problem 3.10) that

$$x_i(t) * h(t) = h(t) * x_i(t) \qquad (3.230)$$

In order to understand the convolution operation we consider the RC circuit of figure 3.45 where now a single square voltage pulse, $v_i(t)$, is applied to its input. The output voltage $v_o(t)$ can be determined according to eqn 3.228. However, we shall evaluate $v_o(t)$ by first approximating the input square pulse by a sum of $(N+1)$ Dirac delta functions, as illustrated in figure 3.46. Now

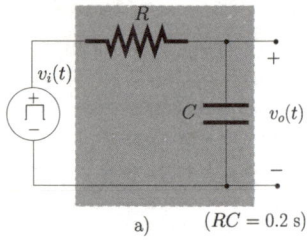
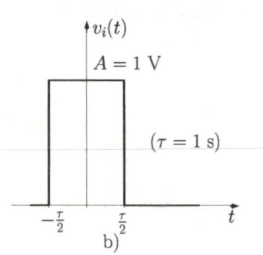

Figure 3.45: *Square voltage pulse, $v_i(t)$, is applied to an RC circuit.*

---

[7]This change of the order of integration is possible whenever the functions are absolutely integrable. The variety of signals of interest and their corresponding spectra obey this requirement. See [2] for more details.

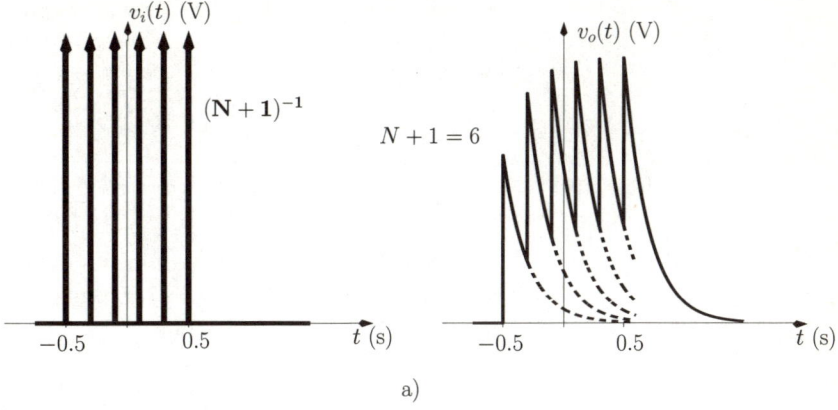

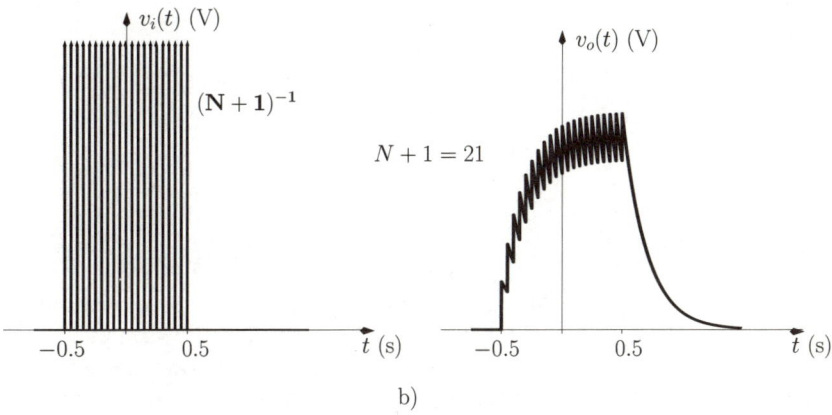

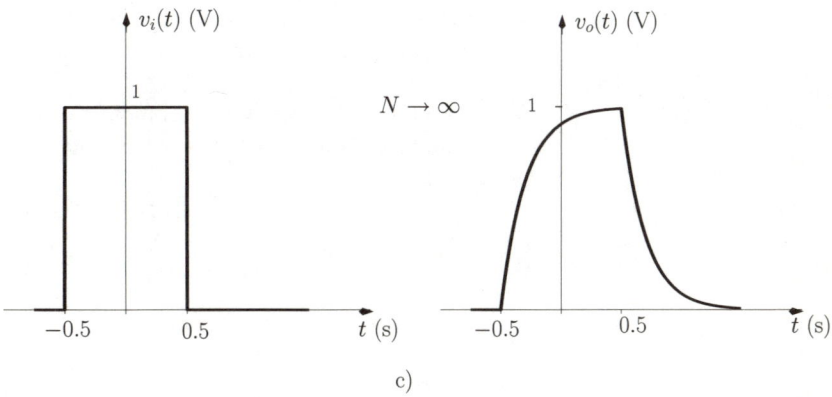

Figure 3.46: *Illustration of the convolution operation with the input voltage signal in the circuit of figure 3.45 being approximated as a sum of $(N+1)$ Dirac delta functions. a) $N+1 = 6$, b) $N+1 = 21$, c) $N \to \infty$.*

$v_i(t)$ is approximated by the following expression:

$$v_i(t) \simeq \frac{A\tau}{N+1} \sum_{k=0}^{N} \delta\left(t + \frac{\tau}{2} - k\frac{\tau}{N}\right)$$

$$= \frac{1}{N+1} \sum_{k=0}^{N} \delta\left(t + \frac{N-2k}{2N}\right) \quad (3.231)$$

Note that the sum of the areas of the $(N+1)$ delta functions is equal to the area of the rectangular pulse, $A\tau = 1$. Using eqn 3.228 the voltage at the output of the RC circuit, $v_o(t)$ is given by

$$v_o(t) = \int_{-\infty}^{\infty} v_i(\lambda) h(t-\lambda) d\lambda \quad (3.232)$$

$$= \frac{1}{N+1} \sum_{k=0}^{N} \int_{-\infty}^{\infty} \delta\left(\lambda + \frac{N-2k}{2N}\right) h(t-\lambda) d\lambda \quad (3.233)$$

where $h(t)$ is given by eqn 3.222 with $RC = 0.2$ s. From eqn 3.193 we can write this as,

$$v_o(t) = \frac{5}{N+1} \sum_{k=0}^{N} e^{-5\left(t + \frac{N-2k}{2N}\right)} u\left(t + \frac{N-2k}{2N}\right) \quad (3.234)$$

This eqn is shown in figure 3.46 with $(N+1) = 6$, $(N+1) = 21$ and $N \to \infty$. From figure 3.46 a) $(N+1=6)$ it is clear that the result of the convolution between $v_i(t)$ and $h(t)$ can be seen as a weighted sum of the impulse response $h(t)$ induced by each of the Dirac delta functions which approximates the input signal $v_i(t)$. By increasing N we increase the number of delta functions and, of course, we increase their density in the time interval τ. If $N \to \infty$ then $v_i(t)$ 'becomes' the rectangular pulse as shown in figure 3.45 c) and $v_o(t)$ is now a smooth waveform. Note the similarity of $v_o(t)$ obtained now, when the input voltage is a single rectangular pulse, with the output voltage when the input voltage is a periodic sequence of rectangular pulses (see also figures 3.23 and 3.26).

Figure 3.47 illustrates the computation of $v_o(t)$ given by eqn 3.232. According to the definition of $h(t)$ we can write

$$h(t-\lambda) = \begin{cases} e^{-\frac{t-\lambda}{RC}} & \text{for } t-\lambda \geq 0 \\ 0 & \text{for } t-\lambda < 0 \end{cases} \quad (3.235)$$

and since $RC = 0.2$ we have

$$h(t-\lambda) = \begin{cases} e^{-5(t-\lambda)} & \text{for } \lambda \leq t \\ 0 & \text{for } \lambda > t \end{cases} \quad (3.236)$$

Figure 3.47 a) illustrates the integrand of eqn 3.232 for $t = -0.75$ s. Note the inversion of $h(-0.75 - \lambda)$ in the λ axis. In this figure it is clear that the

3. Frequency domain electrical signal and circuit analysis

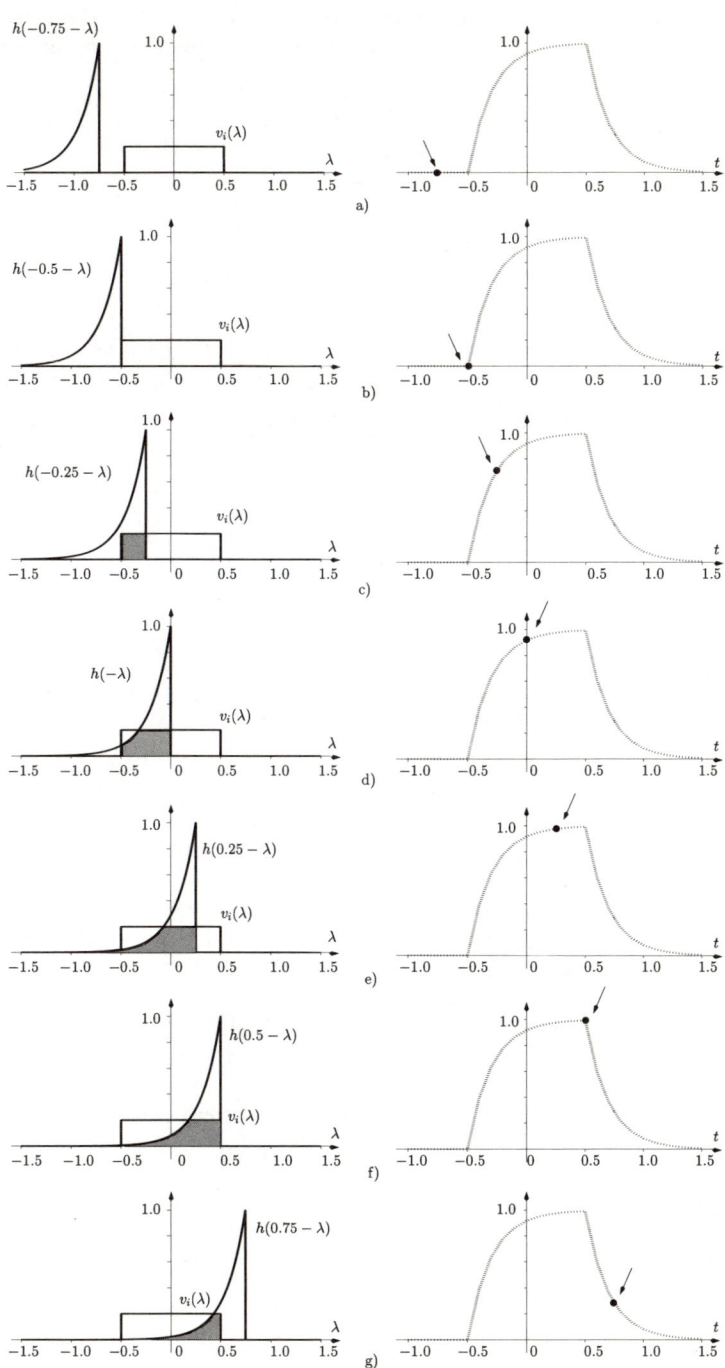

Figure 3.47: *Illustration of mathematical convolution.*

product of $h(-0.75 - \lambda)$ with $v_i(\lambda)$ is zero and, accordingly, $v_o(0.75) = 0$. In fact, the output voltage is zero until $t > -0.5$ s as illustrated by figure 3.47 b). For $-0.5 < t < 0.5$ the output voltage can be obtained from the following expression:

$$v_o(t) = 5 \int_{-0.5}^{t} e^{-5(t-\lambda)} d\lambda , \qquad -0.5 \leq t < 0.5 \qquad (3.237)$$

$$= 5 \left[\frac{1}{5} e^{-5(t-\lambda)} \right]_{-0.5}^{t} , \qquad -0.5 \leq t < 0.5$$

$$= 1 - e^{-5(t+0.5)} , \qquad -0.5 \leq t < 0.5 \qquad (3.238)$$

Figures 3.47 c), d) and e) illustrate the calculation of eqn 3.237 for $t = -0.25$, $t = 0$ and $t = 0.25$, respectively. For $t \geq 0.5$ the output voltage can be obtained from the expression indicated below (see figures 3.47 f) and 3.47 g)):

$$v_o(t) = 5 \int_{-0.5}^{0.5} e^{-5(t-\lambda)} d\lambda , \qquad t \geq 0.5$$

$$= 5 \left[\frac{1}{5} e^{-5(t-\lambda)} \right]_{-0.5}^{0.5} , \qquad t \geq 0.5$$

$$= e^{-5(t-0.5)} - e^{-5(t+0.5)} , \qquad t \geq 0.5 \qquad (3.239)$$

From the above we can write $v_o(t)$ as follows;

$$v_o(t) = \begin{cases} 0 & \text{for } t < -0.5 \\ 1 - e^{-5(t+0.5)} & \text{for } -0.5 \leq t < 0.5 \\ e^{-5(t-0.5)} - e^{-5(t+0.5)} & \text{for } t \geq 0.5 \end{cases} \qquad (3.240)$$

Example 3.3.12 Determine the waveform resulting from the convolution of two identical rectangular waveforms $x_1(t)$ and $x_2(t)$ with amplitude $A = 1$ and width $T = 1$ s.

Solution: According to the definition of a rectangular waveform (see also eqn 3.114) we can write $x_1(\lambda)$ as

$$x_1(\lambda) = \begin{cases} A , & \frac{-1}{2} < \frac{\lambda}{T} < \frac{1}{2} \\ 0 , & \text{elsewhere} \end{cases} \qquad (3.241)$$

that is

$$x_1(\lambda) = \begin{cases} 1 , & \frac{-1}{2} < \lambda < \frac{1}{2} \\ 0 , & \text{elsewhere} \end{cases} \qquad (3.242)$$

and $x_2(t - \lambda)$ can be written as

$$x_2(t - \lambda) = \begin{cases} A , & \frac{-1}{2} < \frac{t-\lambda}{T} < \frac{1}{2} \\ 0 , & \text{elsewhere} \end{cases} \qquad (3.243)$$

3. Frequency domain electrical signal and circuit analysis

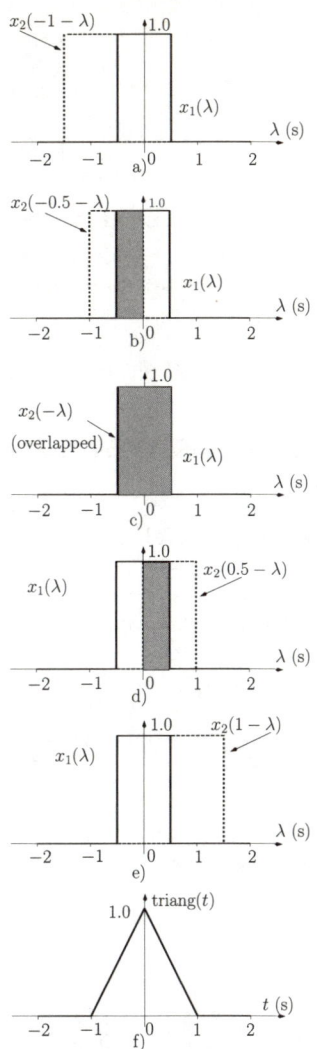

Figure 3.48: *Convolution of two identical rectangular waveforms.*

that is

$$x_2(t - \lambda) = \begin{cases} 1, & t - \tfrac{1}{2} < \lambda < t + \tfrac{1}{2} \\ 0, & \text{elsewhere} \end{cases} \quad (3.244)$$

The convolution of $x_1(t)$ and $x_2(t)$ is given by

$$y(t) = \int_{-\infty}^{\infty} x_1(\lambda)\, x_2(t - \lambda)\, d\lambda \quad (3.245)$$

Figure 3.48 a) shows the functions whose product forms the integrand of eqn 3.245 for $t = -1$ s, that is, this figure shows $x_1(\lambda)$ and $x_2(-1 - \lambda)$. From this figure it is clear that the product of these two functions is zero and so is the result of its integration. Note that for $t \leq -1$ the product of $x_1(\lambda)$ with $x_2(t - \lambda)$ is zero. Figures 3.48 b), c) and d) indicate that for the time interval $-1 < t \leq 1$ the two functions overlap. This overlap is maximum for $t = 0$ as shown by figure 3.48 c). For the time interval, $-1 < t \leq 0$, we can write eqn 3.245 as follows:

$$\begin{aligned} y(t) &= \int_{-0.5}^{t+0.5} d\lambda, & -1 < t \leq 0 \\ &= t + 1, & -1 < t \leq 0 \end{aligned} \quad (3.246)$$

For the time interval $0 < t < 1$ the overlap of the two functions decreases as illustrated by figure 3.48 d) for $t = 0.5$. For this time interval we can write eqn 3.245 as follows:

$$\begin{aligned} y(t) &= \int_{t-0.5}^{0.5} d\lambda, & 0 < t < 1 \\ &= 1 - t, & 0 < t < 1 \end{aligned} \quad (3.247)$$

For $t \geq 1$ there is no overlap between $x_1(\lambda)$ and $x_2(-1 - \lambda)$ and $y(t)$ is again zero. From the above we can write $y(t)$ as

$$y(t) = \begin{cases} 0 & \text{if } t \leq -1 \\ 1 + t & \text{if } -1 < t \leq 0 \\ 1 - t & \text{if } 0 < t < 1 \\ 0 & \text{if } t \geq 1 \end{cases} \quad (3.248)$$

Figure 3.48 f) shows that $y(t)$ represents a triangle. In fact eqn. 3.248 defines the triangular function, triang(t).

The discussion presented above reveals, once again, the advantage of analysing circuits and signals in the frequency domain. While time domain analysis involves the calculation of convolution integrals using the circuit impulse response and the time domain signal, the frequency domain involves the multiplication of the circuit transfer functions with the signal spectrum (or signal Fourier transform) which is, by far, a more simple mathematical operation.

This is a consequence of the *convolution theorems*:

$$x(t) * y(t) \overset{\mathfrak{F}}{\Longleftrightarrow} X(f) \times Y(f) \tag{3.249}$$

$$x(t) \times y(t) \overset{\mathfrak{F}}{\Longleftrightarrow} X(f) * Y(f) \tag{3.250}$$

These two theorems state that the convolution of two functions in the time domain corresponds to multiplication of its Fourier transforms in the frequency domain while multiplication of two functions in the time domain corresponds to convolution of its Fourier transforms in the frequency domain.

3.4 Bibliography

1. K.C.A. Smith and R.E. Alley, *Electrical Circuits, an Introduction*, 1992 (Cambridge University Press).

2. A.V. Oppenheim and A.S. Willsky, *Signals and Systems*, 1996 (Prentice Hall Signal Processing Series), 2nd edition.

3. C. Chen, *System and Signal Analysis*, 1994 (Saunders College Publishing), 2nd edition.

4. A.B. Carlson, P.B. Crilly, J.C. Rutledge, *Communication Systems: An Introduction to Signals and Noise in Electrical Communication*, 2001 (McGraw-Hill Series in Electrical Engineering), 4th edition.

5. M.J. Roberts, *Signals and Systems: Analysis using Transform Methods and Matlab®*, 2003, (McGraw-Hill International Editions).

3.5 Problems

3.1 Determine the effective (or RMS) value of each periodic waveform shown in figure 3.49. Consider $V_A = V_B = V_C = 2$ V and $T = 1$ ms. Figures a) and c) represent sections of sinusoidal waves.

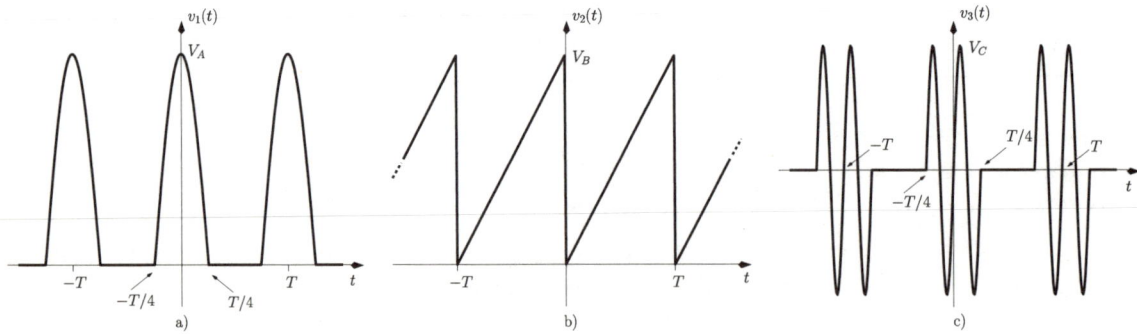

Figure 3.49: *Waveforms of problem 3.1.*

3.2 Sown that the average power dissipated by an impedance Z_L is $P_{AV} = 0.5\,\text{Real}\,[V_A\,I_A^*]$ where V_A is the phasor representing the voltage across Z_L and I_A is the phasor representing the current through Z_L.

3.3 Using phasor analysis, calculate all voltages across and currents through each passive element of the circuit of figure 3.50. $v(t) = 10\,\cos(\omega t + \pi/4)$ V and $i(t) = 0.15\,\cos(\omega t + \pi/3)$ A. The angular frequency ω is 30 krad/s.

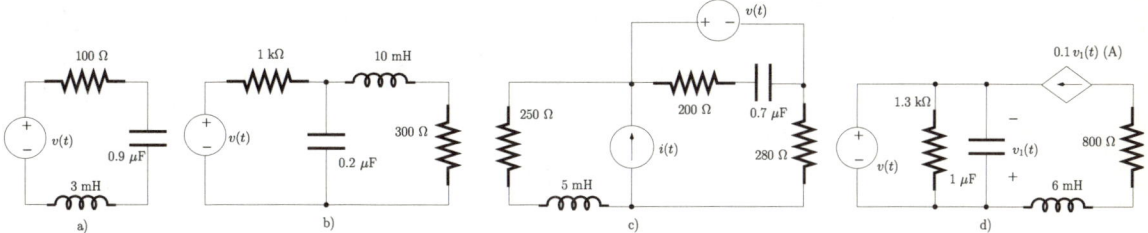

Figure 3.50: *Circuits of problem 3.3.*

3.4 For the circuit of figure 3.51 find the load Z_L for which maximum power transfer occurs at $f = 35$ kHz.

3.5 Consider again the waveforms of figure 3.49. Determine the Fourier series of each waveform and sketch the correspondent line spectrum.

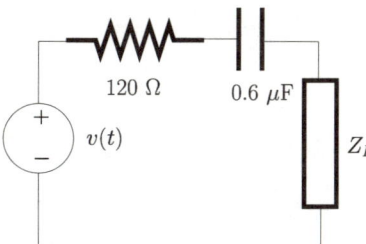

Figure 3.51: *Circuit of problem 3.4.*

3.6 Show that the transfer function, $H(f) = V_o/V_s$, of the circuit of figure 3.29 can be expressed by eqn 3.151.

3.7 Determine the output voltage, $v_o(t)$, for the circuits of figure 3.52. $v_s(t)$ is the periodic sequence of triangular pulses as shown in figure 3.49. Consider $T = 0.1$ ms and $V_B = 2.5$ V.

3.8 Determine the transfer functions, $H(f)$, and the 3 dB bandwidth of the circuits of figure 3.53. For the circuit of figure 3.53 a) determine the quality factor.

3.9 Determine the Fourier transforms of the signals shown in figure 3.54.

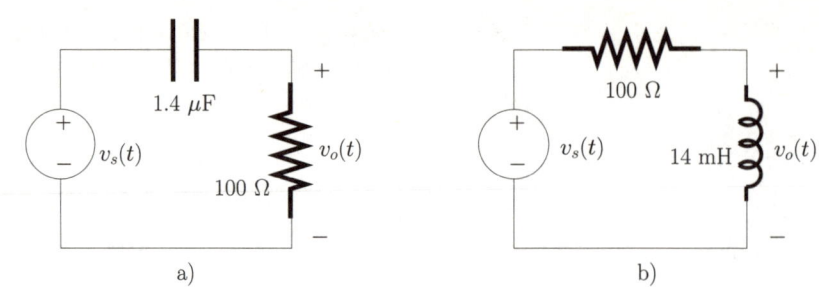

Figure 3.52: *Circuits of problem 3.7.*

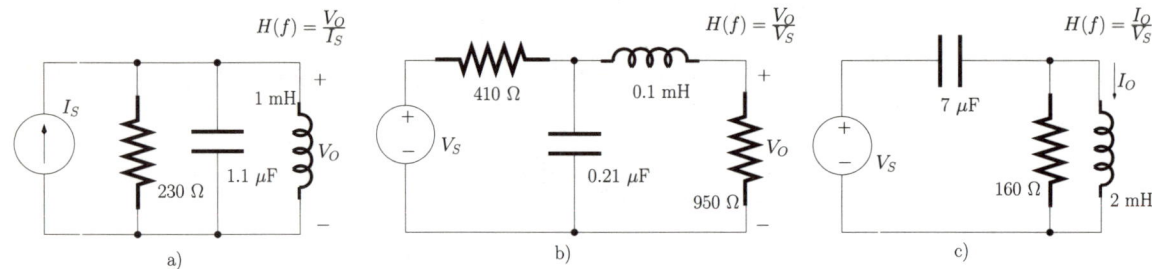

Figure 3.53: *Circuits of problem 3.8.*

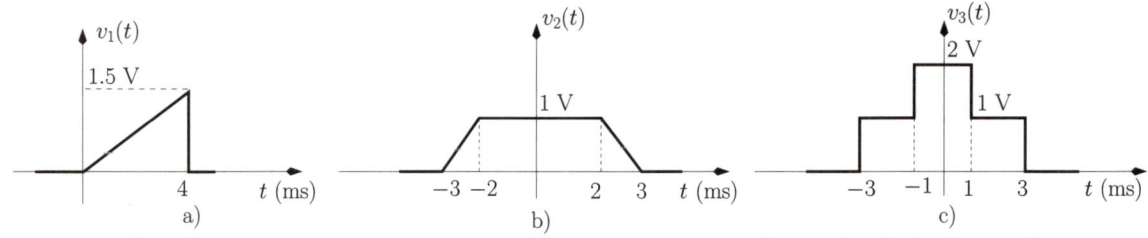

Figure 3.54: *Signals of problem 3.9.*

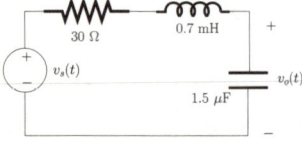

Figure 3.55: *Circuit of problem 3.11.*

3.10 Show that the mathematical convolution satisfies the commutative property expressed by eqn 3.230.

3.11 Find the impulse response of the circuit of figure 3.55.

3.12 Consider the circuit of problem 3.11 driven by a rectangular pulse; $v_i = V_a \text{rect}(t/T_a - 1/2)$. $V_a = 2$ V, $T_a = 0.5$ ms. Determine the output voltage using the convolution operation.

4 Natural and forced responses circuit analysis

4.1 Introduction

In this chapter we discuss the natural and the forced responses of passive electrical circuits. The natural response of a circuit is its response when the circuit is not been driven by any signal source and it usually describes how the energies stored in the capacitors and inductors are transferred and finally dissipated in resistive elements.

The forced response of an electrical circuit is the combination of the steady-state response, studied in detail in the last chapter, with the transient response and is a consequence of the application of a voltage or current signal to a circuit.

The time domain expressions for the voltages or currents in the circuit can be derived by using Fourier transform techniques, discussed in the last chapter, or by using the Laplace transform.

We start by presenting the analysis of a circuit with non-zero initial conditions driven by a signal source which is switched-on at a specific instant of time (usually $t = 0$). Then we show that this analysis can be effected using Fourier transform techniques. In section 4.4 we present the Laplace transform and in section 4.5 we use this tool to analyse the natural response and the step forced responses of RC, RL, LC and RLC circuits.

4.2 Time domain analysis

In the last chapter we saw that the choice of the time origin is not relevant for the study of the steady-state behaviour of electrical circuits. However, to address both the natural and the transient responses, the definition of a time origin, usually associated with the start of an event such as the switching-on of a circuit, is obviously of fundamental importance. Signals which have a time origin, that is, they are zero for $t < 0$, are called *causal* signals since they have a physical origin or *cause*. Figure 4.1 shows two examples of causal voltage signals which can be expressed as follows:

$$v_{S_1}(t) = \frac{V_s}{2}\left[1 - \cos(\omega\, t)\right] u(t) \qquad (4.1)$$

$$v_{S_2}(t) = V_s \sum_{k=0}^{\infty} \left[u(t - kT) - u\left(t - kT - \frac{T}{2}\right)\right] \qquad (4.2)$$

where V_s represents the amplitude of the signal and $\omega = 2\pi/T$ is the angular frequency where T is the period. $u(t)$ is the unit-step function as described in the last chapter (eqn 3.200). The multiplication by $u(t)$ in eqn 4.1 guarantees

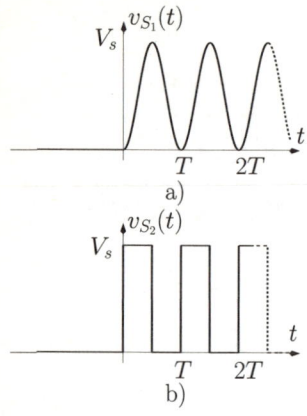

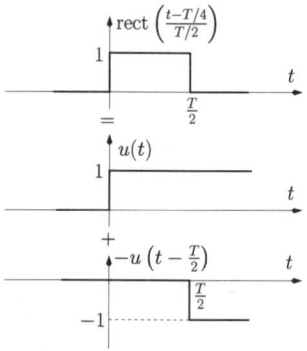

Figure 4.1: *Causal signals. Examples.*

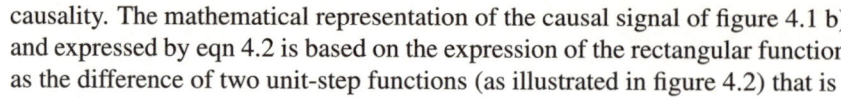

Figure 4.2: *Rectangular signal as the sum to two unit-step functions.*

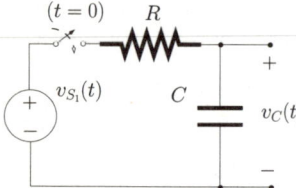

Figure 4.3: *RC circuit.*

causality. The mathematical representation of the causal signal of figure 4.1 b) and expressed by eqn 4.2 is based on the expression of the rectangular function as the difference of two unit-step functions (as illustrated in figure 4.2) that is

$$\text{rect}\left(\frac{t-\frac{T}{4}}{\frac{T}{2}}\right) = u(t) - u\left(t - \frac{T}{2}\right)$$

Circuits with non-zero initial conditions

In section 3.3.7, we presented the concept of the impulse response of a circuit. This, together with the convolution operation, allows us to obtain the output response of a circuit in the time domain when a particular input signal is driving the circuit. However, it is important to note that this type of analysis assumed that no capacitors or inductors were storing energy prior to the application of the input signal. That is, this type of analysis is valid when all the *initial conditions* associated with the capacitors and inductors are zero.

In order to address analysis with non-zero initial conditions driven by causal signal sources we consider the circuit of figure 4.3 driven by the voltage source described by eqn 4.1. We want to determine the voltage across the capacitor, $v_C(t)$, in the time domain. This circuit incorporates a switch which is open for $t < 0$. We assume that for $t < 0$ the capacitor has a voltage across its terminals equal to V_{co}[1]. The switch is closed at $t = 0$ applying $v_{S_1}(t)$ to the RC circuit. Applying Kirchhoff's current law to the RC circuit of figure 4.3 we can write, for $t \geq 0$, the following eqn:

$$\frac{v_{S_1}(t) - v_C(t)}{R} = C\frac{dv_C(t)}{dt} \qquad (4.3)$$

This can also be written as

$$v_{S_1}(t) = v_C(t) + RC\frac{dv_C(t)}{dt}$$

or

$$\frac{V_s}{2}[1 - \cos(\omega t)]u(t) = v_C(t) + \tau\frac{dv_C(t)}{dt} \qquad (4.4)$$

where $\tau = RC$. The differential equation defined by 4.4 has a general solution, for $t \geq 0$, which can be written as follows:

$$v_C(t) = \underbrace{\left(\frac{V_s}{2} - \frac{V_s}{2}\frac{\omega^2\tau^2}{1+\omega^2\tau^2}e^{-\frac{t}{\tau}} - \frac{V_s}{2}\left[\frac{1}{1+\omega^2\tau^2}\cos(\omega t)\right.\right.}_{1^{\text{st term}}}$$

$$\underbrace{\left.\left. + \frac{\omega\tau}{1+\omega^2\tau^2}\sin(\omega t)\right]\right)u(t)}_{1^{\text{st term (cont.)}}} + \underbrace{V_{co}\,e^{-\frac{t}{\tau}}u(t)}_{2^{\text{nd term}}} \qquad (4.5)$$

---

[1] Note that this corresponds to a stored energy $CV_{co}^2/2$ (see also eqn 1.25).

4. Natural and forced responses circuit analysis

We observe that the first term is related to the application of the input signal $v_{S_1}(t)$ to the RC circuit. In fact, it can be shown that this first term can be obtained from the convolution operation between the input voltage $v_{S_1}(t)$ and the circuit impulse response (see example 4.2.1). However, the second term of eqn 4.5 depends only on the initial voltage of the capacitor, V_{co}.

By using trigonometric identities (appendix A) and taking into consideration the initial condition ($t<0$) eqn 4.5 can be written, for all time t, as follows

$$v_C(t) = \underbrace{\frac{V_s}{2}\left(1 - \frac{1}{\sqrt{1+\omega^2\tau^2}}\sin\left[\omega t + \tan^{-1}\left(\frac{1}{\omega\tau}\right)\right]\right)u(t)}_{\text{Contribution by the steady-state response (Fig. 4.4 b)}}$$

$$+ \underbrace{\frac{-V_s}{2}\frac{\omega^2\tau^2}{1+\omega^2\tau^2}e^{-\frac{t}{\tau}}u(t)}_{\text{Contribution by the transient response (Fig. 4.4 c)}}$$

$$+ \underbrace{V_{co}e^{-\frac{t}{\tau}}u(t)}_{\text{Natural response (Fig. 4.4 d)}} + \underbrace{V_{co}u(-t)}_{\text{Initial condition (Fig. 4.4 d)}} \qquad (4.6)$$

where we can identify all the different contributions to $v_C(t)$. The steady-state response represents the voltage $v_C(t)$ when all transient phenomena of the circuit have vanished, as extensively discussed in the previous chapter.

To illustrate the different contributions we plot $v_C(t)$ and its constituent parts in figure 4.4. Figure 4.4 a) shows $v_C(t)$, given by eqn 4.6, assuming $\tau = 10/\omega$ and $V_{co} = V_s/6$. Figures 4.4 b) and 4.4 c) show the contributions by the steady-state and transient regimes, respectively. Figure 4.4 d) shows the initial condition and the natural response associated with $v_C(t)$. The transient response corresponds to the response of the circuit between the time when the signal is applied to the circuit and the time where the circuit is considered to be in steady-state. The combination of the steady-state response with the transient response is called the 'forced response' since it is caused by the forcing signal source applied to the circuit. The natural response associated with $v_C(t)$ corresponds to the voltage across the capacitor, for $t \geq 0$, that would be obtained if the voltage source $v_{S_1}(t)$ was replaced by a short-circuit at $t=0$. In fact, this natural response can be seen as the contribution of the voltage V_{co} to the overall voltage across the capacitor for $t \geq 0$. It is interesting to note that this contribution can also be obtained by applying the superposition theorem to the circuit of figure 4.3 resulting in the circuit of figure 4.5. For this circuit we can write

$$C\frac{dv_C(t)}{dt} + \frac{v_C(t)}{R} = 0 \qquad (4.7)$$

which is a homogeneous first order differential equation with a general solution given by

$$v_C(t) = V_{co}\,e^{-\frac{t}{\tau}} \qquad (4.8)$$

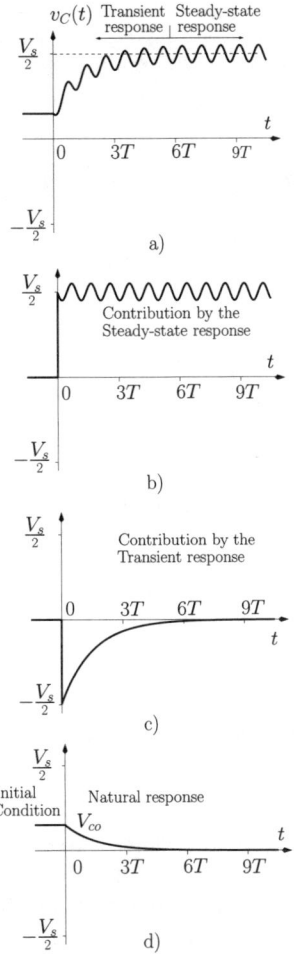

Figure 4.4: a) Time domain voltage across the capacitor of figure 4.3. b) Contribution by the steady-state regime. c) Contribution by the transient response. d) Natural response.

Since $v_C(t)$ is also the voltage across the resistance R we can obtain the current $i(t)$ as

$$i(t) = \frac{V_{co}}{R} e^{-\frac{t}{\tau}} \qquad (4.9)$$

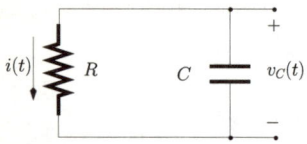

Figure 4.5: *Equivalent circuit for the calculation of the natural response ($t \geq 0$) of the RC circuit of figure 4.3.*

Example 4.2.1 Consider the RC circuit of figure 4.3 driven by the voltage source described by eqn 4.1. Determine the voltage $v_C(t)$ obtained by calculating the convolution between $v_{S_1}(t)$ and the circuit impulse response and show that this result is equal to the first term of eqn 4.5.

Solution: The RC circuit impulse response, $h(t)$, was derived in Chapter 3 and is given by eqn 3.222. The convolution between $v_{S_1}(t)$ and $h(t)$ can be determined as

$$
\begin{aligned}
v_C(t) &= \int_{-\infty}^{\infty} v_{S_1}(\lambda) \, h(t-\lambda) \, d\lambda \\
&= \int_{-\infty}^{\infty} \frac{V_s}{2} [1 - \cos(\omega\lambda)] \, u(\lambda) \, e^{-\frac{t-\lambda}{\tau}} \, u(t-\lambda) \, d\lambda \\
&= \int_0^t \frac{V_s}{2} [1 - \cos(\omega\lambda)] \, e^{-\frac{t-\lambda}{\tau}} \, d\lambda \\
&= \frac{V_s}{2} e^{-\frac{t-\lambda}{\tau}} \frac{1 + \omega^2 \tau^2 - \cos(\omega\lambda) - \omega\tau \sin(\omega\lambda)}{1+\omega^2\tau^2} \bigg|_0^t \\
&= \frac{V_s}{2} \left(1 - \frac{1}{1+\omega^2\tau^2} \cos(\omega t) - \frac{\omega\tau}{1+\omega^2\tau^2} \sin(\omega t) \right. \\
&\quad \left. - \frac{\omega^2 \tau^2}{1+\omega^2\tau^2} e^{-\frac{t}{\tau}} \right), \qquad (t \geq 0) \qquad (4.10)
\end{aligned}
$$

Comparing this eqn with the first term of eqn 4.5 we observe that they are identical.

Example 4.2.2 Consider again the RC circuit of figure 4.3 but now driven by the voltage source described by eqn 4.2. This voltage source corresponds to the periodic sequence of square pulses, as illustrated in figure 4.1 b). Consider a period T equal to $\tau/3$ ($\tau = RC$). Determine the voltage $v_C(t)$.

Solution: According to eqn 4.4, and for $t \geq 0$, the voltage $v_C(t)$ satisfies the following eqn:

$$V_s \sum_{k=0}^{\infty} \left[u(t-kT) - u\left(t-kT-\frac{T}{2}\right) \right] = v_C(t) + \tau \frac{dv_C(t)}{dt} \qquad (4.11)$$

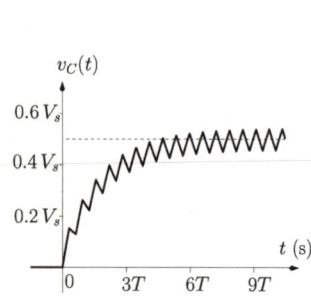

Figure 4.6: *The voltage across the capacitor given by eqn 4.12. $V_{co} = 0$ V.*

This differential equation has a general solution which can be written as follows:

$$
\begin{aligned}
v_C(t) &= V_{co} \, e^{-\frac{t}{\tau}} u(t) + V_s \sum_{k=0}^{\infty} \left(1 - e^{-\frac{t-kT}{\tau}} \right) u(t-kT) \\
&\quad - V_s \sum_{n=0}^{\infty} \left(1 - e^{-\frac{t-nT-T/2}{\tau}} \right) u\left(t - nT - \frac{T}{2}\right) \qquad (4.12)
\end{aligned}
$$

where the first term represents the natural response and the last two terms represent the forced response (given by the sum of the transient and steady-state responses). Figure 4.6 shows $v_C(t)$ given by eqn 4.12 where we assume that V_{co} is zero. Note the similarity between this waveform, when is in its steady-state condition, and that shown in figure 3.26 a) of section 3.3.3. In both situations the cut-off frequency, $f_c = (2\pi\tau)^{-1}$, of the RC circuit is smaller than $1/T$ causing the output signal, $v_C(t)$, to approximate a triangular waveform as discussed in example 3.3.4.

4.3 Transient analysis using Fourier transforms

The Fourier transform, studied in detail in the previous chapter, can also be used to analyse the application of causal signals to electrical circuits. In order to demonstrate this application we introduce two important theorems which allow us to take into account the initial conditions of electrical circuits; the differentiation and the integration theorems for causal signals.

4.3.1 Differentiation theorem

This theorem states that if a causal signal $x(t)$ has a Fourier transform $X(f)$ then the Fourier transform of the time derivative of $x(t)$ is given by:

$$\mathfrak{F}\left[\frac{d\,x(t)}{dt}\right] \;=\; j\,2\pi\,f\,X(f) - x(0) \qquad (4.13)$$

where $\mathfrak{F}[\cdot]$ designates the Fourier transform operation defined by eqn 3.178. In order to prove this theorem we consider the Fourier transform of $x(t)$;

$$X(f) \;=\; \int_{-\infty}^{\infty} x(t)\,e^{-j\,2\pi\,f\,t}\,dt$$

Since $x(t)$ is a causal signal the last eqn can be written as:

$$X(f) \;=\; \int_{0}^{\infty} x(t)\,e^{-j\,2\pi\,f\,t}\,dt \qquad (4.14)$$

This eqn can be evaluated by integrating (by parts) using the following equality:

$$\int w\,dz \;=\; w\,z - \int z\,dw$$

with

$$\begin{aligned} w &= x(t) \\ dz &= e^{-j\,2\pi\,f\,t}\,dt \end{aligned}$$

Thus, we have

$$\begin{aligned} dw &= \frac{d\,x(t)}{dt}\,dt \\ z &= \frac{-1}{j\,2\pi\,f}\,e^{-j\,2\pi\,f\,t} \end{aligned}$$

Now eqn 4.14 can be written as:

$$\int_0^\infty x(t)\, e^{-j2\pi f t}\, dt = x(t)\, \frac{-1}{j2\pi f} e^{-j2\pi f t}\Big|_0^\infty$$
$$+ \frac{1}{j2\pi f} \int_0^\infty \frac{dx(t)}{dt} e^{-j2\pi f t}\, dt$$

that is:

$$X(f) = \frac{1}{j2\pi f} x(0) + \frac{1}{j2\pi f} \int_0^\infty \frac{dx(t)}{dt} e^{-j2\pi f t}\, dt$$
$$= \frac{1}{j2\pi f} x(0) + \frac{1}{j2\pi f} \mathfrak{F}\left[\frac{dx(t)}{dt}\right]$$

where it is assumed that:

$$\lim_{t\to\infty} \frac{x(t)}{j2\pi f} e^{-j2\pi f t} = 0$$

Finally, we have the required result

$$\mathfrak{F}\left[\frac{dx(t)}{dt}\right] = j2\pi f\, X(f) - x(0) \qquad (4.15)$$

4.3.2 Integration theorem

This theorem states that if a causal signal $x(t)$ has a Fourier transform $X(f)$ then the Fourier transform of the time integration of $x(t)$ is given by:

$$\mathfrak{F}\left[\int_0^t x(\lambda)\, d\lambda\right] = \frac{1}{j2\pi f} X(f) + \frac{X(f)}{2}\delta(f) \qquad (4.16)$$

The proof of this theorem can be performed by noting first that

$$\int_0^t x(\lambda)\, d\lambda = \int_{-\infty}^\infty x(\lambda)\, u(t-\lambda)\, d\lambda$$
$$= x(t) * u(t)$$

Hence, from eqn 3.249, we can write (see also eqn 3.205)

$$\mathfrak{F}\left[\int_0^t x(\lambda)\, d\lambda\right] = X(f)\, U(f)$$
$$= \frac{1}{j2\pi f} X(f) + \frac{X(f)}{2}\delta(f) \qquad (4.17)$$

4.3.3 I–V characteristics for passive elements

These two theorems have a significant influence on the frequency domain current–voltage characteristics of electrical components that are capable of storing energy and when the voltage or the current applied to these elements are causal

4. Natural and forced responses circuit analysis

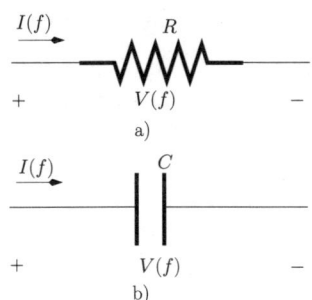

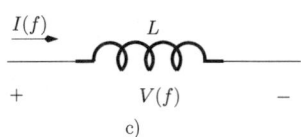

Figure 4.7: *I–V* characteristics for a) resistances, b) capacitances, c) inductances.

signals. Recall that for those electrical components capable of storing energy the *I–V* relationships for causal voltage and current signals accommodate the initial conditions. Thus, there will be a voltage associated with the energy stored in a capacitor and a current associated with the energy stored in an inductor.

Resistance

For resistances the *I–V* characteristic obeys the linear relationship between $V(f)$ and $I(f)$

$$V(f) = R\,I(f) \qquad (4.18)$$

where $V(f)$ and $I(f)$ are the Fourier transforms of the causal voltage and current applied to the resistance as indicated in figure 4.7 a).

Capacitance

The current through a capacitor can be related to the voltage across its terminals according to the following eqn:

$$i(t) = C\,\frac{d\,v(t)}{d\,t} \qquad (4.19)$$

Applying the Fourier transform to this eqn we obtain (see also eqn 4.13)

$$I(f) \;=\; j\,2\pi f\,C\,V(f) - C\,v(0) \qquad (4.20)$$

where $v(0)$ is the voltage across the capacitor terminals at $t = 0$. $V(f)$ and $I(f)$ represent the Fourier transforms of the voltage across and the current through the capacitor, respectively. Solving to obtain $V(f)$ we get

$$\begin{aligned}
V(f) &= \frac{I(f)}{j\,2\pi f\,C} + \frac{v(0)}{j\,2\pi f} \\
&= I(f)\,Z_C(f) + \frac{v(0)}{j\,2\pi f}
\end{aligned} \qquad (4.21)$$

with

$$Z_C(f) \;=\; \frac{1}{j\,2\pi f\,C}$$

Note that $Z_C(f)$ represents the complex impedance associated with the capacitor C as discussed in the last chapter.

Inductance

The voltage across an inductor can be related to the current flowing through it according to the following eqn:

$$v(t) = L\,\frac{d\,i(t)}{d\,t} \qquad (4.22)$$

Applying the Fourier transform to this eqn we obtain (see also eqn 4.13)

$$V(f) = j2\pi f L I(f) - L i(0) \quad (4.23)$$

where $i(0)$ is the current flowing through the inductor at $t = 0$. $V(f)$ and $I(f)$ represent the Fourier transforms of the voltage across and the current through the inductor, respectively. Eqn 4.23 can also be written as:

$$V(f) = Z_L(f) I(f) - L i(0) \quad (4.24)$$

with

$$Z_L(f) = j2\pi f L$$

where $Z_L(f)$ is the complex impedance of the inductor L.

Solving to obtain $I(f)$ we get

$$I(f) = \frac{V(f)}{j2\pi f L} + \frac{i(0)}{j2\pi f} \quad (4.25)$$

Example 4.3.1 Use the Fourier transform method to determine the voltage across the capacitor of the RC circuit of figure 4.3 when driven by the voltage source described by eqn 4.2.

Solution: Figure 4.8 shows the equivalent circuit for $t \geq 0$. Since $I(f)$ flows through both the resistor and the capacitor we can write (see also eqns 4.18 and 4.20):

$$\frac{V_{S_2}(f) - V_C(f)}{R} = j2\pi f C V_C(f) - C V_{co}$$

where $V_{S_2}(f)$ is the Fourier transform of $v_{S_2}(t)$. Solving this to obtain $V_C(f)$ we have:

$$V_C(f) = \frac{V_{S_2}(f)}{1 + j2\pi f \tau} + \frac{\tau V_{co}}{1 + j2\pi f \tau} \quad (4.26)$$

$V_{S_2}(f)$ can be obtained, according to eqns 3.188 and 3.205, as follows:

$$V_{S_2}(f) = \sum_{k=-\infty}^{\infty} \left(e^{-j2\pi f kT} - e^{-j2\pi f (kT+T/2)} \right)$$

$$\times \left(\frac{1}{2}\delta(f) + \frac{1}{j2\pi f} \right)$$

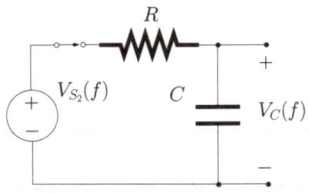

Figure 4.8: *RC equivalent circuit in the frequency domain.*

Now, eqn 4.26 can be written as

$$V_C(f) = \frac{\tau V_{co}}{1 + j2\pi f \tau} + \sum_{k=-\infty}^{\infty} \left(e^{-j2\pi f kT} - e^{-j2\pi f (kT+T/2)} \right)$$

$$\times \underbrace{\left(\frac{1}{1 + j2\pi f \tau} \frac{1}{2}\delta(f) + \frac{1}{(1 + j2\pi f \tau) j2\pi f} \right)}_{\frac{H(f)}{2}\delta(f) + \frac{1}{j2\pi f} H(f)} \quad (4.27)$$

where

$$H(f) = \frac{1}{1 + j2\pi f \tau}$$

Taking into account eqns 3.214, 4.16 and 3.188, the inverse Fourier transform of eqn 4.27, $v_C(t)$, for $t \geq 0$, can be calculated as

$$v_C(t) = V_{co} e^{-\frac{t}{\tau}} u(t) + \sum_{k=-\infty}^{\infty} \left(\int_0^{t-kT} \frac{1}{\tau} e^{-\frac{\lambda}{\tau}} u(\lambda) \, d\lambda \right.$$
$$\left. - \int_0^{t-kT-T/2} \frac{1}{\tau} e^{-\frac{\lambda}{\tau}} u(\lambda) \, d\lambda \right) \quad (4.28)$$

that is, $v_C(t)$ for $t \geq 0$ is given by

$$v_C(t) = V_{co} e^{-\frac{t}{\tau}} u(t) + V_s \sum_{k=0}^{\infty} \left(1 - e^{-\frac{t-kT}{\tau}}\right) u(t - kT)$$
$$- V_s \sum_{n=0}^{\infty} \left(1 - e^{-\frac{t-nT-T/2}{\tau}}\right) u\left(t - nT - \frac{T}{2}\right) \quad (4.29)$$

Note that this expression for $v_C(t)$ is the same as that obtained in example 4.2.2 and is shown in figure 4.6 for $T = \tau/3$.

4.4 The Laplace transform

We have shown above that the Fourier transform can be used to analyse the transient behaviour of electrical circuits. Here we introduce the Laplace transform which is highly suited for the analysis of circuits driven by causal signals.

Definition

The unilateral Laplace transform of a causal signal, $x(t)$ is defined as[2]:

$$X(s) = \int_0^\infty x(t) e^{-st} \, dt \quad (4.30)$$

where s is a complex number which is called the Laplace transform variable. As an example we calculate the Laplace transform of the unit-step function which, according to eqn 4.30, is given by

$$U(s) = \int_0^\infty u(t) e^{-st} \, dt$$
$$= \int_0^\infty 1 \times e^{-st} \, dt$$
$$= \left. \frac{-1}{s} e^{-st} \right|_0^\infty$$

---

[2]There is also the bilateral Laplace transform which is defined from $-\infty$ to ∞. Such a transform has special applications and is not used here. In this book when we refer to the Laplace transform we always mean the unilateral one.

$$= \frac{1}{s}\left(1 - \lim_{t \to \infty} e^{-st}\right) \quad (4.31)$$

$$= \frac{1}{s} \quad \text{for Real}(s) > 0 \quad (4.32)$$

We emphasise that the Laplace transform of the unit-step function exists only for values of s such that their real part is greater than zero, otherwise the limit in eqn 4.31 does not converge. The range of allowed values for s defines the so-called Region Of Convergence (ROC) of the Laplace transform. For the unit-step the ROC is indicated in figure 4.9.

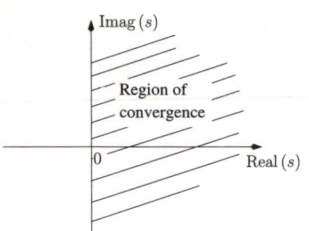

Figure 4.9: *Region of convergence of the Laplace transform of the unit-step.*

Example 4.4.1 Find the Laplace transform of the following functions

1. $x_1(t) = e^{-t/\tau}\, u(t)$
2. $x_2(t) = e^{-j\alpha t}\, u(t)$
3. $x_3(t) = e^{-\gamma t}\, u(t)$
4. $x_4(t) = \cos(2\pi f_o t)\, u(t)$

where τ, α and f_o are positive real numbers and γ is a complex number.

Solution:

1. The Laplace transform of the causal, real exponential is

$$\begin{aligned}
X_1(s) &= \int_0^\infty e^{-t/\tau}\, u(t)\, e^{-st}\, dt \\
&= -\frac{\tau}{s\tau + 1} e^{-t\frac{1+s\tau}{\tau}} \bigg|_0^\infty \\
&= \left(1 - \lim_{t \to \infty} e^{-t\frac{1+s\tau}{\tau}}\right) \\
&= \frac{\tau}{s\tau + 1} \quad \text{for Real}(s) > -1/\tau \quad (4.33)
\end{aligned}$$

2. Similarly, the Laplace transform of the causal, complex exponential can be expressed as

$$\begin{aligned}
X_2(s) &= \int_0^\infty e^{-j\alpha t}\, u(t)\, e^{-st}\, dt \\
&= \frac{1}{s + j\alpha} \quad \text{for Real}(s) > 0 \quad (4.34)
\end{aligned}$$

3. The Laplace transform of this causal, complex exponential can be expressed as

$$\begin{aligned}
X_3(s) &= \int_0^\infty e^{-\gamma t}\, u(t)\, e^{-st}\, dt \\
&= \frac{1}{s + \gamma} \quad \text{for Real}(s) > -\text{Real}(\gamma) \quad (4.35)
\end{aligned}$$

4. The Laplace transform of the causal cosine waveform can be expressed as

$$\begin{aligned} X_4(s) &= \int_0^\infty \frac{e^{j\,2\pi f_o t} + e^{-j\,2\pi f_o t}}{2} u(t)\, e^{-st}\, dt \\ &= \frac{-1}{2}\left(\frac{1}{j\,2\pi f_o - s} + \frac{1}{-j\,2\pi f_o - s}\right) \quad \text{for Real}(s) > 0 \\ &= \frac{s}{(2\pi f_o)^2 + s^2} \quad \text{for Real}(s) > 0 \end{aligned} \qquad (4.36)$$

The inverse Laplace transform

The inverse Laplace transform of $X(s)$ is defined by

$$x(t) = \frac{1}{j\,2\pi} \int_{c-j\infty}^{c+j\infty} X(s)\, e^{st}\, ds \quad \text{for } t \geq 0 \qquad (4.37)$$

where c is a real number which sets the path of integration in the complex domain. This path of integration must be defined within the ROC of $X(s)$. The solution of this type of integral is out of the scope of this book and is not discussed further[3]. In practice, eqn 4.37 is rarely used and we use instead the method of partial fractions. In addition, Laplace transform tables like those presented in appendix A are also used.

4.4.1 Theorems of the Laplace transform

Before presenting the partial fractions expansion method, we discuss some important theorems associated with the Laplace transform. The proofs of these are left as an exercise for the reader. All the signals expressed in the time domain are assumed to be zero for $t < 0$.

Linearity

Let us consider two causal signals $x_1(t)$ and $x_2(t)$ with Laplace transforms given by $X_1(s)$ and $X_2(s)$, respectively. If $z(t) = \alpha_1 x_1(t) + \alpha_2 x_2(t)$, where α_1 and α_2 are real constants, then its Laplace transform can be written as:

$$Z(s) = \alpha_1 X_1(s) + \alpha_2 X_2(s) \qquad (4.38)$$

Time delay

If $z(t) = x(t - \tau)$ where τ represents a positive time delay then the Laplace transform of $z(t)$ is given by:

$$Z(s) = e^{-s\tau} X(s) \qquad (4.39)$$

---

[3]For a detailed discussion on this subject see, for example, [1].

Time differentiation

If $z(t)$ is a signal resulting from the time differentiation of a causal signal $x(t)$, such that

$$z(t) = \frac{d}{dt} x(t) \qquad (4.40)$$

then the Laplace transform of $z(t)$ is given by:

$$Z(s) = s X(s) - x(0) \qquad (4.41)$$

The proof of this theorem is similar to that presented in the context of Fourier transforms in section 4.3.1.

Example 4.4.2 Find the Laplace transform of $h(t) = \sin(2\pi f_o t) u(t)$.

Solution: Taking into account that

$$\sin(2\pi f_o t) u(t) = \frac{-1}{2\pi f_o} \frac{d}{dt} \cos(2\pi f_o t) u(t) \qquad (4.42)$$

then using eqns 4.38, 4.36 and 4.41 we have

$$H(s) = \frac{-1}{2\pi f_o} \left(s \frac{s}{(2\pi f_o)^2 + s^2} - 1 \right)$$

$$= \frac{2\pi f_o}{(2\pi f_o)^2 + s^2} \qquad (4.43)$$

Time integration

If $z(t)$ is a signal resulting from the time integration of a causal signal $x(t)$, such that

$$z(t) = \int_0^t x(\lambda) \, d\lambda \qquad (4.44)$$

then the Laplace transform of $z(t)$ is given by:

$$Z(s) = \frac{X(s)}{s} \qquad (4.45)$$

The proof of this theorem is similar to that presented in the context of Fourier transforms in section 4.3.2.

Example 4.4.3 Find the inverse Laplace transform of

$$Z(s) = \frac{1}{s(1 + s\tau)}$$

Solution: $Z(s)$ can be expressed as

$$Z(s) = \frac{X(s)}{s} \tag{4.46}$$

with

$$X(s) = \frac{1}{1+s\tau} \tag{4.47}$$

The inverse Laplace transform of $X(s)$ is, according to eqn 4.33, equal to

$$x(t) = \frac{1}{\tau} e^{-t/\tau} u(t) \tag{4.48}$$

Now, using eqn 4.44 we have

$$\begin{aligned} z(t) &= \int_0^t \frac{1}{\tau} e^{-t/\tau} u(t)\, d\tau \\ &= \left. -e^{-t/\tau} u(t) \right|_0^t \\ &= \left(1 - e^{-t/\tau}\right) u(t) \end{aligned} \tag{4.49}$$

Time scaling

If $z(t) = x(\alpha t)$ then the Laplace transform of $z(t)$ is given by:

$$Z(s) = \frac{1}{\alpha} X\left(\frac{s}{\alpha}\right) \tag{4.50}$$

Convolution

If $z(t)$ results from the convolution of $x(t)$ with $y(t)$ then the Laplace transform of $z(t)$ is given by:

$$Z(s) = X(s)\, Y(s) \tag{4.51}$$

Shift in the s domain

Let us consider a causal signal $x(t)$ with Laplace transform $X(s)$. If $Z(s) = X(s + \alpha)$ then

$$z(t) = x(t)\, e^{-\alpha t} \tag{4.52}$$

Differentiation in the s domain

Let us consider a causal signal $x(t)$ with Laplace transform $X(s)$. If $Z(s)$ is such that

$$Z(s) = \frac{d^n}{ds^n} X(s) \tag{4.53}$$

then
$$z(t) = (-1)^n \, t^n \, x(t) \tag{4.54}$$

Example 4.4.4 Find the inverse Laplace transform of
$$G(s) = \frac{1}{(s+a)^{n+1}} \tag{4.55}$$
where $n = 0, 1, 2, \ldots$.
Solution: Since,
$$\frac{1}{(s+a)^{n+1}} = \frac{(-1)^n}{n!} \frac{d^n}{ds^n} \frac{1}{s+a} \tag{4.56}$$
then, by using eqn 4.33, with $a = 1/\tau$, together with eqn 4.54 we can write
$$g(t) = \frac{t^n}{n!} e^{-ta} u(t) \tag{4.57}$$

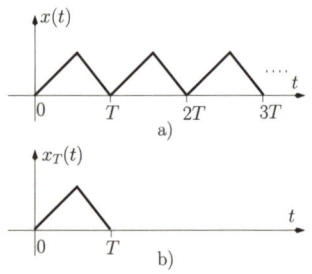

Figure 4.10: *Periodic waveform. a) $x(t)$. b) $x_T(t)$.*

Time periodicity

Let us consider $x(t)$ to be a causal periodic waveform with period T as illustrated in figure 4.10. This can be written as
$$x(t) = \sum_{k=0}^{\infty} x_T(t - kT) \tag{4.58}$$
where $x_T(t)$ is zero outside the time interval $[0, T]$. The Laplace transform of $x(t)$ is given by:
$$X(s) = \frac{1}{1 - e^{-sT}} X_T(s) \tag{4.59}$$
with
$$X_T(s) = \int_0^{\infty} x_T(t) e^{-st} \, dt \tag{4.60}$$

Example 4.4.5 Find the Laplace transform of the waveform of figure 4.1 b).
Solution: According to eqn 4.60 we can write:
$$V_{S_T}(s) = \int_0^{T/2} V_s \, e^{-st} \, dt$$
$$= \frac{V_s}{s} \left(1 - e^{-sT/2}\right) \tag{4.61}$$
Using eqn 4.59 we have
$$V_{S_2}(s) = \frac{V_s}{s} \frac{1 - e^{-sT/2}}{1 - e^{-sT}} \tag{4.62}$$

4.4.2 Partial-fraction expansion

The partial-fraction expansion method is used to obtain the inverse Laplace transform if the s-domain function can be expressed as a ratio of polynomials in s. To illustrate the application of this method we consider the partial-fraction expansion of the following

$$X(s) = \frac{s^2 a_2 + s a_1 + a_0}{(s+p_2)(s+p_1)(s+p_0)} \qquad (4.63)$$

where each p_i is a distinct pole of $X(s)$.

This last eqn can also be written in the partial-fraction form as follows:

$$X(s) = \frac{K_2}{s+p_2} + \frac{K_1}{s+p_1} + \frac{K_0}{s+p_0} \qquad (4.64)$$

where each K_i is a constant to be determined by first equating eqns 4.63 and 4.64 as indicated below:

$$\frac{s^2 a_2 + s a_1 + a_0}{(s+p_2)(s+p_1)(s+p_0)} = $$
$$\frac{K_2(s+p_1)(s+p_0) + K_1(s+p_2)(s+p_0) + K_0(s+p_2)(s+p_1)}{(s+p_2)(s+p_1)(s+p_0)} \qquad (4.65)$$

that is

$$\begin{aligned} s^2 a_2 + s a_1 + a_0 &= (K_0 + K_1 + K_2) s^2 \\ &+ [K_0(p_1 + p_2) + K_1(p_0 + p_2) + K_2(p_1 + p_0)]s \\ &+ K_0 p_1 p_2 + K_1 p_0 p_2 + K_2 p_1 p_0 \end{aligned} \qquad (4.66)$$

By equating the coefficients of the powers of s in the last eqn we obtain the following set of eqns

$$\begin{cases} a_2 = K_0 + K_1 + K_2 \\ a_1 = K_0(p_1 + p_2) + K_1(p_0 + p_2) + K_2(p_1 + p_0) \\ a_0 = K_0 p_1 p_2 + K_1 p_0 p_2 + K_2 p_1 p_0 \end{cases} \qquad (4.67)$$

Finally, solving this set of eqns to obtain K_0, K_1 and K_2 we get

$$K_0 = \frac{p_0^2 a_2 - a_1 p_0 + a_0}{(p_2 - p_0)(p_1 - p_0)}$$

$$K_1 = \frac{p_1^2 a_2 - a_1 p_1 + a_0}{(p_2 - p_1)(p_0 - p_1)}$$

$$K_2 = \frac{p_2^2 a_2 - a_1 p_2 + a_0}{(p_1 - p_2)(p_0 - p_2)}$$

Taking the inverse Laplace of eqn 4.64 using eqns 4.35 and 4.38 we can determine $x(t)$ as

$$x(t) = \left(K_0 e^{-t p_0} + K_1 e^{-t p_1} + K_2 e^{-t p_2} \right) u(t)$$

Example 4.4.6 Determine the inverse Laplace transform of the following function which occurs in the analysis of a damped simple harmonic oscillator.

$$Y(s) = \frac{\omega_n^2}{s^2 + 2\eta\omega_n s + \omega_n^2} \qquad (4.68)$$

ω_n is called the natural frequency while η is called the damping factor. Consider the following situations: $\eta < 1$, $\eta = 1$ and $\eta > 1$.

Solution:

1. $\eta < 1$. First we express the denominator of $Y(s)$ as a product of the two poles which can be obtained solving the following quadratic eqn

$$s^2 + 2\eta\omega_n s + \omega_n^2 = 0 \qquad (4.69)$$

that is (see also eqn 2.36)

$$s = -\eta\omega_n \pm j\omega_n \sqrt{1-\eta^2}, \qquad (4.70)$$

Now, eqn 4.68 can be written as

$$Y(s) = \frac{\omega_n^2}{(s + \eta\omega_n + j\omega_n\sqrt{1-\eta^2})(s + \eta\omega_n - j\omega_n\sqrt{1-\eta^2})}$$

and $Y(s)$ can then be written in the partial-fraction form as

$$Y(s) = \frac{K_0}{s + \eta\omega_n + j\omega_n\sqrt{1-\eta^2}} + \frac{K_1}{s + \eta\omega_n - j\omega_n\sqrt{1-\eta^2}}$$

Equating the last two eqns we have

$$\omega_n^2 = (K_0 + K_1)s + K_0\left(\eta\omega_n - j\omega_n\sqrt{1-\eta^2}\right) \\ + K_1\left(\eta\omega_n + j\omega_n\sqrt{1-\eta^2}\right)$$

By equating the coefficients of the powers of s we obtain the following set of eqns

$$\begin{cases} 0 = K_1 + K_0 \\ \omega_n^2 = K_0\left(\eta\omega_n - j\omega_n\sqrt{1-\eta^2}\right) + K_1\left(\eta\omega_n + j\omega_n\sqrt{1-\eta^2}\right) \end{cases} \qquad (4.71)$$

Solving to obtain K_0 and K_1 we get

$$K_0 = j\frac{\omega_n}{2\sqrt{1-\eta^2}}$$

$$K_1 = -j\frac{\omega_n}{2\sqrt{1-\eta^2}}$$

4. Natural and forced responses circuit analysis

Using eqns 4.35 and 4.38 we can determine $y(t)$ as

$$y(t) = \left(j\frac{\omega_n}{2\sqrt{1-\eta^2}} e^{-t\left(\eta\omega_n + j\omega_n\sqrt{1-\eta^2}\right)} \right.$$

$$\left. - j\frac{\omega_n}{2\sqrt{1-\eta^2}} e^{-t\left(\eta\omega_n - j\omega_n\sqrt{1-\eta^2}\right)} \right) u(t)$$

It is left to the reader to show that the last eqn can be written as

$$y(t) = \frac{\omega_n}{\sqrt{1-\eta^2}} \sin\left(\omega_n\sqrt{1-\eta^2}\,t\right) e^{-t\eta\omega_n} u(t) \quad (4.72)$$

2. $\eta = 1$. For this situation we can write $Y(s)$ as follows:

$$Y(s) = \frac{\omega_n^2}{(s+\omega_n)^2}$$

Using eqn 4.57 we have

$$y(t) = \omega_n^2 \, t \, e^{-\omega_n t} u(t) \quad (4.73)$$

3. $\eta > 1$. In this situation $Y(s)$ has two real poles;

$$s_i = -\eta\omega_n \pm \omega_n\sqrt{\eta^2 - 1}, \quad i = 0, 1 \quad (4.74)$$

Hence, $Y(s)$ can be written in the partial-fraction form as

$$Y(s) = \frac{K_0}{s + \eta\omega_n + \omega_n\sqrt{\eta^2 - 1}} + \frac{K_1}{s + \eta\omega_n - \omega_n\sqrt{\eta^2 - 1}}$$

Equating this with eqn 4.68 we have

$$\omega_n^2 = (K_0 + K_1)s + K_0\left(\eta\omega_n - \omega_n\sqrt{\eta^2 - 1}\right)$$
$$+ K_1\left(\eta\omega_n + \omega_n\sqrt{\eta^2 - 1}\right)$$

From this we can write

$$\begin{cases} 0 = K_1 + K_0 \\ \omega_n^2 = K_0\left(\eta\omega_n - \omega_n\sqrt{\eta^2-1}\right) + K_1\left(\eta\omega_n + \omega_n\sqrt{\eta^2-1}\right) \end{cases} \quad (4.75)$$

and solving to obtain K_0 and K_1 we get

$$K_0 = -\frac{\omega_n}{2\sqrt{\eta^2 - 1}}$$

$$K_1 = \frac{\omega_n}{2\sqrt{\eta^2 - 1}}$$

Using eqns 4.33 and 4.38 we can determine $y(t)$ as

$$y(t) = \left(\frac{-\omega_n}{2\sqrt{\eta^2-1}} e^{-t\left(\eta\omega_n + \omega_n\sqrt{\eta^2-1}\right)} \right.$$

$$\left. + \frac{\omega_n}{2\sqrt{\eta^2-1}} e^{-t\left(\eta\omega_n - \omega_n\sqrt{\eta^2-1}\right)} \right) u(t)$$

It is left to the reader to show that $y(t)$ can also be written as

$$y(t) = \frac{\omega_n}{\sqrt{\eta^2-1}} e^{-t\eta\omega_n} \sinh\left(\omega_n\sqrt{\eta^2-1}\,t\right) u(t) \quad (4.76)$$

The application of this analysis to RLC circuits is discussed later in this chapter.

The discussion presented above applies to s-functions without repeated poles. In practice this is the most frequently encountered situation. However, let us consider an s-function which has $n+1$ poles including n identical ones as shown below:

$$W(s) = \frac{1}{(s+p_1)^n (s+p_0)} \quad (4.77)$$

This function can be written in a partial-fraction form as follows:

$$W(s) = \frac{K_{1_n}}{(s+p_1)^n} + \frac{K_{1_{n-1}}}{(s+p_1)^{n-1}} + \ldots + \frac{K_{1_2}}{(s+p_1)^2} + \frac{K_{1_1}}{s+p_1}$$

$$+ \frac{K_0}{s+p_0} \quad (4.78)$$

Now, applying a procedure similar to that described above we can determine the coefficients $K_{1_{n-k}}$ and K_0. Using eqns 4.35 and 4.57 the inverse Laplace transform of $W(s)$ can be obtained as

$$w(t) = \sum_{k=1}^{n} \frac{K_{1_k} t^{k-1}}{(k-1)!} e^{-t p_1} u(t) + K_0 e^{-t p_0} u(t) \quad (4.79)$$

Relationship between Laplace and Fourier transforms

The Laplace transform can be seen as the transformation of time functions into a sum of generic exponentials of complex arguments, $\exp(s\,t)$ with $s = \sigma + j\,2\pi\,f$, instead of the sum of exponentials with purely imaginary arguments, $\exp(j\,2\pi\,f\,t)$, obtained from the Fourier transform. For any causal time function $x(t)$ whose Laplace transform Region Of Convergence (ROC) includes the imaginary axis, $j\,2\,\pi\,f$, the Fourier transform, $X_{\mathcal{F}}(f)$ can be obtained from the Laplace transform, $X_{\mathcal{L}}(s)$ as follows:

$$X_{\mathcal{F}}(f) = X_{\mathcal{L}}(s) \quad (4.80)$$

with s replaced by $j\,2\,\pi\,f$.

4. Natural and forced responses circuit analysis

Example 4.4.7 Use eqn 4.80 to obtain the Fourier transform of $x(t) = \exp(-t/\tau)\,u(t)$ where τ is a positive real number.

Solution: According to eqn 4.33 the Laplace transform of $x(t)$ is

$$X(s) = \frac{\tau}{s\tau + 1} \quad \text{for Real}(s) > -1/\tau \tag{4.81}$$

Since the ROC includes the imaginary axis the Fourier transform of $x(t)$ is

$$X(f) = \left.\frac{\tau}{s\tau + 1}\right|_{s \to j2\pi f}$$
$$= \frac{\tau}{j2\pi f\tau + 1} \tag{4.82}$$

Note that this result is the same as that given by eqn 3.214 replacing σ by $1/\tau$.

4.5 Analysis using Laplace transforms

We are now in position to employ the Laplace transform to analyse the natural and forced responses of electrical circuits. Again we use the circuit of figure 4.3 driven by the periodic sequence of square pulses of figure 4.1 b) to illustrate this procedure.

4.5.1 Solving differential equations

The differential equation expressed by eqn 4.11 can be solved in the Laplace domain, taking into account the time differentiation theorem expressed by eqn 4.41, as follows:

$$V_{S_2}(s) = V_C(s) + \tau\,[s\,V_C(s) - V_{co}]$$
$$= V_C(f)(1 + s\tau) - V_{co}\,\tau \tag{4.83}$$

$V_{S_2}(s)$ and $V_C(s)$ are the Laplace transforms of $v_{S_2}(t)$ and $v_C(t)$, respectively. Equation 4.83 can be solved to obtain $V_C(s)$ as follows:

$$V_C(s) = \frac{V_{S_2}(s)}{1 + s\tau} + \frac{V_{co}\,\tau}{1 + s\tau} \tag{4.84}$$

Since $V_{S_2}(s)$ has been determined in example 4.4.5, we can write $V_C(s)$ as

$$V_C(s) = \frac{V_{s2}}{s\,(1 + s\tau)}\,\frac{1 - e^{-sT/2}}{1 - e^{-sT}} + \frac{V_{co}\,\tau}{1 + s\tau} \tag{4.85}$$

The solution of the differential equation expressed by eqn 4.11 can now be obtained after calculating the inverse Laplace transform of $V_C(s)$ given by the last eqn. From eqn 4.33 we have

$$\frac{V_{co}\,\tau}{1 + s\tau} \quad \overset{\mathcal{L}}{\Longleftrightarrow} \quad V_{co}\,e^{-t/\tau}\,u(t)$$

where $\overset{\mathcal{L}}{\Longleftrightarrow}$ denotes a Laplace transform pair. From eqn 4.49 we have

$$\frac{V_s}{s(1+s\tau)} \overset{\mathcal{L}}{\Longleftrightarrow} V_s \left(1 - e^{-t/\tau}\right) u(t)$$

Using eqn 4.39 we have

$$\frac{V_s}{s(1+s\tau)} \left(1 - e^{-sT/2}\right) \overset{\mathcal{L}}{\Longleftrightarrow} V_s \left(1 - e^{-t/\tau}\right) u(t)$$
$$- V_s \left(1 - e^{-(t-T/2)/\tau}\right) u\left(t - \frac{T}{2}\right)$$

Finally, from eqn 4.59 we get

$$\frac{V_s}{s(1+s\tau)} \frac{1 - e^{-sT/2}}{1 - e^{-sT}} \overset{\mathcal{L}}{\Longleftrightarrow} V_s \sum_{k=0}^{\infty} \left[\left(1 - e^{-\frac{t-kT}{\tau}}\right) u(t)\right.$$
$$\left. - \left(1 - e^{-\frac{t-kT-T/2}{\tau}}\right) u\left(t - kT - \frac{T}{2}\right)\right]$$

From the above we can write $v_C(t)$, for $t \geq 0$, as

$$v_C(t) = V_{co} e^{-\frac{t}{\tau}} u(t) + V_s \sum_{k=0}^{\infty} \left(1 - e^{-\frac{t-kT}{\tau}}\right) u(t - kT)$$
$$- V_s \sum_{n=0}^{\infty} \left(1 - e^{-\frac{t-nT-T/2}{\tau}}\right) u\left(t - nT - \frac{T}{2}\right) \quad (4.86)$$

which is the same expression as that obtained in examples 3.3.4 and 4.2.2.

4.5.2 I–V characteristics for passive elements

The discussion above illustrates the usefulness of the Laplace transform in solving linear integral-differential equations in the context of electrical circuit and signal analysis. Hence, any circuit can be analysed by first applying Kirchhoff's laws, using the time domain relationships between current and voltage for each electrical element discussed in Chapter 1, and then solving the circuits equations in the Laplace domain. The voltage across or current through any circuit element, in the time domain, can of course be obtained by determining the corresponding inverse Laplace transform.

Although this procedure already provides a significant simplification of the analysis of circuits it would be useful to apply the Laplace transform directly to the circuit. This can, in fact, be done by determining the current–voltage relationships of the passive elements in the Laplace domain.

Resistance

Applying the Laplace transform to Ohm's law we obtain a linear relationship between $V(s)$ and $I(s)$

$$V(s) = R\,I(s) \quad (4.87)$$

where $V(s)$ and $I(s)$ represent the Laplace transforms of the voltage across and the current through the resistance, respectively.

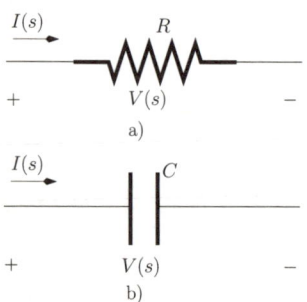
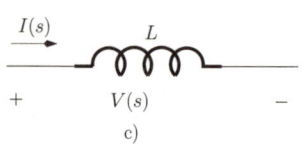

Figure 4.11: *Laplace domain I–V characteristic for a) resistance, b) capacitance, c) inductance.*

Capacitance

Applying Laplace transforms to eqn 4.19 we obtain (see also eqn 4.41)

$$I(s) = sCV(s) - Cv(0) \qquad (4.88)$$

where $v(0)$ is the voltage across the capacitor terminals at $t = 0$. $V(s)$ and $I(s)$ represent the Laplace transforms of the voltage across and the current through the capacitor, respectively. Solving eqn 4.88 in order to obtain $V(s)$ we get

$$V(s) = \frac{I(s)}{sC} + \frac{v(0)}{s} \qquad (4.89)$$

$$= I(s)Z_C(s) + \frac{v(0)}{s} \qquad (4.90)$$

with

$$Z_C(s) = \frac{1}{sC}$$

where $Z_C(s)$ represents the complex impedance, in the Laplace domain, associated with the capacitor C. Note that eqn 4.89 can also be obtained by finding the Laplace transform of $v(t)$ in eqn 1.24.

Inductance

Applying Laplace transforms to eqn 4.22 we obtain (see also eqn 4.41)

$$V(s) = sLI(s) - Li(0) \qquad (4.91)$$

where $i(0)$ is the current flowing through the inductor at $t = 0$. $V(s)$ and $I(s)$ represent the Laplace transforms of the voltage across and the current through the inductor, respectively. Eqn 4.91 can also be written as:

$$V(s) = Z_L(s)I(s) - Li(0) \qquad (4.92)$$

with

$$Z_L(s) = sL$$

where $Z_L(s)$ represents the complex impedance of L, in the Laplace domain. Solving eqn 4.91 we get

$$I(s) = \frac{V(s)}{sL} + \frac{i(0)}{s} \qquad (4.93)$$

Note that this last eqn can also be obtained by applying Laplace transforms to eqn 1.27.

The generalised impedance in the s domain

There is a straightforward relationship between the generalised impedance provided by the phasor analysis $Z_\mathcal{F}(f)$, presented in the last chapter, and the generalised impedance in the Laplace domain associated with each of the passive elements, $Z_\mathcal{L}(s)$:

- for a resistance; $Z_\mathcal{F}(f) = Z_\mathcal{L}(s) = R$;
- for a capacitance; $Z_\mathcal{F}(f) = Z_\mathcal{L}(s)|_{s=j\,2\,\pi\,f} = (j\,2\,\pi\,f\,C)^{-1}$
- for an inductance; $Z_\mathcal{F}(f) = Z_\mathcal{L}(s)|_{s=j\,2\,\pi\,f} = j\,2\,\pi\,f\,L$

Note that these relationships are consistent with the relationship between the Laplace and Fourier transforms (see eqn 4.80).

Circuit analysis

Figure 4.12 is used to illustrate a partial analysis of the RC circuit in the Laplace domain. Applying Kirchhoff's current law we can write (see also eqn 4.88):

$$\frac{V_{S_2}(s) - V_C(s)}{R} = s\,C\,V(s) - C\,v_{co} \qquad (4.94)$$

Solving this eqn in order to obtain $V_C(s)$ we can write

$$V_C(s) = \frac{V_{S_2}(s)}{1+s\tau} + \frac{V_{co}\,\tau}{1+s\tau} \qquad (4.95)$$

with $\tau = RC$ as before. Note that this last eqn, obtained by applying Kirchhoff's current law in the Laplace domain, is the same as eqn 4.85 which is obtained applying the Laplace transform to the time domain differential equation expressed by eqn 4.11.

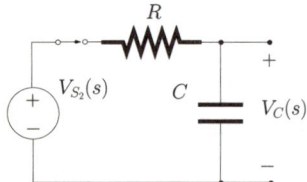

Figure 4.12: *Analysis of the RC circuit in the Laplace domain ($t \geq 0$).*

4.5.3 Natural response

RL and RC circuits

We have studied the natural response of the RC circuit in section 4.2. Let us consider now the RL circuit of figure 4.13 a). The switch is closed for $t < 0$ and it is open for $t \geq 0$. The equivalent circuit for $t < 0$ is shown in figure 4.13 b). Because the inductor is conducting a constant current the voltage across its terminals is zero. This means that the voltage across R_1 is zero and, therefore the voltage V is applied to R_2. It follows that the DC current that flows through R_2 and the inductor is

$$I_{lo} = \frac{V}{R_2}$$

For $t < 0$ the inductor stores energy equal to $L\,I_{lo}^2/2$ (see eqn 1.28). From the above, it is clear that the initial condition associated with the current through

4. Natural and forced responses circuit analysis

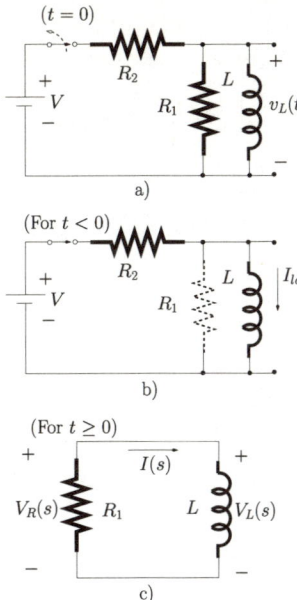

Figure 4.13: *a) RL circuit. b) Equivalent circuit for $t < 0$. c) Equivalent circuit for $t \geq 0$.*

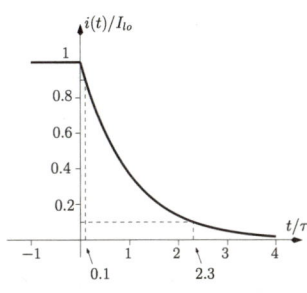

Figure 4.14: $i(t)$ *normalised to I_{lo} versus the time normalised to τ.*

the inductor, when the switch is closed at $t = 0$, is I_{lo}. Figure 4.13 c) shows the equivalent circuit for $t \geq 0$ (the switch is open). Now the energy stored by the inductor will be dissipated by the resistance R_1 and we can write

$$V_L(s) = V_R(s)$$

Using the expressions for $V_L(s)$ and $V_{R_1}(s)$, given by eqns 4.91 and 4.87, respectively,

$$s L I(s) - L I_{lo} = -R_1 I(s)$$

Solving the last eqn in order to obtain $I(s)$ we have

$$I(s) = \frac{\tau I_{lo}}{1 + s \tau}$$

with $\tau = L/R_1$. Finally, using eqn 4.33 we obtain

$$i(t) = I_{lo} e^{-t/\tau} u(t)$$

Note that the natural response of this circuit is similar to that discussed for the RC circuit. Figure 4.14 shows $i(t)$ normalised to I_{lo} versus the time normalised to τ. From this figure we observe that the time necessary for the current to go from 90% of I_{lo} to 10% of I_{lo}, is about $2.2 \times \tau$. It is interesting to find the voltage $v_L(t) = v_{R_1}(t)$ developed across the resistor/inductor parallel combination of figure 4.13 c). This can be obtained either by using eqn 4.22 or simply by finding $R_1 \times i(t)$. This gives

$$v_L(t) = -R_1 I_{lo} e^{-t/\tau} u(t)$$

A plot of $v_L(t)$ normalised to $R_1 I_{lo}$ versus time normalised to τ is shown in figure 4.15. Note that the voltage developed across the inductor is negative at $t = 0$ and tends toward zero. Such a voltage is known as the inductor's back emf (electro-motive force).

You might like to note that if $R_1 \to \infty$ then $v_L(0) \to \infty$. This is the large emf produced when $i_L(t)$ is suddenly reduced to zero and is the basis of the traditional ignition coil in a car. Modern car ignition systems use capacitive discharge together with sophisticated electronic circuitry.

Example 4.5.1 Consider the circuit shown in figure 4.16 a). Determine the voltage across the capacitor and the current flowing through R_3 for $t \geq 0$.

Solution: The equivalent circuit for $t < 0$ is shown in figure 4.16 b). In this situation the capacitor is not conducting and the voltage drop across R_2 is zero. This means that the voltage across the capacitor terminals is equal to V_A where:

$$V_A = V \frac{R'}{R' + R4} \qquad (4.96)$$

with $R' = R_3 || R_1$. The capacitor stores energy equal to $C V_A^2 / 2$ and the initial condition is V_{co} being equal to V_A.

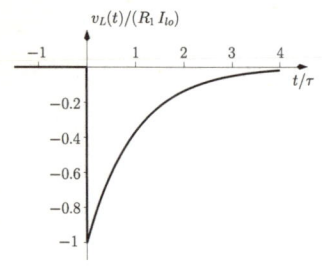

Figure 4.15: $v_L(t)$ normalised to $(R_1 I_{lo})$ versus the time normalised to τ.

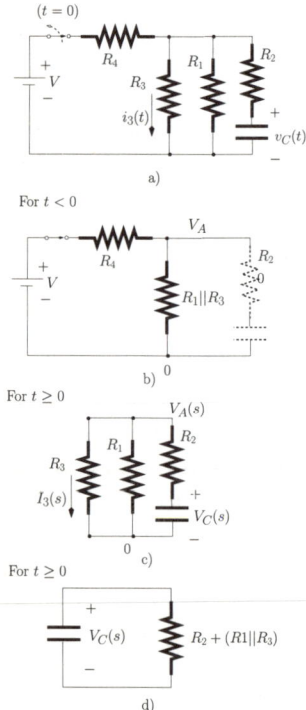

Figure 4.16: *a)* RC circuit. *b)* Equivalent circuit for $t < 0$. *c)* Equivalent circuit for $t \geq 0$. *d)* Equivalent resistance 'seen' by C for $t \geq 0$.

Figure 4.16 c) shows the equivalent circuit for $t \geq 0$, that is when the switch is closed. For this situation, the energy stored by the capacitor will be dissipated by the resistances R_1, R_2 and R_3.

For this circuit we can write after applying Kirchhoff's current law, the following eqns:

$$\frac{V_A(s) - V_C(s)}{R_2} = s\,C\,V_C(s) - C\,V_{co} \quad (4.97)$$

$$\frac{V_A(s) - V_C(s)}{R_2} = -\frac{V_A(s)}{R'} \quad (4.98)$$

Solving these two eqns in order to obtain $V_C(s)$ we get:

$$V_C(s) = \frac{C\,(R' + R_2)\,V_{co}}{1 + s\,C\,(R' + R_2)} \quad (4.99)$$

Using eqn 4.33 we have

$$v_C(t) = V_{co}\,e^{-t/\tau'}\,u(t) \quad (4.100)$$

where $\tau' = C\,(R' + R_2)$. Note the similarity of the expression for the natural response given by the last eqn and that obtained for the circuit of figure 4.3 (see also eqns 4.6 and 4.8). The effective resistance seen across the capacitor terminals is given by $R_2 + R' = R_2 + (R_1 \| R_3)$ as shown in figure 4.16 d).

In order to obtain the current flowing through R_3, $I_3(s)$, we consider again eqn 4.98 from which we can write

$$V_A(s) = \frac{R'}{R_2 + R'}\,V_C(s)$$

$$= \frac{C\,R'\,V_{co}}{1 + s\,C\,(R' + R_2)} \quad (4.101)$$

and the current flowing through R_3 is

$$I_3(s) = \frac{V_A(s)}{R_3}$$

$$= \frac{C\,R'\,V_{co}}{R_3[1 + s\,C\,(R' + R_2)]} \quad (4.102)$$

again using eqn 4.33 we obtain

$$i_3(t) = \frac{R'\,V_{co}}{R_3\,(R' + R_2)}\,e^{-t/\tau'}\,u(t) \quad (4.103)$$

Note that the behaviour of the current $i_3(t)$ is similar to that of the voltage across the capacitor.

LC circuits

Let us consider now the LC circuit of figure 4.17 a). For $t < 0$ the switch S_1 is closed while the switch S_2 is open. Hence, the DC voltage applied to the capacitor, V_{co}, is V. There is no voltage across the inductor. Hence, for $t < 0$ the capacitor stores energy while the inductor is not storing any energy.

At $t = 0$ the switch S_1 is opened and the switch S_2 is closed. Figure 4.17 b) shows the equivalent circuit for $t \geq 0$. Now the voltage across the capacitor is the same as the voltage across the inductor; V_{LC}. Hence, we can write

$$s L I(s) = -\frac{I(s)}{sC} + \frac{V_{co}}{s}$$

that is

$$I(s) = \frac{C V_{co}}{s^2 L C + 1}$$

The voltage can be determined as

$$V_{LC}(s) = \frac{s L C V_{co}}{s^2 L C + 1}$$

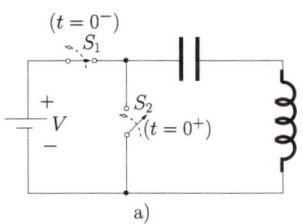

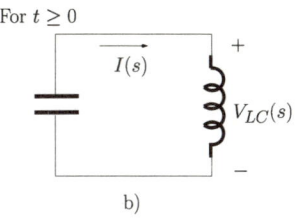

Figure 4.17: *a) LC circuit. b) Equivalent circuit for $t \geq 0$.*

Using eqns 4.36 and 4.43 we can determine the current and the voltage in the time domain

$$i(t) = -C V_{co} \sin\left(\frac{1}{LC} t\right) u(t)$$

$$v_{LC}(t) = V_{co} \cos\left(\frac{1}{LC} t\right) u(t)$$

These two last equations indicate that the current flows in a sinusoidal manner between the capacitor and the inductor, with the electrical energy being transferred periodically between electrostatic energy in the capacitor to magnetic energy in the inductor. The frequency of this energy transfer (oscillation) is $f = (2\pi\sqrt{LC})^{-1}$. This LC circuit has a simple hydraulic analogue where a water pipe, with its section covered by an elastic membrane, is connected to a flywheel forming a closed circuit as shown in figure 4.18 (see also section 1.3). We assume that this circuit is completely filled with water. If we rotate the fly-wheel manually the membrane will be stretched due to the water pressure created in one of its sides. This procedure is 'equivalent' to storing energy in the capacitor of the LC circuit. If we release the flywheel there will be an oscillatory motion of water flow. If we assume that there are no losses in this circuit this oscillation will carry on indefinitely. We note that LC circuits form the basis of analogue oscillators used in various telecommunication circuits.

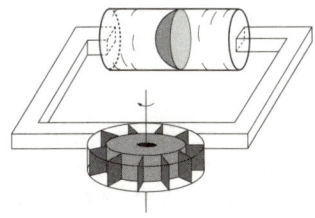

Figure 4.18: *Hydraulic analogue of the LC circuit of figure 4.17.*

RLC circuits

We consider now the RLC circuit of figure 4.19 a). For $t < 0$ the switch is closed and for $t \geq 0$ the switch is open. For $t < 0$ the DC voltage V is simultaneously applied to the resistance R and to the capacitor C. The voltage across the capacitor, V_{co}, is equal to the DC source voltage V. Figure 4.19 b) shows the equivalent circuit for $t \geq 0$. From this circuit we can write the following eqn

$$V_R(s) = V_C(s) + V_L(s)$$

that is:

$$R\,I(s) = -\frac{I(s)}{sC} + \frac{V_{co}}{s} - sL\,I(s)$$

Solving in order to get $I(s)$ we have

$$I(s) = \frac{C\,V_{co}}{s^2\,LC + s\,RC + 1}$$

This last eqn can be written as follows:

$$\begin{aligned} I(s) &= \frac{V_{co}}{L}\,\frac{1}{s^2 + s\frac{R}{L} + \frac{1}{LC}} \\ &= C\,V_{co}\,\frac{\omega_n^2}{s^2 + 2\,\eta\,\omega_n\,s + \omega_n^2} \end{aligned}$$

with

$$\omega_n = \frac{1}{\sqrt{LC}} \qquad (4.104)$$

$$\eta = \frac{1}{2}R\sqrt{\frac{C}{L}} \qquad (4.105)$$

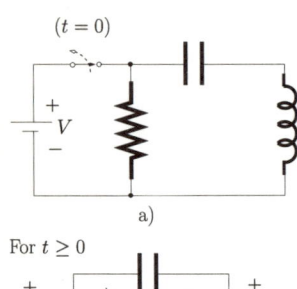
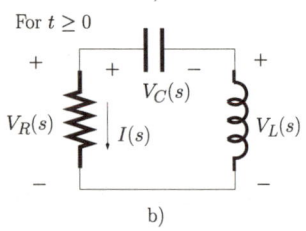

Figure 4.19: a) RLC. b) Equivalent circuit for $t \geq 0$.

The expression for $I(s)$ is similar to that expressed by eqn 4.68 and discussed in example 4.4.6. Therefore, by using eqns 4.72, 4.73, 4.76 we can obtain the current of the RLC circuit, for $t \geq 0$, as follows:

$$i(t) = C\,V_{co} \times \begin{cases} \dfrac{\omega_n}{\sqrt{1-\eta^2}}\sin\left(\omega_n\sqrt{1-\eta^2}\,t\right)e^{-t\eta\omega_n}\,u(t) & \text{if } \eta < 1 \\ \omega_n^2\,t\,e^{-\omega_n t}\,u(t) & \text{if } \eta = 1 \\ \dfrac{\omega_n}{\sqrt{\eta^2-1}}\sinh\left(\omega_n\sqrt{\eta^2-1}\,t\right)e^{-t\eta\omega_n}\,u(t) & \text{if } \eta > 1 \end{cases}$$

(4.106)

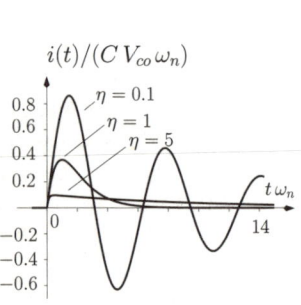

Figure 4.20: $i(t)$ given eqn 4.106 normalised to $C\,V_{co}\,\omega_n$ versus the time normalised to ω_n^{-1}.

Figure 4.20 shows $i(t)$ normalised to $(C\,V_{co}\,\omega_n)$ versus the time normalised to ω_n^{-1} for three different values of η. We observe that, in all situations, the current tends to zero. This is expected since the energy initially stored by the capacitor is constantly dissipated by the resistance. We also observe that the value of η influences the behaviour of the current. For $\eta < 1$ the current exhibits an oscillatory behaviour and the circuit is said to be *underdamped*. In this situation the circuit natural response is dominated by the LC combination. For $\eta \geq 1$ this oscillatory behaviour does not occur. For $\eta = 1$ the circuit is said to be *critically damped* and for $\eta > 1$ the circuit is said to be *overdamped*.

For values of $\eta > 1$, the transient behaviour of the circuit is dominated by the combination of the resistance with the capacitance. In fact, if $\eta > 2$ then the current $i(t)$, given by eqn 4.106, can first be approximated as follows:

$$i(t) \simeq CV_{co}\frac{\omega_n\, 2\eta}{2\eta^2-1}\,\frac{e^{-\frac{t\omega_n}{2\eta}} - e^{\frac{t\omega_n}{2\eta}-2\eta\omega_n t}}{2}u(t)$$

$$= CV_{co}\frac{\omega_n\, 2\eta}{2\eta^2-1}\, e^{-\frac{t\omega_n}{2\eta}}\,\frac{1 - e^{-t\omega_n\frac{2\eta^2-1}{\eta}}}{2}u(t) \qquad (4.107)$$

where we use the following approximation:

$$\sqrt{\eta^2 - 1} \simeq \eta - \frac{1}{2\eta} \quad \text{if } \eta > 2$$

However, eqn 4.107 can be further simplified as

$$i(t) \simeq CV_{co}\frac{\omega_n}{2\eta}e^{-\frac{t\omega_n}{2\eta}}u(t)$$

$$= \frac{V_{co}}{R}e^{-\frac{t}{RC}}u(t) \qquad (4.108)$$

after assuming that

$$2\eta^2 \gg 1$$

and

$$e^{-2\omega_n \eta t} \ll 1$$

Equation 4.108 is equal to eqn 4.9 regarding the natural response associated with the current of the RC circuit of figure 4.5. Note that increasing η is equivalent to increasing the value of R, assuming that ω_n is kept constant. The hydraulic equivalent for this circuit is similar to that shown in figure 4.18 where now we consider losses in the hydraulic pipes. These losses attenuate the oscillatory movement of the the water flow. If the resistance to water flow is very high (very thin pipes) then there will be no oscillatory water movement at all ($\eta > 1$).

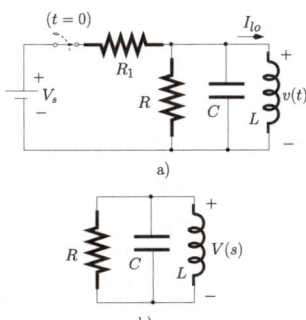

Figure 4.21: *a) RLC. b) Equivalent circuit for $t \geq 0$.*

Example 4.5.2 Consider the RLC circuit of figure 4.21 a). For $t < 0$ the switch is closed. Determine the voltage $v(t)$.

Solution: Since the DC voltage source has been applied to the inductor for a long time ($t < 0$) the voltage $v(t)$ is zero. Therefore, V_s appears across R_1. The current flowing through this resistance, R_1, also flows through the inductor and it is given by

$$I_{lo} = \frac{V_s}{R_1}$$

Note that the voltage across the capacitor is zero. For $t \geq 0$ the switch is open, resulting in the equivalent circuit shown in 4.21 b) for which we can write:

$$\frac{V(s)}{R} + sCV(s) + \frac{V(s)}{sL} + \frac{I_{lo}}{s} = 0$$

Solving this last eqn in order to obtain $V(s)$ we have

$$V(s) = \frac{-L\,I_{lo}}{s^2\,LC + s\,L/R + 1}$$

This can also be written in terms of its natural frequency ω_n and damping factor, η, as follows:

$$V(s) = -L\,I_{lo}\,\frac{\omega_n^2}{s^2 + 2\,\eta\,\omega_n + \omega_n^2}$$

with

$$\omega_n = \frac{1}{\sqrt{LC}} \qquad (4.109)$$

and

$$\eta = \frac{1}{2}\frac{1}{R}\sqrt{\frac{L}{C}} \qquad (4.110)$$

Again, by using eqns 4.72, 4.73, 4.76 we can obtain the voltage $v(t)$, for $t \geq 0$, as follows:

$$v(t) = -L\,I_{lo} \times \begin{cases} \frac{\omega_n}{\sqrt{1-\eta^2}}\sin\left(\omega_n\sqrt{1-\eta^2}\,t\right)e^{-t\,\eta\,\omega_n}\,u(t) & \text{if } \eta < 1 \\ \omega_n^2\,t\,e^{-\omega_n\,t}\,u(t) & \text{if } \eta = 1 \\ \frac{\omega_n}{\sqrt{\eta^2-1}}\sinh\left(\omega_n\sqrt{\eta^2-1}\,t\right)e^{-t\,\eta\,\omega_n}\,u(t) & \text{if } \eta > 1 \end{cases}$$
(4.111)

Note the similarity of the natural response for the voltage $v(t)$ for this circuit and the natural response of the current, $i(t)$ of the previous RLC shown in figure 4.19 a). However, it should be noted that now the resistance R plays an opposite role to that played by the resistance of the circuit of figure 4.19. Now, if we want to decrease the damping factor, η, we have to increase the value of R. This is reasonable since, as the resistance R tends to infinity (open circuit), the circuit of figure 4.19 tends to the lossless circuit of figure 4.17.

4.5.4 Response to the step function

We consider now the response of various passive circuits to the step function. The circuits are driven by either a step voltage or current source.

RC circuits

Figure 4.22 shows the RC circuit driven by a step voltage source. Assuming that the capacitor is discharged at $t = 0$ we can write the following eqns for

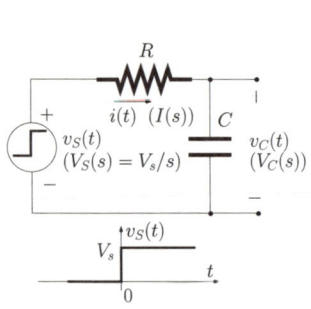

Figure 4.22: *RC driven by a step voltage source.*

this circuit in the Laplace domain, as follows:

$$V_C(s) = \frac{I(s)}{sC} \qquad (4.112)$$

$$V_S(s) = RI(s) + \frac{I(s)}{sC} \qquad (4.113)$$

Since the Laplace transform for $v_S(t)$ is (see eqn 4.32)

$$V_S(s) = \frac{V_s}{s} \qquad (4.114)$$

we can determine $V_C(s)$ and $I(s)$ as indicated below

$$V_C(s) = \frac{V_s}{s(s\tau+1)} \qquad (4.115)$$

$$I(s) = \frac{V_s}{R}\frac{\tau}{s\tau+1} \qquad (4.116)$$

with $\tau = RC$. Using eqns 4.49 and 4.33 we can obtain the time domain expression for the voltage across and the current through the capacitor:

$$v_C(t) = V_s\left(1 - e^{-\frac{t}{\tau}}\right)u(t) \qquad (4.117)$$

$$i(t) = \frac{V_s}{R}e^{-\frac{t}{\tau}}u(t) \qquad (4.118)$$

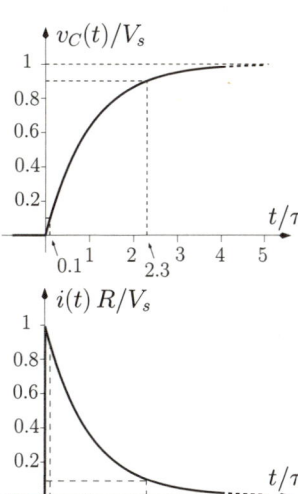

Figure 4.23: a) $v_C(t)$ normalised to V_s versus the time normalised to τ. b) $i(t)$ normalised to V_s/R versus the time normalised to τ.

Figure 4.23 a) shows $v_C(t)$ normalised to V_s versus the time normalised to τ and figure 4.23 b) shows $i(t)$ normalised to V_s/R versus the time normalised to τ. Note that, during the transient response the voltage across the capacitor terminals increases in an exponential manner towards V_s while the current $i(t)$ tends to zero. In this figure we also show that the time required for the voltage to go from 10% to 90% of its final value is about $2.2\,\tau$ (see figure 4.23 a)). This time is called the *rise-time*, t_r. On the other hand, the *fall-time*, t_f, is defined as the time taken by a signal to fall from 90% to 10% of its peak value as is the case for the current in this example (see figure 4.23 b)). Note that the current fall-time is equal to the voltage rise-time; $2.2\,\tau$.

Example 4.5.3 The rise time of an RC low-pass filter was measured to be $t_r = 50\ \mu s$. Determine its bandwidth. Also determine the delay time of the circuit, t_d, defined as the time taken for the signal to reach 50% of its peak value.

Solution: The bandwidth of an RC low-pass filter (see section 3.3.5) is:

$$BW = \frac{1}{2\pi\tau}$$

with $\tau = RC$. This eqn can be written as

$$BW = \frac{2.2}{2\pi t_r} \simeq \frac{0.35}{t_r}$$
$$= 7\text{ kHz}$$

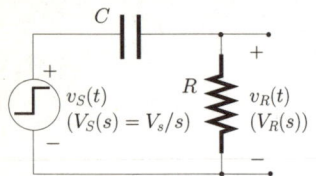

Figure 4.24: CR circuit driven by a step voltage source.

To find the delay time we solve eqn 4.117 for $v_C(t_d) = V_s/2$. This gives

$$t_d = 0.7\,\tau = \frac{0.11}{BW}$$
$$= 15.7\ \mu s$$

Example 4.5.4 Consider the CR circuit of figure 4.24. Determine the output voltage $v_R(t)$.

Solution: We can write the following eqn

$$sC[V_S(s) - V_R(s)] - CV_{co} = \frac{V_R(s)}{R} \qquad (4.119)$$

Assuming that the initial charge on the capacitor is zero ($V_{co} = 0$) we have:

$$V_R(s) = \frac{\tau s}{1+\tau s} V_S(s)$$
$$= V_s \frac{\tau}{1+\tau s} \qquad (4.120)$$

with $\tau = RC$. Using eqn 4.33 we can write

$$v_R(t) = V_s\, e^{-\frac{t}{\tau}} u(t) \qquad (4.121)$$

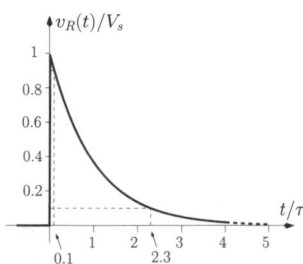

Figure 4.25: $v_R(t)$ given by eqn 4.121 normalised to V_s versus the time normalised to τ.

Figure 4.25 shows the voltage $v_R(t)$ normalised to V_s versus the time normalised to τ. From this figure we observe the fast rise of the voltage to its peak and then its exponential decay to zero. Note that the time taken for the voltage to fall from 90% to 10% of its peak value (fall-time) is about 2.2 τ.

RL circuits

Figure 4.26 shows an RL circuit driven by a step voltage source. For this circuit we can write:

$$\frac{V_S(s) - V_L(s)}{R} = \frac{V_L(s)}{sL} + \frac{I_{lo}}{s} \qquad (4.122)$$

Assuming that the initial condition of the inductor is zero ($I_{lo} = 0$) we can solve this eqn in order to obtain $V_L(s)$ as

$$V_L(s) = \frac{s\tau}{s\tau+1} V_S(s) = \frac{\tau}{s\tau+1} V_s$$

where $\tau = L/R$. Using eqn 4.33 we can write

$$v_L(t) = V_s\, e^{-\frac{t}{\tau}} u(t) \qquad (4.123)$$

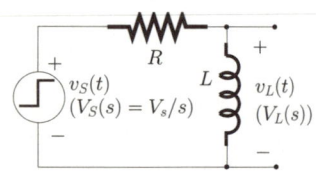

Figure 4.26: RL circuit driven by a step voltage source.

Note the similarity between this last eqn and eqn 4.121. In fact, the time (and frequency) domain behaviour of this circuit is similar to the CR circuit shown in figure 4.24.

4. Natural and forced responses circuit analysis

Example 4.5.5 Consider the LR circuit of figure 4.27 a). Determine the output voltage $v_R(t)$. Assume $I_{lo} = 0$.

Solution: For this circuit we can write:

$$\frac{V_S(s) - V_R(s)}{sL} = \frac{V_R(s)}{R} \qquad (4.124)$$

solving this eqn in order to obtain $V_R(s)$ we have

$$V_R(s) = \frac{V_s}{s(s\tau + 1)} \qquad (4.125)$$

where $\tau = L/R$. Using eqns 4.49 and 4.34 we can write the time domain voltage across the resistance

$$v_R(t) = V_s \left(1 - e^{-\frac{t}{\tau}}\right) u(t) \qquad (4.126)$$

Note that this expression is similar to the voltage across the capacitor (see eqn 4.117) of the RC circuit of figure 4.22.

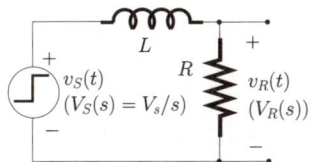

Figure 4.27: LR circuit driven by a step voltage source.

RLC circuits

Figure 4.28 shows an RLC circuit driven by a step-function voltage source. Assuming that all initial conditions of the circuit are zero we can write:

$$I(s) = \frac{V_S(s)}{Z_{eq}(s)} \qquad (4.127)$$

with $Z_{eq}(s)$ representing the equivalent impedance of the series combination of the inductor, resistance and capacitance. $Z_{eq}(s)$ is given by

$$Z_{eq}(s) = R + sL + \frac{1}{sC}$$

and $1/Z_{eq}(s)$ can be written as

$$\frac{1}{Z_{eq}} = C \frac{s\omega_n^2}{s^2 + 2\eta\omega_n + \omega_n^2} \qquad (4.128)$$

where ω_n and η are given by eqns 4.104 and 4.105 respectively. Hence, the current $I(s)$ is given by:

$$I(s) = V_S(s) \frac{sC\omega_n^2}{s^2 + 2\eta\omega_n + \omega_n^2}$$

$$= V_s C \frac{\omega_n^2}{s^2 + 2\eta\omega_n + \omega_n^2} \qquad (4.129)$$

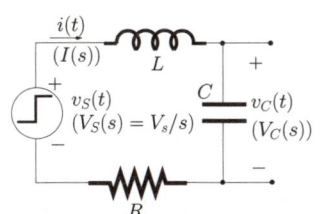

Figure 4.28: RLC circuit driven by a step voltage source.

Again we use eqns 4.72, 4.73, 4.76 to obtain the current of the circuit in the time domain, for $t \geq 0$, as follows:

$$i(t) = C V_s \times \begin{cases} \frac{\omega_n}{\sqrt{1-\eta^2}} \sin\left(\omega_n \sqrt{1-\eta^2}\, t\right) e^{-t\eta\omega_n} u(t) & \text{if } \eta < 1 \\ \omega_n^2 \, t \, e^{-\omega_n t} u(t) & \text{if } \eta = 1 \\ \frac{\omega_n}{\sqrt{\eta^2-1}} \sinh\left(\omega_n \sqrt{\eta^2-1}\, t\right) e^{-t\eta\omega_n} u(t) & \text{if } \eta > 1 \end{cases}$$
(4.130)

The voltage across the capacitor terminals, $v_C(t)$, can be obtained from eqn 1.24, that is

$$v_C(t) = V_s \times \begin{cases} \left[1 - \frac{e^{-t\eta\omega_n}}{\sqrt{1-\eta^2}} \sin\left(\sqrt{1-\eta^2}\,\omega_n t + \phi\right)\right] u(t) & \text{if } \eta < 1 \\ \left[1 - (t\omega_n + 1) e^{-t\omega_n}\right] u(t) & \text{if } \eta = 1 \\ \left(1 - e^{-t\eta\omega_n}\left[\cosh\left(\omega_n \sqrt{\eta^2-1}\, t\right) \right. \right. & \\ \left. \left. + \frac{\eta}{\sqrt{\eta^2-1}} \sinh\left(\omega_n \sqrt{\eta^2-1}\, t\right)\right]\right) u(t) & \text{if } \eta > 1 \end{cases}$$
(4.131)

with

$$\phi = \tan^{-1}\left(\frac{\sqrt{1-\eta^2}}{\eta}\right) \tag{4.132}$$

Figure 4.29 a) shows the current $i(t)$, given by eqn 4.130, normalised to $C V_s \omega_n$ versus the time normalised to ω_n^{-1} considering $\eta = 0.1, 0.3, 0.7, 1$ and 3. From this figure we observe that, regardless of the value of η, the current in the RLC circuit eventually tends, as expected, to zero. As with the natural response we observe that for values of $\eta < 1$ (underdamped circuit) the current exhibits oscillations and its transient behaviour is dominated by the LC combination. On the other hand, for $\eta \geq 1$ (critically damped and overdamped circuit) there is no oscillatory behaviour. Figure 4.29 b) shows the voltage across the capacitor terminals $v_C(t)$, given by eqn 4.131, normalised to V_s versus the time normalised to ω_n^{-1}. We observe that for all values of η the voltage across the capacitor terminals tends to V_s; the voltage of the DC source. Very much like the current, we observe that for values of $\eta < 1$ the voltage overshoots its final value exhibiting an oscillatory behaviour. From this figure it is clear that the amount of overshoot depends on the value of η; the smaller the value of η the greater the amount of overshoot. The first overshoot is called the *peak overshoot* and can be determined by differentiating eqn 4.131 ($\eta < 1$) with respect to the time and by setting this derivative to zero, that is

$$\frac{d v_C(t)}{dt} = V_s \frac{\eta \omega_n e^{-t\eta\omega_n}}{\sqrt{1-\eta^2}} \sin\left(\sqrt{1-\eta^2}\,\omega_n t + \phi\right)$$

4. Natural and forced responses circuit analysis

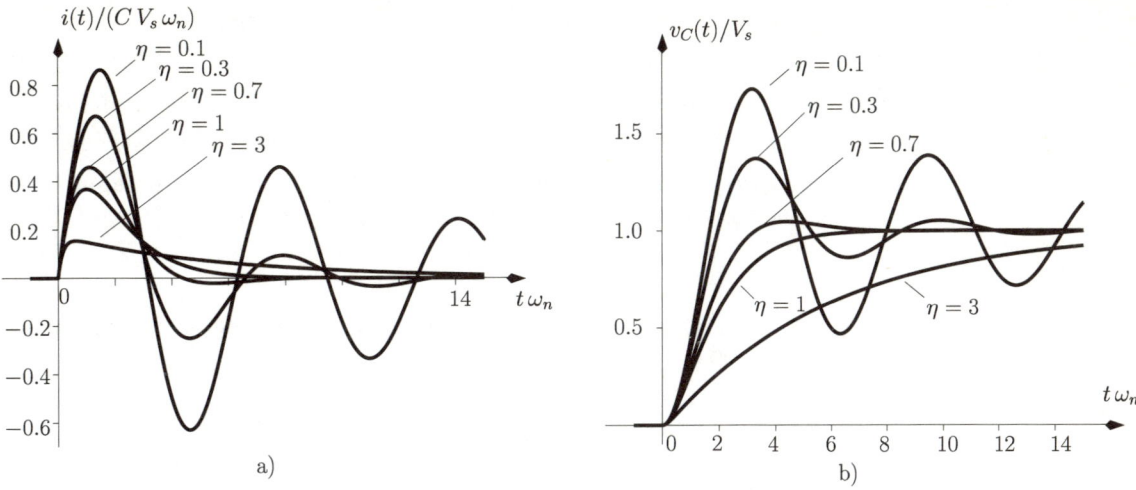

Figure 4.29: a) $i(t)$, given by eqn 4.130, normalised to $CV_s\omega_n$ versus the time normalised to ω_n^{-1}. b) $v_C(t)$ given by eqn 4.131, normalised to V_s versus the time normalised to ω_n^{-1}.

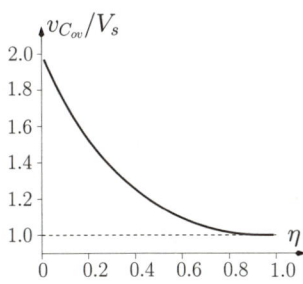

Figure 4.30: *Peak overshoot normalised to V_s versus η.*

$$- V_s w_n e^{-t\eta w_n} \cos\left(\sqrt{1-\eta^2}\omega_n t + \phi\right) = 0 \quad (4.133)$$

The last eqn is zero for $(\sqrt{1-\eta^2}\,\omega_n\,t) = 0,\,\pi,\,2\pi,\,3\pi,\ldots$.

The peak overshoot occurs at

$$t_{ov} = \frac{\pi}{\omega_n\sqrt{1-\eta^2}} \quad (4.134)$$

Note that this value applies only for zero initial conditions. Using this value of t_{ov} in eqn 4.131 ($\eta < 1$) we obtain the value of the peak overshoot of

$$v_{C_{ov}} = V_s\left[1 + \exp\left(\frac{-\pi\eta}{\sqrt{1-\eta^2}}\right)\right] \quad (4.135)$$

Figure 4.30 shows the peak overshoot normalised to V_s versus the damping factor η.

The *settling time*, t_S, is defined as the time that a waveform takes to attain (and to stay within the limits of) a percentage of its final value. This percentage is usually taken as 2% or 5%. It is interesting to note that for values of $\eta < 1$ the rise time of the waveform is relatively fast but the settling time can be quite large. Figure 4.31 shows the settling time of $v_C(t)$ (normalised to ω_n^{-1}) versus η with the percentage of the final value being $\pm 2\%$ and $\pm 5\%$. From this figure it can be seen that for $\pm 2\%$ the value of η which minimises t_S is $\eta = 0.78$. For this situation the minimum settling time is $t_S = 3.6/\omega_n$. If the percentage which defines t_S is $\pm 5\%$, the value of η which minimises t_S is $\eta = 0.69$ and the minimum settling time is $t_S = 2.9/\omega_n$.

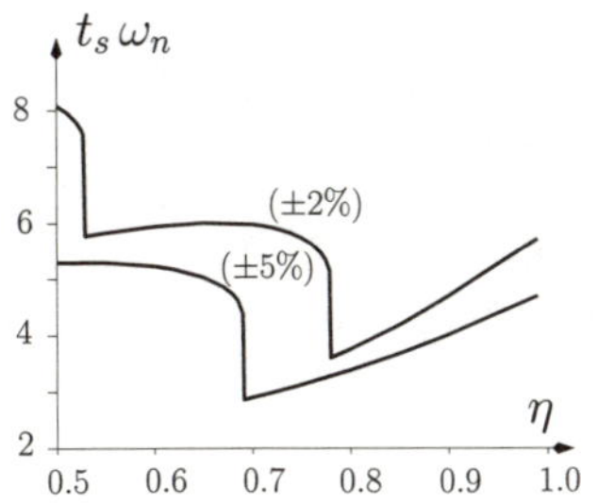

Figure 4.31: *Settling time, normalised to ω_n^{-1}, versus η.*

The rise, fall and settling time values and the overshoot characteristics are used to describe the transient behaviour of different passive and active circuits

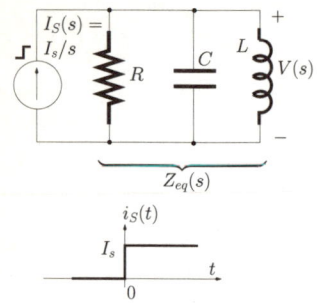

and are of great importance, for example, in filter design.

Example 4.5.6 Consider the RLC circuit of figure 4.32. Determine the current in L. Assume that all initial conditions are zero and that $R > \sqrt{L/(4C)}$.

Solution: Since all the initial conditions of the circuit are zero we can determine the voltage $V(s)$ across the circuit elements as follows:

$$V(s) = I_S(s)\, Z_{eq}(s) \qquad (4.136)$$

where $Z_{eq}(s)$ is the equivalent impedance corresponding to the parallel combination of R with sL and with $(sC)^{-1}$, that is

$$Z_{eq}(s) = \frac{sLR}{s^2 RLC + sL + R} \qquad (4.137)$$

Now eqn 4.136 can be written as

$$V(s) = \frac{I_s}{C}\,\frac{1}{s^2 + 2\eta\omega_n s + \omega_n^2} \qquad (4.138)$$

with η and ω_n given by eqns 4.110 and 4.109, respectively. Since R is greater than $\sqrt{L/(4C)}$ we have $\eta < 1$. Using eqn 4.72 we can write

$$v(t) = \frac{I_s}{C}\,\frac{1}{\omega_n\sqrt{1-\eta^2}}\,\sin\left(\omega_n\sqrt{1-\eta^2}\,t\right)\,e^{-t\eta\omega_n}\,u(t) \qquad (4.139)$$

The current in the inductor $i_L(t)$ can be obtained from eqn 1.27, that is

$$i_L(t) = I_s\left[1 - \frac{e^{-t\eta\omega_n}}{\sqrt{1-\eta^2}}\sin\left(\sqrt{1-\eta^2}\,\omega_n t + \phi\right)\right]u(t) \qquad (4.140)$$

with ϕ given by eqn 4.132. Figure 4.33 shows the current given by eqn 4.140, normalised to I_s, versus the time normalised to ω_n^{-1} with $\eta = 0.1, 0.4, 0.7$ and 0.9. For large values of t the current source behaves as a DC source. Since the inductor behaves as a short-circuit for DC sources then, for large t, the voltage across the resistor and across the capacitor tends to zero and the inductor conducts I_s.

Figure 4.32: RLC circuit driven by a step current source.

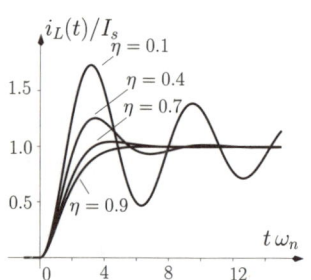

Figure 4.33: $i_L(t)$ given by eqn 4.140, normalised to I_s, versus the time normalised to ω_n.

Example 4.5.7 Consider the RLC circuit of figure 4.34. Determine the voltage across C for all time t. Take $V_{S_1} = 3$ V and $V_{S_2} = 7$ V.

Solution: Figure 4.35 a) shows the equivalent circuit for $t < 0$. Note that for $t < 0$ the inductor behaves as a short-circuit while the capacitor behaves as an open-circuit. The voltage across the terminals of the capacitor is the voltage across R_1 which can be calculated as

$$\begin{aligned}V_{co} &= V_{R_1}\\ &= -V_{S_2}\frac{R_1}{R_1 + R_3}\\ &= -3.5\text{ V}\end{aligned}$$

4. Natural and forced responses circuit analysis

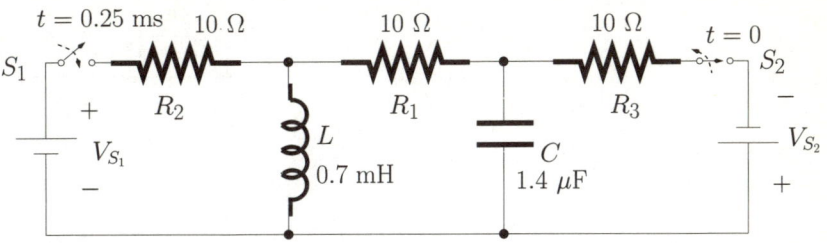

Figure 4.34: *RLC circuit.*

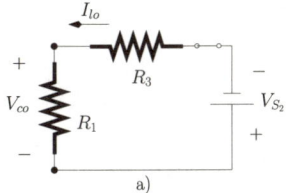

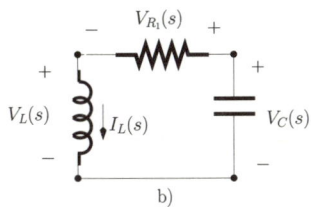

Figure 4.35: *Equivalent circuit for; a) $t < 0$; b) $0 \leq t < 0.25$ ms.*

The current flowing through the inductor also flows through R_1 and R_3. This current can be expressed as

$$I_{lo} = \frac{-V_{S_2}}{R_1 + R_3}$$
$$= -0.35 \text{ A}$$

At $t = 0$ the switch S_2 is open and S_1 remains open. Figure 4.35 b) shows the equivalent circuit for $0 \leq t < 0.25$ ms. For this circuit we can write:

$$V_C(s) = V_{R_1}(s) + V_L(s)$$

that is:

$$-\frac{I_L(s)}{sC} + \frac{V_{co}}{s} = R_1 I_L(s) + s L I_L(s) - L I_{lo}$$

where V_{co} and I_{lo} are the initial conditions (at $t = 0$) associated with the capacitor and inductor, respectively. Solving this last eqn in order to obtain $I_L(s)$ we get:

$$I_L(s) = C V_{co} \frac{\omega_n^2}{s^2 + 2\eta \omega_n s + \omega_n^2}$$
$$+ I_{lo} \frac{s}{s^2 + 2\eta \omega_n s + \omega_n^2}$$

with

$$\omega_n = \frac{1}{\sqrt{LC}}$$
$$= 32 \text{ krad/s}$$
$$\eta = \frac{1}{2} R_1 \sqrt{\frac{C}{L}}$$
$$= 0.22$$

Taking the inverse Laplace transform ($\eta < 1$) we obtain the current flowing through the inductor, $i_L(t)$ for the time interval $0 \leq t < 0.25$ ms, that is,

$$i_L(t) = \underbrace{C V_{co} \frac{\omega_n}{\sqrt{1-\eta^2}} \sin\left(\omega_n \sqrt{1-\eta^2}\, t\right) e^{-t\eta\omega_n}}_{\text{Contribution from } V_{co}}$$

$$+ I_{lo} \frac{1}{\sqrt{1-\eta^2}} \cos\left(\omega_n \sqrt{1-\eta^2}\, t + \phi\right) e^{-t\eta\omega_n}$$
$$\underbrace{\hspace{6cm}}_{\text{Contribution from } I_{lo}}$$

with

$$\phi = \tan^{-1}\left(\frac{\eta}{\sqrt{1-\eta^2}}\right) \tag{4.141}$$

The voltage across the capacitor, for $0 \leq t < 0.25$ ms, is given by:

$$V_C(s) = -\frac{I_L(s)}{sC} + \frac{V_{co}}{s}$$
$$= -V_{co}\frac{\omega_n^2}{s(s^2 + 2\eta\omega_n s + \omega_n^2)}$$
$$- \frac{I_{lo}}{C}\frac{1}{s^2 + 2\eta\omega_n s + \omega_n^2} + \frac{V_{co}}{s}$$

taking the inverse Laplace transform we obtain

$$v_C(t) = \frac{V_{co}}{\sqrt{1-\eta^2}} \cos\left(\omega_n \sqrt{1-\eta^2}\, t - \phi\right) e^{-t\eta\omega_n}$$
$$\underbrace{\hspace{7cm}}_{\text{Contribution from } V_{co}}$$
$$- \frac{I_{lo}}{C}\frac{1}{\omega_n\sqrt{1-\eta^2}} \sin\left(\omega_n \sqrt{1-\eta^2}\, t\right) e^{-t\eta\omega_n}$$
$$\underbrace{\hspace{7cm}}_{\text{Contribution from } I_{lo}}$$

Figure 4.36 a) shows the voltage across the capacitor while figure 4.36 b) shows the current through the inductor. From these figures we observe that at $t = 0.25$ ms the voltage across the capacitor is 1.17 V while the current through the inductor is -17.8 mA. These values are the initial conditions associated with the capacitor and inductor when the switch S_1 is closed at $t = 0.25$ ms, that is, $V_{co'} = 1.17$ V and $I_{lo'} = -17.8$ mA.

Figure 4.37 shows the equivalent circuit for $t \geq 0.25$ ms. Note that when the switch S_1 is closed at $t = 0.25$ ms V_{S1} is applied to the circuit as a step-forcing voltage, that is, $v_{S_1}(t) = 3\, u(t - t_o)$ V with $t_o = 0.25$ ms. Using nodal analysis we can write the following eqns:

$$\begin{cases} I_{R_2}(s) = I_L(s) + I_{R_1}(s) \\ I_{R_1}(s) = I_C(s) \end{cases} \tag{4.142}$$

that is:

$$\begin{cases} \dfrac{V_{S_1}(s) - V_L(s)}{R_2} = \dfrac{V_L(s)}{sL} + \dfrac{I_{lo'}}{s} e^{-st_o} + \dfrac{V_L(s) - V_C(s)}{R_1} \\ \dfrac{V_L(s) - V_C(s)}{R_1} = sC V_C(s) - C V_{co'} e^{-st_o} \end{cases} \tag{4.143}$$

4. Natural and forced responses circuit analysis

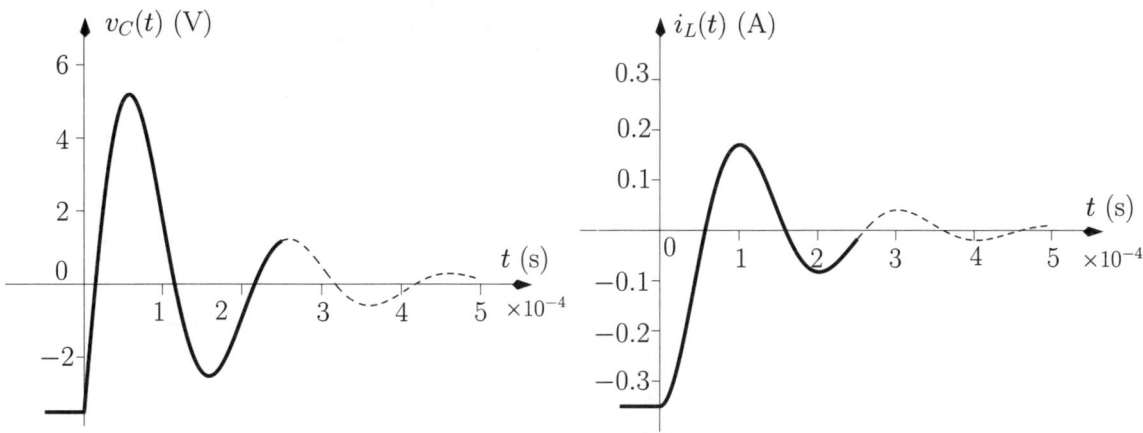

Figure 4.36: *a) Voltage across the capacitor. b) Current through the inductor.*

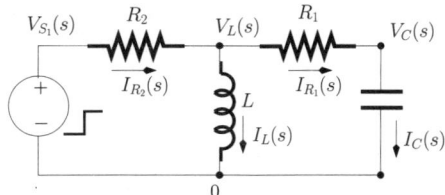

Figure 4.37: *Equivalent circuit for $t \geq 0.25$ ms.*

with

$$V_{S_1}(s) = \frac{V_s}{s} e^{-s t_o}$$

$V_s = 3$ V. Solving eqn 4.143 to obtain $V_C(s)$ we get:

$$\begin{aligned} V_C(s) &= V_{S_1}(s) \frac{s L \omega_n'^2}{R_2 \left(s^2 + 2\eta' \omega_n' s + \omega_n'^2 \right)} \\ &+ V_{co'} e^{-s t_o} \frac{s + R_1 C \omega_n'^2}{s^2 + 2\eta' \omega_n' s + \omega_n'^2} \\ &- I_{lo'} e^{-s t_o} \frac{L \omega_n'^2}{s^2 + 2\eta' \omega_n' s + \omega_n'^2} \end{aligned}$$

with

$$\begin{aligned} \omega_n' &= \sqrt{\frac{R_2}{R_1 + R_2} \frac{1}{LC}} \\ &= 22.6 \text{ krad/s} \\ \eta' &= \frac{1}{2} R_1 \sqrt{\frac{R_2}{R_1 + R_2} \frac{C}{L}} \\ &= 0.16 \end{aligned}$$

taking the inverse Laplace transform ($\eta < 1$) we obtain the voltage across the capacitor for $t > 0.25$ ms, that is,

$$V_C(t) = \underbrace{\frac{V_s L}{R_2} \frac{\omega'_n}{\sqrt{1-\eta'^2}} \sin\left(\omega'_n \sqrt{1-\eta'^2}\,(t-t_o)\right) e^{-(t-t_o)\eta'\omega'_n}}_{\text{Contribution from } v_{S_1}(t)}$$

$$+ \underbrace{V_{co'} R_1 C \frac{\omega'_n}{\sqrt{1-\eta'^2}} \sin\left(\omega'_n \sqrt{1-\eta'^2}\,(t-t_o)\right) e^{-(t-t_o)\eta'\omega'_n}}_{\text{Contribution from } V_{co'}}$$

$$+ \underbrace{\frac{V_{co'}}{\sqrt{1-\eta'^2}} \cos\left(\omega'_n \sqrt{1-\eta'^2}\,(t-t_o) + \phi'\right) e^{-(t-t_o)\eta'\omega'_n}}_{\text{Contribution from } V_{co'}}$$

$$- \underbrace{I_{lo'} L \frac{\omega'_n}{\sqrt{1-\eta'^2}} \sin\left(\omega'_n \sqrt{1-\eta'^2}\,(t-t_o)\right) e^{-(t-t_o)\eta'\omega'_n}}_{\text{Contribution from } I_{lo'}}$$

with ϕ' given by

$$\phi' = \tan^{-1}\left(\frac{\eta'}{\sqrt{1-\eta'^2}}\right)$$

Figure 4.38 shows the voltage across the capacitor for $0 < t < 1$ ms.

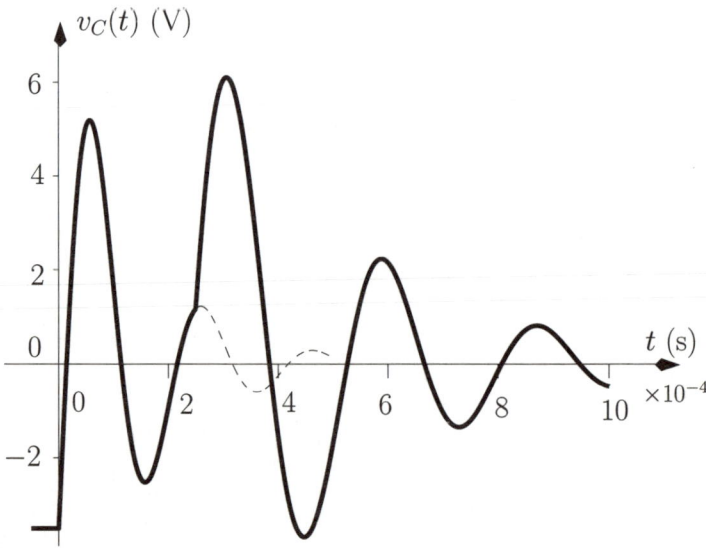

Figure 4.38: *Voltage across the capacitor.*

4.6 Bibliography

1. C.R. Wylie and L.C. Barrett, *Advanced Engineering Mathematics*, 1995 (McGraw-Hill International Editions), 6th edition.

2. C. Chen, *System and Signal Analysis*, 1994 (Saunders College Publishing), 2nd edition.

3. J.J. D'Azzo and C.H. Houpis, *Linear Control System, Analysis and Design*, 1995 (McGraw-Hill International Editions), 4th edition.

4. M.J. Roberts, *Signals and Systems: Analysis using Transform Methods and Matlab®*, 2003 (McGraw-Hill International Editions).

4.7 Problems

4.1 Consider example 4.2.2. Plot $v_C(t)$ when the period T is longer than τ (for example $T = 3\tau$). Draw comments in relation to the time taken to reach steady-state.

4.2 Consider the causal signals of figure 4.39. Determine their Laplace transforms and indicate their ROC.

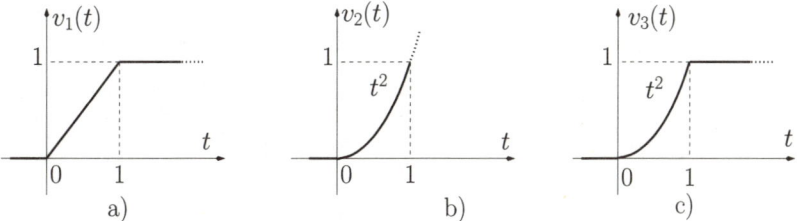

Figure 4.39: *Signals of problem 4.2.*

4.3 Using the partial fraction expansion method determine the inverse Laplace transform of the following s functions:

$$X_1(s) = \frac{1}{(s-a)^2 + b^2}$$

$$X_2(s) = \frac{s}{(s+a)^2 (s+b)^2}$$

$$X_3(s) = \frac{a}{s^2 - a^2}$$

$$X_4(s) = \frac{s}{s^2 - a^2}$$

4.4 Consider the circuits of figure 4.40. Using first Fourier transforms calculate the voltage across R_1 for all time t. Then use Laplace transforms to determine the same voltage. Draw conclusions.

4.5 Consider the circuits of figure 4.41. Determine the current through the inductor for all time t.

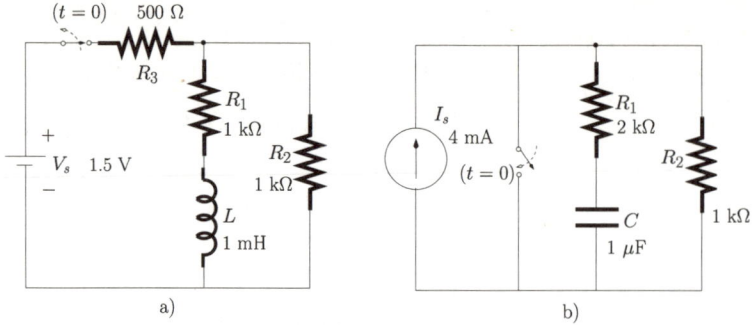

Figure 4.40: *Circuits of problem 4.4.*

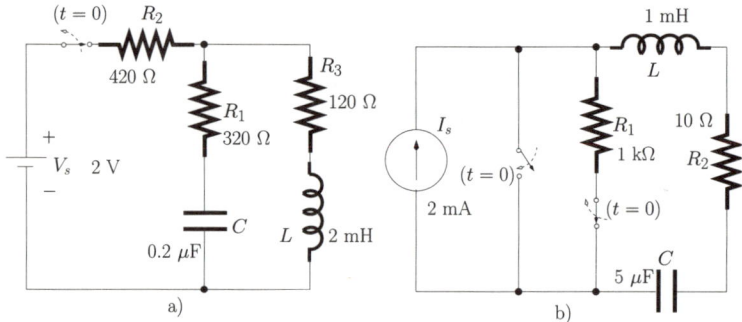

Figure 4.41: *Circuits of problem 4.5.*

4.6 Consider the circuits of figure 4.42. Determine the voltage across the capacitor C_1 for all time t. Take $v_S(t) = 3\,u(t)$ volts. Assume zero initial conditions.

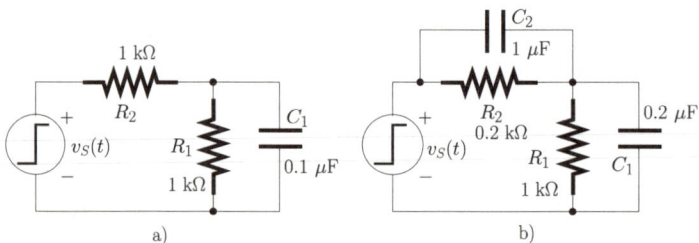

Figure 4.42: *Circuits of problem 4.6.*

4.7 Consider the circuits of figure 4.43. Determine the current through the inductor for all time t. Take $v_S(t) = 4\,u(t)$ volts. Assume zero initial conditions.

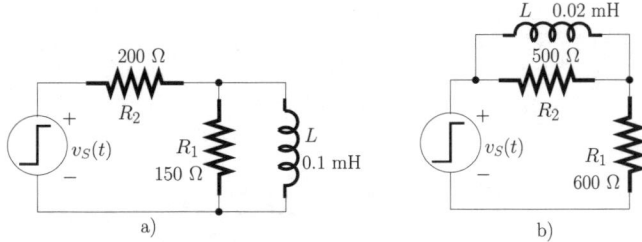

Figure 4.43: *Circuits of problem 4.7.*

4.8 Consider the circuit of figure 4.44. Determine the current through the resistance, the capacitor and the inductor for all time t. Take $i_S(t) = u(t)$ mA. Take $R = 20$ Ω. Assume zero initial conditions.

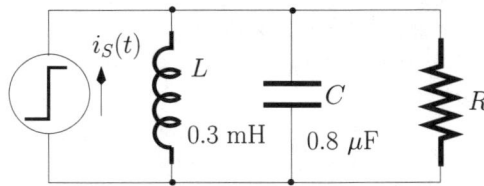

Figure 4.44: *Circuit of problem 4.8 and problem 4.9.*

4.9 Consider the circuit of figure 4.44. Determine the value of the resistance such that the damping factor is 1.

5 Electrical two-port network analysis

5.1 Introduction

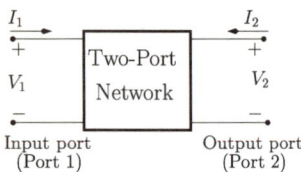

Figure 5.1: *Electrical two-port network.*

Two-port circuit techniques are usually employed to analyse and characterise linear electrical and electronic circuits. As its name suggests, a two-port circuit or network is a circuit with an input port and an output port, as shown in figure 5.1. Hence, two-port circuit analysis techniques can be used to analyse and characterise circuits ranging from a simple resistive voltage divider to very complex electronic amplifiers.

The theory of two-port circuits relates the voltage and the current variables at the ports. Depending upon which two of these four variables (V_1, V_2, I_1, or I_2) are chosen as the independent variables a different set of parameters can be defined each of which completely characterises the network. Each set of parameters is defined here as an electrical representation. In this chapter we consider the impedance, admittance and chain (or ABCD) representations[1]. The electrical representations (or parameters) are obtained in the frequency domain by means of phasor analysis. In section 5.3 we show how these electrical representations are suited for computer-based electrical analysis by means of a systematic analysis approach.

5.2 Electrical representations

As mentioned previously a two-port network can be characterised by a set of parameters which relates current and voltage values at the ports. Figure 5.1 shows the convention for the voltages across each port as well as the direction of the current that might be present at each port.

5.2.1 Electrical impedance representation

The electrical impedance representation uses the currents I_1 and I_2 as the excitation signals and the voltages V_1 and V_2 as the responses to these excitations. Hence, the relevant parameters are impedances, the Z-parameters, and these satisfy the two following eqns:

$$V_1 = Z_{11} I_1 + Z_{12} I_2 \tag{5.1}$$
$$V_2 = Z_{21} I_1 + Z_{22} I_2 \tag{5.2}$$

These eqns can be written in a matrix form:

$$[V] = [Z][I] \tag{5.3}$$

---

[1] There are other types of two-port representation not discussed here. The most important of these is the S-parameter representation discussed in detail in Chapter 7.

5. Electrical two-port network analysis

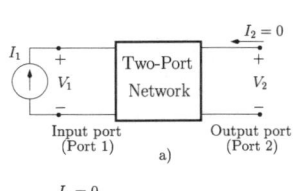

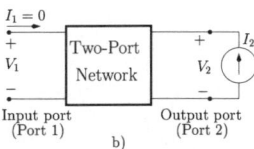

Figure 5.2: Set-up for the measurement of the Z-parameters. a) Z_{11} and Z_{21}. b) Z_{12} and Z_{22}.

with

$$[\boldsymbol{I}] = \begin{bmatrix} I_1 \\ I_2 \end{bmatrix} \tag{5.4}$$

$$[\boldsymbol{V}] = \begin{bmatrix} V_1 \\ V_2 \end{bmatrix} \tag{5.5}$$

$$[\boldsymbol{Z}] = \begin{bmatrix} Z_{11} & Z_{12} \\ Z_{21} & Z_{22} \end{bmatrix} \tag{5.6}$$

The matrix $[\boldsymbol{Z}]$ is called here the 'electrical impedance representation' of the two-port network. The evaluation of the coefficients Z_{ij}, which have dimensions of impedance, can be effected by performing measurements at the two-port circuit terminals as illustrated in figure 5.2. Figure 5.2 a) shows the set-up for the measurements of Z_{11} and of Z_{21}. In this situation the output port is an open-circuit so that $I_2 = 0$ and a current source I_1 drives the input port. Measuring the voltages V_1 and V_2 we can determine Z_{11} and Z_{21} as follows:

$$Z_{11} = \left.\frac{V_1}{I_1}\right|_{I_2=0} \qquad Z_{21} = \left.\frac{V_2}{I_1}\right|_{I_2=0} \tag{5.7}$$

Z_{11} is the input impedance (see also section 1.4.2) while Z_{21} is called the forward transimpedance gain. It is clear that Z_{11} and Z_{21} can be calculated from eqns 5.1 and 5.2 by setting $I_2 = 0$. Figure 5.2 b) shows the set-up for the measurements of Z_{12} and of Z_{22}. Now the input port is an open-circuit so that $I_1 = 0$ and a current source I_2 excites the circuit in port 2. Measuring again the voltages V_1 and V_2 we can determine Z_{12} and Z_{22} as follows:

$$Z_{12} = \left.\frac{V_1}{I_2}\right|_{I_1=0} \qquad Z_{22} = \left.\frac{V_2}{I_2}\right|_{I_1=0} \tag{5.8}$$

Z_{22} is the output impedance while Z_{12} is called the reverse transimpedance gain. Note that Z_{12} and Z_{22} can be calculated from eqns 5.1 and 5.2 by setting $I_1 = 0$.

It is important to note that Z_{21} and Z_{12} also represent impedance transfer functions (see section 3.3.7).

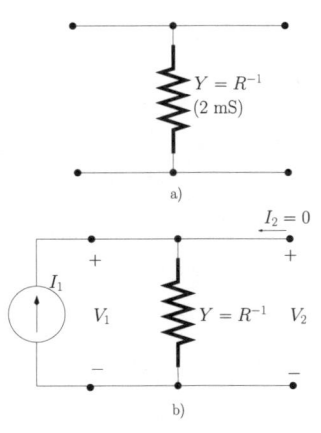

Figure 5.3: a) Shunt admittance. b) Calculation of Z_{11} and of Z_{21}.

Example 5.2.1 Determine the Z-parameters of the shunt admittance of figure 5.3 a).

<u>Solution</u>: Figure 5.3 b) shows the equivalent circuit for the calculation of Z_{11} and of Z_{21}. Since I_1 flows through Y we have

$$\begin{aligned} Z_{11} &= \frac{V_1}{I_1} \\ &= \frac{1}{Y} \\ &= R \\ &= 500 \ \Omega \end{aligned}$$

Since $V_2 = V_1$ we have

$$\begin{aligned} Z_{21} &= \frac{V_2}{I_1} \\ &= Z_{11} = R \end{aligned}$$

From symmetry considerations we have:

$$\begin{aligned} Z_{12} &= Z_{11} = R \\ Z_{22} &= Z_{11} = R \end{aligned}$$

Example 5.2.2 Determine the Z-parameters of the T-network of figure 5.4 a).

<u>Solution</u>: Figure 5.4 b) shows the equivalent circuit for the calculation of Z_{11} and Z_{21}. Since I_1 flows through Z_1 and Z_3 we can write:

$$\begin{aligned} Z_{11} &= Z_1 + Z_3 \\ &= j\omega L_1 + \frac{1}{j\omega C} \\ &= \frac{1 - \omega^2 L_1 C}{j\omega C} \\ &= \frac{1 - \omega^2 6 \times 10^{-15}}{j\omega 3 \times 10^{-9}}\, \Omega \end{aligned}$$

Z_{21} is calculated as follows:

$$\begin{aligned} Z_{21} &= Z_3 \\ &= \frac{1}{j\omega C} \\ &= \frac{1}{j\omega 3 \times 10^{-9}}\, \Omega \end{aligned}$$

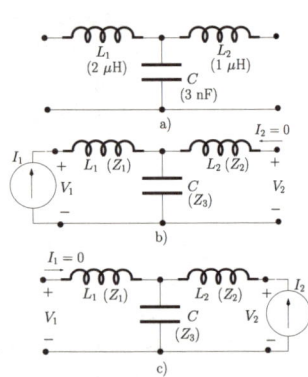

Figure 5.4: *a) T-network. b) Calculation of Z_{11} and of Z_{21}. c) Calculation of Z_{12} and of Z_{22}.*

Figure 5.4 c) shows the equivalent circuit for the calculation of Z_{22} and of Z_{12}. Since I_2 flows through Z_2 and Z_3 we can write:

$$\begin{aligned} Z_{22} &= Z_2 + Z_3 \\ &= \frac{1 - \omega^2 L_2 C}{j\omega C} \\ &= \frac{1 - \omega^2 3 \times 10^{-15}}{j\omega 3 \times 10^{-9}}\, \Omega \end{aligned}$$

Z_{12} is calculated as follows:

$$\begin{aligned} Z_{12} &= Z_3 \\ &= \frac{1}{j\omega 3 \times 10^{-9}}\, \Omega \end{aligned}$$

5. Electrical two-port network analysis

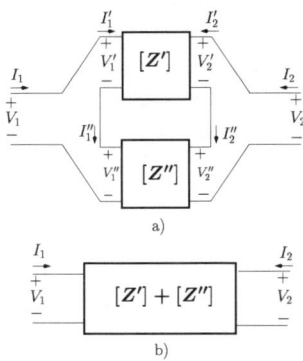

Figure 5.5: *a) Series connection of two-port networks. b) Equivalent two-port network.*

Series connection

Let us consider the interconnection of a two-port network, characterised by an impedance representation $[Z']$, with another two-port network, characterised by an impedance representation $[Z'']$, as illustrated in figure 5.5 a). Such an interconnection is called a series connection as the two networks share the same input current and the same output current[2].

The equivalent two-port network resulting from this interconnection can be characterised by an equivalent impedance representation, $[Z_{eq}]$, which can be determined according to eqns 5.7 and 5.8, that is;

$$Z_{eq_{11}} = \left.\frac{V_1}{I_1}\right|_{I_2=0} \tag{5.9}$$

Since port 2 is an open-circuit then $I_2 = I_2' = I_2'' = 0$ and $I_1 = I_1' = I_1''$. In addition, $V_1 = V_1' + V_1''$. Therefore, we can write the last eqn as follows:

$$\begin{aligned}Z_{eq_{11}} &= \left.\frac{V_1'}{I_1'}\right|_{I_2'=0} + \left.\frac{V_1''}{I_1''}\right|_{I_2''=0}\\ &= Z_{11}' + Z_{11}''\end{aligned} \tag{5.10}$$

Similarly, it can be shown (see problem 5.2) that:

$$Z_{eq_{12}} = Z_{12}' + Z_{12}'' \tag{5.11}$$
$$Z_{eq_{21}} = Z_{21}' + Z_{21}'' \tag{5.12}$$
$$Z_{eq_{22}} = Z_{22}' + Z_{22}'' \tag{5.13}$$

that is

$$[Z_{eq}] = [Z'] + [Z''] \tag{5.14}$$

5.2.2 Electrical admittance representation

The electrical admittance representation considers the voltages (V_1 and V_2) to be the excitation signals and the currents (I_1 and I_2) as the responses to these excitations. Hence, the Y-parameters satisfy the two following equations:

$$I_1 = Y_{11} V_1 + Y_{12} V_2 \tag{5.15}$$
$$I_2 = Y_{21} V_1 + Y_{22} V_2 \tag{5.16}$$

Equations 5.15 and 5.16 can be written in a matrix form, that is;

$$[I] = [Y][V] \tag{5.17}$$

where $[I]$ and $[V]$ are defined in eqns 5.4 and 5.5, respectively.

$$[Y] = \begin{bmatrix} Y_{11} & Y_{12} \\ Y_{21} & Y_{22} \end{bmatrix} \tag{5.18}$$

---

[2]The series connection of two-port networks should not be confused with the series connection of impedances which are *one*-port-terminal networks.

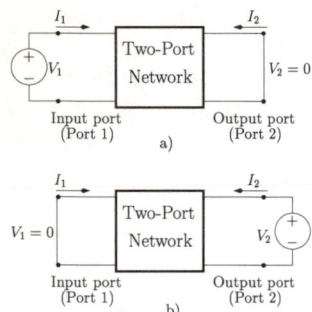

Figure 5.6: *Set-up for the Y-parameters measurement. a) Y_{11} and Y_{21}. b) Y_{12} and Y_{22}.*

The matrix $[Y]$ is the electrical admittance representation (or the admittance parameters) of the two-port network.

We can interpret the coefficients Y_{ij}, all of which have dimensions of admittance, in terms of measurements effected at the two-port circuit terminals as illustrated in figure 5.6. Figure 5.6 a) shows the set-up for the measurements of Y_{11} and of Y_{21}. In this situation the output port is short-circuited so that $V_2 = 0$ and a voltage source V_1 drives the input port. Measuring the currents I_1 and I_2 gives Y_{11} and Y_{21} as follows:

$$Y_{11} = \left.\frac{I_1}{V_1}\right|_{V_2=0} \qquad Y_{21} = \left.\frac{I_2}{V_1}\right|_{V_2=0} \qquad (5.19)$$

Y_{11} is the input admittance while Y_{21} is the forward transconductance gain. Figure 5.6 b) shows the set-up for the measurements of Y_{12} and Y_{22}. Now the input port is short-circuited so that $V_1 = 0$ and a voltage source V_2 excites the circuit in port 2. Measuring again the currents I_1 and I_2 we can determine Y_{12} and Y_{22} as:

$$Y_{12} = \left.\frac{I_1}{V_2}\right|_{V_1=0} \qquad Y_{22} = \left.\frac{I_2}{V_2}\right|_{V_1=0} \qquad (5.20)$$

Y_{22} is the output admittance while Y_{12} is the reverse transconductance gain.

Again, we note that Y_{21} and Y_{12} also represent admittance transfer functions (see section 3.3.7).

Example 5.2.3 Determine the admittance representation of the Π-network of figure 5.7 a).

Solution: Figure 5.7 b) shows the circuit for the calculation of Y_{11} and Y_{21}. The admittances associated with C_1, C_2 and L can be written as:

$$\begin{aligned} Y_1 &= j\omega C_1 \\ Y_2 &= j\omega C_2 \\ Y_3 &= \frac{1}{j\omega L} \end{aligned}$$

Since Y_2 is short-circuited we observe that Y_1 is in parallel with Y_3. Since V_1 is applied to the admittance resulting from the parallel connection of Y_1 with Y_3, we can write:

$$I_1 = V_1(Y_1 + Y_3)$$

that is

$$\begin{aligned} Y_{11} &= \frac{I_1}{V_1} \\ &= Y_1 + Y_3 \\ &= \frac{1 - \omega^2 L C_1}{j\omega L} \\ &= \frac{1 - \omega^2 3 \times 10^{-15}}{j\omega\, 10^{-6}} \text{ S} \end{aligned}$$

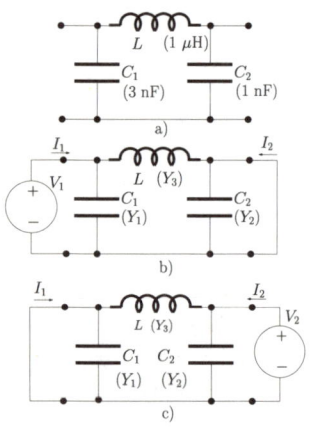

Figure 5.7: *a) Π-network. b) Calculation of Y_{11} and Y_{21}. c) Calculation of Y_{12} and Y_{22}.*

I_2 is the current that flows through Y_3. Since V_1 is directly applied to this admittance we can write:

$$I_2 = -Y_3 V_1$$

that is

$$\begin{aligned} Y_{21} &= \frac{I_2}{V_1} \\ &= -Y_3 \\ &= -\frac{1}{j\omega L} \\ &= -\frac{1}{j\omega\, 10^{-6}} \text{ S} \end{aligned}$$

Figure 5.7 shows the equivalent circuit for the calculation of Y_{12} and Y_{22}. V_2 is applied to the admittance resulting from the parallel connection of Y_2 with Y_3. We can write:

$$I_2 = V_2(Y_2 + Y_3)$$

that is

$$\begin{aligned} Y_{22} &= \frac{I_2}{V_2} \\ &= Y_2 + Y_3 \\ &= \frac{1 - \omega^2 L C_2}{j\omega L} \\ &= \frac{1 - \omega^2\, 10^{-15}}{j\omega\, 10^{-6}} \text{ S} \end{aligned}$$

I_1 is the current that flows through Y_3. Since V_2 is directly applied to this admittance we can write:

$$\begin{aligned} Y_{12} &= -Y_3 \\ &= Y_{21} \end{aligned}$$

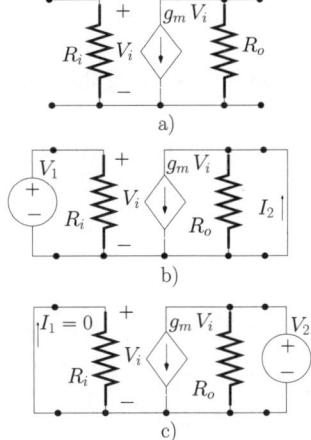

Figure 5.8: a) Two-port circuit. b) Calculation of Y_{11} and of Y_{21}. c) Calculation of Y_{12} and of Y_{22}.

Example 5.2.4 The circuit in figure 5.8 a) describes a basic representation of an electronic amplifier. This amplifier has an input resistance $R_i = 2.5$ kΩ and an output resistance $R_o = 5$ kΩ. The gain of the amplifier is modelled by the transconductance $g_m = 50$ mS. Determine its Y-parameters.

Solution: Figure 5.8 b) shows the circuit for the calculation of Y_{11} and Y_{21}. Since V_1 is the voltage across R_i we can write:

$$\begin{aligned} Y_{11} &= \frac{I_1}{V_1} \\ &= \frac{1}{R_i} \\ &= 0.4 \text{ mS} \end{aligned}$$

Also, we have that $I_2 = g_m V_i$ and that $V_1 = V_i$. Hence,

$$\begin{aligned} Y_{21} &= \frac{I_2}{V_1} \\ &= \frac{g_m V_i}{V_i} \\ &= g_m \\ &= 50 \text{ mS} \end{aligned}$$

Figure 5.8 c) shows the circuit for the calculation of Y_{12} and Y_{22}. Since $V_1 = V_i = 0$ the voltage controlled current source presents an infinite impedance. Hence we have:

$$\begin{aligned} Y_{12} &= 0 \\ Y_{22} &= \frac{1}{R_o} \\ &= 0.2 \text{ mS} \end{aligned}$$

Parallel connection of two-port networks

Let us consider the connection of a two-port network, characterised by an admittance representation $[Y']$, with another two-port network, characterised by an admittance representation $[Y'']$, as illustrated in figure 5.9 a). Such a connection is the parallel combination of the two networks as they have the same input and output voltages[3].

The equivalent two-port network resulting from this connection can be characterised by an equivalent admittance representation, $[Y_{eq}]$, which can be determined according to eqns 5.19–5.20, that is;

$$Y_{eq_{11}} = \left.\frac{I_1}{V_1}\right|_{V_2=0} \tag{5.21}$$

Since port 2 is short-circuited $V_2 = V_2' = V_2'' = 0$. In addition we have $V_1 = V_1' = V_1''$ and $I_1 = I_1' + I_1''$. Therefore, we can rewrite the last eqn as follows:

$$\begin{aligned} Y_{eq_{11}} &= \left.\frac{I_1'}{V_1'}\right|_{V_2'=0} + \left.\frac{I_1''}{V_1''}\right|_{V_2''=0} \\ &= Y_{11}' + Y_{11}'' \end{aligned} \tag{5.22}$$

Similarly, it can be shown (see problem 5.4) that:

$$Y_{eq_{12}} = Y_{12}' + Y_{12}'' \tag{5.23}$$
$$Y_{eq_{21}} = Y_{21}' + Y_{21}'' \tag{5.24}$$
$$Y_{eq_{22}} = Y_{22}' + Y_{22}'' \tag{5.25}$$

that is

$$[Y_{eq}] = [Y'] + [Y''] \tag{5.26}$$

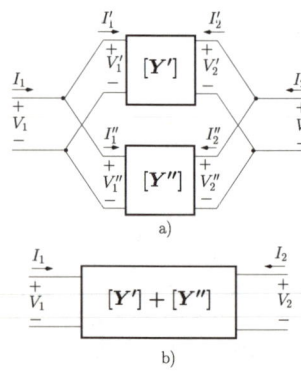

Figure 5.9: *a) Parallel connection of two-port networks. b) Equivalent two-port network.*

---
[3]Once again, the parallel connection of two-port networks should not be confused with the parallel connection of admittances or impedances which are *one*-port-terminal networks.

5.2.3 Electrical chain representation

The electrical chain (or cascade) representation (some times referred to as ABCD parameters) is a very attractive one when analysis is required for cascaded circuits, that is when the output port of one circuit is connected to the input port of another. In this representation the input current I_1 and the input voltage V_1 are the excitation signals and the output current I_2 and the output voltage V_2 are the responses to these excitations. The chain parameters satisfy the two following equations:

$$V_1 = A_{11} V_2 - A_{12} I_2 \tag{5.27}$$
$$I_1 = A_{21} V_2 - A_{22} I_2 \tag{5.28}$$

These eqns can be written in a matrix form as follows:

$$\begin{bmatrix} V_1 \\ I_1 \end{bmatrix} = \begin{bmatrix} A_{11} & A_{12} \\ A_{21} & A_{22} \end{bmatrix} \begin{bmatrix} V_2 \\ -I_2 \end{bmatrix} \tag{5.29}$$

$$[\boldsymbol{A}] = \begin{bmatrix} A_{11} & A_{12} \\ A_{21} & A_{22} \end{bmatrix} \tag{5.30}$$

where the matrix $[\boldsymbol{A}]$ is called here the chain electrical representation of the two-port network. As with the two previous representations the coefficients A_{ij}, can be interpreted in terms of measurements effected at the two-port circuit terminals as illustrated in figure 5.10. Figure 5.10 a) shows the set-up for the measurement of A_{11} where the output port is an open-circuit. A_{11} represents the inverse of the forward voltage gain and can be determined as follows:

$$A_{11} = \left. \frac{V_1}{V_2} \right|_{I_2=0} \tag{5.31}$$

Figure 5.10 b) shows the set-up for the measurement of A_{12} where the output port is now short-circuited. A_{12} represents the inverse of the forward transconductance gain and can be determined as indicated below:

$$A_{12} = \left. \frac{V_1}{-I_2} \right|_{V_2=0} \tag{5.32}$$

Figure 5.10 c) shows the set-up for the measurement of A_{21} where the output port is an open-circuit. A_{21} represents the inverse of the forward transimpedance gain and can be determined as follows:

$$A_{21} = \left. \frac{I_1}{V_2} \right|_{I_2=0} \tag{5.33}$$

Figure 5.10 d) shows the set-up for the measurement of A_{22} where the output port is now short-circuited. A_{22} represents the inverse of the current gain and can be determined according to the following equation:

$$A_{22} = \left. \frac{I_1}{-I_2} \right|_{V_2=0} \tag{5.34}$$

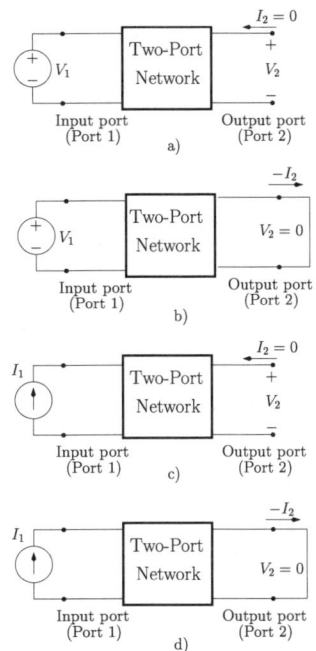

Figure 5.10: *Set-up for the chain parameters measurement. a) A_{11}. b) A_{12}. c) A_{21}. d) A_{22}.*

Note that A_{11} and A_{22} are dimensionless whilst A_{12} and A_{21} have dimensions of impedance and admittance, respectively.

Example 5.2.5 Determine the chain-parameters of the two-port network of figure 5.11 a) which represents a simplified model of an electronic amplifier. $R_i = 10$ kΩ, $R_o = 1$ kΩ and $g_m = 40$ mS.

Solution: Figure 5.11 b) shows the equivalent circuit for the calculation of A_{11}. From eqn 5.31 we can write:

$$A_{11} = \frac{V_i}{-g_m V_i R_o}$$
$$= -(g_m R_o)^{-1}$$
$$= -(40)^{-1}$$

A_{11}^{-1} is the voltage gain of the amplifier and is equal to -40.

Figure 5.11 c) shows the equivalent circuit for the calculation of A_{12}. From eqn 5.32 we can write:

$$A_{12} = \frac{V_i}{-g_m V_i}$$
$$= -(g_m)^{-1}$$
$$= -(40 \times 10^{-3})^{-1} \; \Omega$$

and the transconductance gain of the amplifier is $A_{12}^{-1} = -40$ mS.

Figure 5.11 d) shows the equivalent circuit for the calculation of A_{21}. From eqn 5.33 we can write:

$$A_{21} = \frac{V_i}{-g_m V_i R_o R_i}$$
$$= -(g_m R_o R_i)^{-1}$$
$$= -(400 \times 10^3)^{-1} \; \text{S}$$

Hence, the transimpedance gain is $A_{21}^{-1} = -400$ kΩ.

Figure 5.11 e) shows the equivalent circuit for the calculation of A_{22}. From eqn 5.34 we can write:

$$A_{22} = \frac{V_i}{-g_m V_i R_i}$$
$$= -(g_m R_i)^{-1}$$
$$= -(400)^{-1}$$

and the current gain is $A_{22}^{-1} = -400$.

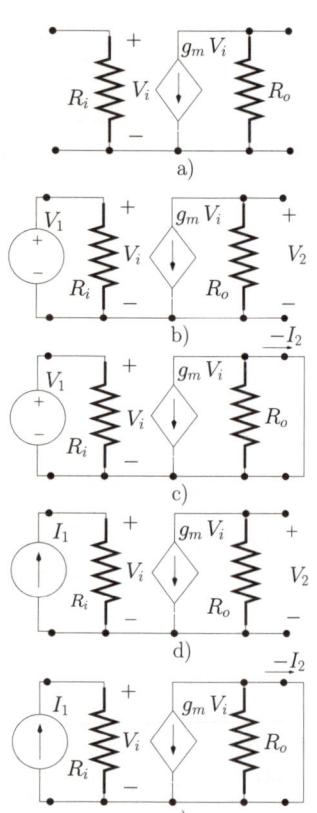

Figure 5.11: *Calculation of the chain parameters of a two-port network (electronic amplifier model).*

Chain/Cascade connection

Let us consider the connection of a two-port network, characterised by a chain representation $[A']$, with another two-port network, characterised by a chain

5. Electrical two-port network analysis

representation $[A'']$, as illustrated in figure 5.12. Such a connection is called a chain or cascade connection of two-port networks.

The equivalent two-port network resulting from this connection can be characterised by an equivalent chain representation, $[A_{eq}]$, which can be determined according to eqns 5.31–5.34, that is;

$$A_{eq_{11}} = \left.\frac{V_1}{V_2}\right|_{I_2=0} \tag{5.35}$$

According to eqn 5.27 we can write:

$$V_1' = A_{11}' V_2' - A_{12}' I_2' \tag{5.36}$$

Since $V_1 = V_1'$, $V_2' = V_1''$, $V_2 = V_2''$ and $I_2' = -I_1''$ we can write eqn 5.35 as follows:

$$A_{eq_{11}} = A_{11}' \left.\frac{V_1''}{V_2''}\right|_{I_2''=0} + A_{12}' \left.\frac{I_1''}{V_2''}\right|_{I_2''=0} \tag{5.37}$$

which, according to eqns 5.31 and 5.33, is:

$$A_{eq_{11}} = A_{11}' A_{11}'' + A_{12}' A_{21}'' \tag{5.38}$$

Similarly, it can be shown (see problem 5.6):

$$A_{eq_{12}} = A_{11}' A_{12}'' + A_{12}' A_{22}'' \tag{5.39}$$
$$A_{eq_{21}} = A_{21}' A_{11}'' + A_{22}' A_{21}'' \tag{5.40}$$
$$A_{eq_{22}} = A_{21}' A_{12}'' + A_{22}' A_{22}'' \tag{5.41}$$

Equations 5.38–5.41 can be recognised as those resulting from the product of matrix $[A']$ with matrix $[A'']$, that is;

$$[A_{eq}] = [A'] \times [A''] \tag{5.42}$$

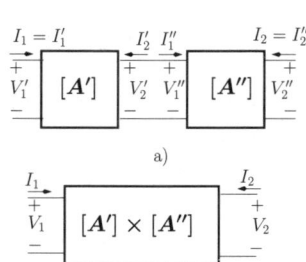

Figure 5.12: *a) Chain connection of two-port networks. b) Equivalent two-port network.*

5.2.4 Conversion between electrical representations

Most two-port circuits can be described by any of the electrical representations discussed above. This means that it is possible to convert between the different electrical representations. The formulae for such conversions can be obtained using elementary matrix algebra. For example, the transformation between the impedance representation and the admittance representation can be obtained as follows:

$$[I] = [Y][V] \tag{5.43}$$

with $[I]$ and $[V]$ described as in eqns 5.4 and 5.5, respectively. $[Y]$ is the electrical admittance representation, as in eqn 5.18. Using elementary matrix algebra we can solve eqn 5.43 to obtain $[V]$

$$[V] = [Y]^{-1}[I] \tag{5.44}$$

Comparing this with eqn 5.3 it is clear that $[Z] = [Y]^{-1}$, that is;

$$[Z] = \begin{bmatrix} \frac{Y_{22}}{|Y|} & \frac{-Y_{12}}{|Y|} \\ \frac{-Y_{21}}{|Y|} & \frac{Y_{11}}{|Y|} \end{bmatrix} \quad (5.45)$$

where $|Y|$ is the determinant of the matrix $[Y]$ defined as:

$$|Y| = Y_{11} Y_{22} - Y_{12} Y_{21} \quad (5.46)$$

Similarly, it can be shown (see problem 5.7) that $[Y] = [Z]^{-1}$.

Example 5.2.6 Determine the electrical chain representation of a two-port network from its electrical impedance representation.

Solution:
$$V_1 = Z_{11} I_1 + Z_{12} I_2 \quad (5.47)$$
$$V_2 = Z_{21} I_1 + Z_{22} I_2 \quad (5.48)$$

Solving eqn 5.48 in order to obtain I_1 we have:

$$I_1 = \frac{1}{Z_{21}} V_2 - \frac{Z_{22}}{Z_{21}} I_2 \quad (5.49)$$

Substituting I_1 in eqn 5.47 we obtain

$$V_1 = \frac{Z_{11}}{Z_{21}} V_2 - \frac{Z_{11} Z_{22} - Z_{12} Z_{21}}{Z_{21}} I_2 \quad (5.50)$$

Comparing eqn 5.49 with eqn 5.28 and comparing eqn 5.50 with eqn 5.27 we can write (see also eqn 5.30)

$$[A] = \begin{bmatrix} \frac{Z_{11}}{Z_{21}} & \frac{Z_{11} Z_{22} - Z_{12} Z_{21}}{Z_{21}} \\ \frac{1}{Z_{21}} & \frac{Z_{22}}{Z_{21}} \end{bmatrix} \quad (5.51)$$

Example 5.2.7 Determine the chain parameters of the shunt admittance of figure 5.3 a).

Solution: The Z-parameters of the shunt admittance are calculated in example 5.2.1. From eqn 5.51 the chain parameters can be written as follows:

$$[A] = \begin{bmatrix} 1 & 0 \\ Y & 1 \end{bmatrix} \quad (5.52)$$

with $Y = 2$ mS.

In appendix C we present tables of the conversions between all electrical representations discussed in this chapter.

5.2.5 Miller's theorem

Miller's theorem is used extensively to simplify the analysis of some two-port network configurations such as that of figure 5.13 a). The theorem states that if the voltage gain, A_v, between port 1 and port 2 is known, then it is possible to obtain a circuit like that shown in figure 5.13 b) which is equivalent to the former circuit in terms of input impedance and forward voltage gain. The extra admittances Y_1 and Y_2, indicated in figure 5.13 b), are given by:

$$Y_1 = Y_f(1 - A_v) \tag{5.53}$$

$$Y_2 = Y_f \frac{A_v - 1}{A_v} \tag{5.54}$$

with the voltage gain, A_v defined as

$$A_v = \left. \frac{V_2}{V_1} \right|_{I_2=0} \tag{5.55}$$

In order to prove this theorem we consider that the two-port circuit a can be characterised by an admittance representation such that[4]:

$$[\boldsymbol{Y_a}] = \begin{bmatrix} Y_{a_{11}} & 0 \\ Y_{a_{21}} & Y_{a_{22}} \end{bmatrix} \tag{5.56}$$

Since $Y_{a_{12}} = 0$ the circuit is called unilateral, in the sense that the output voltage and current are influenced by but *do not* influence the input voltage and current. The series admittance Y_f of figure 5.13 a) can be considered as the single element two-port network of figure 5.13 c) with the admittance representation given below (see problem 5.8):

$$[\boldsymbol{Y_f}] = \begin{bmatrix} Y_f & -Y_f \\ -Y_f & Y_f \end{bmatrix} \tag{5.57}$$

Since these two-port networks are in parallel then, from eqn 5.26, the circuit of figure 5.13 c) can be characterised by an impedance representation $[\boldsymbol{Y_M}]$ given by:

$$[\boldsymbol{Y_M}] = \begin{bmatrix} Y_{a_{11}} + Y_f & -Y_f \\ Y_{a_{21}} - Y_f & Y_{a_{22}} + Y_f \end{bmatrix} \tag{5.58}$$

The forward voltage gain, A_v, the input impedance, Z_{in}, and the output impedance, Z_{out}, can be determined from eqn 5.58 as follows (see also eqns 5.15 and 5.16):

$$A_v = \left. \frac{V_2}{V_1} \right|_{I_2=0}$$

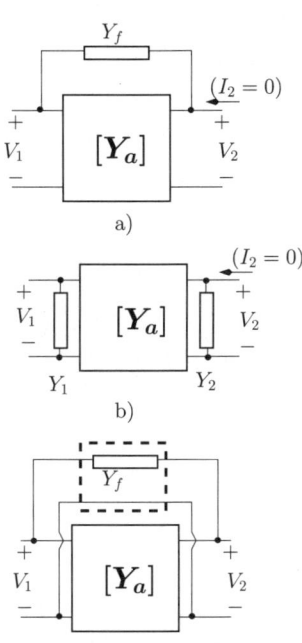

Figure 5.13: *a) Parallel connection of the two-port network* (a) *with a series admittance Y_f. b) Application of Miller's theorem. c) Equivalent circuit for the parallel connection of a two-port network with a series admittance.*

---

[4]For clarity of derivation we take $Y_{a_{12}}$ to be zero. In other words, the reverse transconductance gain is determined by Y_f alone.

$$= -\frac{Y_{a_{21}} - Y_f}{Y_{a_{22}} + Y_f} \tag{5.59}$$

$$Z_{in} = \left.\frac{V_1}{I_1}\right|_{I_2=0}$$

$$= \left[Y_{a_{11}} + Y_f \left(1 + \frac{Y_{a_{21}} - Y_f}{Y_{a_{22}} + Y_f}\right)\right]^{-1}$$

$$= [Y_{a_{11}} + Y_f (1 - A_v)]^{-1} \tag{5.60}$$

$$Z_{out} = \left.\frac{V_2}{I_2}\right|_{I_1=0}$$

$$= \left[Y_{a_{22}} + Y_f \left(1 + \frac{Y_{a_{21}} - Y_f}{Y_{a_{11}} + Y_f}\right)\right]^{-1} \tag{5.61}$$

On the other hand it is possible to show (see problem 5.9) that the circuit of figure 5.13 b) can be characterised by an admittance representation $\left[\boldsymbol{Y}'_M\right]$ given by:

$$[\boldsymbol{Y}'_M] = \begin{bmatrix} Y_{a_{11}} + Y_f(1 - A_v) & 0 \\ Y_{a_{21}} & Y_{a_{22}} + Y_f - \frac{Y_f}{A_v} \end{bmatrix} \tag{5.62}$$

with A_v given by eqn 5.59. The forward voltage gain, A_v, the input impedance, Z_{in}, and the output impedance, Z_{out}, can be determined from eqn 5.62 as follows:

$$A_v = \left.\frac{V_2}{V_1}\right|_{I_2=0}$$

$$= -\frac{Y_{a_{21}} - Y_f}{Y_{a_{22}} + Y_f} \tag{5.63}$$

$$Z_{in} = \left.\frac{V_1}{I_1}\right|_{I_2=0}$$

$$= [Y_{a_{11}} + Y_f (1 - A_v)]^{-1} \tag{5.64}$$

$$Z_{out} = \left.\frac{V_2}{I_2}\right|_{I_1=0}$$

$$= \left[Y_{a_{22}} + Y_f \left(1 - \frac{1}{A_v}\right)\right]^{-1}$$

$$= \left[Y_{a_{22}} + Y_f \left(1 - \frac{Y_{a_{22}} + Y_f}{Y_{a_{21}} - Y_f}\right)\right]^{-1} \tag{5.65}$$

Comparing eqn 5.59 with eqn 5.63, and eqn 5.60 with eqn 5.64 we recognise that the admittance representation resulting from Miller's theorem provides the same forward voltage gain and the same input impedance as those obtained from applying the standard circuit theory. However, comparing eqn 5.61 with eqn 5.65 we observe that the output impedance provided by Miller's theorem is not the same as that provided by the circuit theory analysis. Hence, this method is not applicable to the determination of the output impedance of a circuit topology like that depicted in figure 5.13 a).

In many practical situations where, for example, voltage amplification is required we have $|Y_{a_{22}}| \gg |Y_f|$ and $|Y_{a_{21}}| \gg |Y_f|$. Under these two conditions we can approximate the voltage gain given by eqn 5.59 by the following eqn:

$$A_v \simeq -\frac{Y_{a_{21}}}{Y_{a_{22}}} \qquad (5.66)$$

The voltage gain given by the last equation is the voltage gain of the two-port network a (without the series admittance Y_f). This approximation turns out to be one of the most attractive advantages of the use of Miller's theorem (see [2] for a more detailed discussion). Miller's theorem is a very important analysis tool which is used in Chapter 6 to analyse and to discuss the high-frequency response of electronic amplifiers.

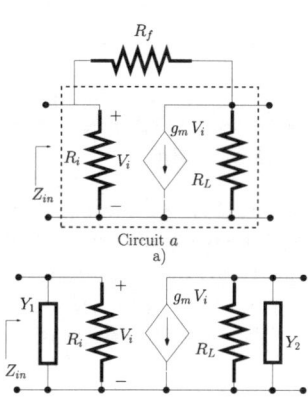

Figure 5.14: *Application of Miller's theorem: example.*

Example 5.2.8 Apply Miller's theorem to the circuit shown in figure 5.14 a) in order to obtain the input impedance, Z_{in}. $g_m = 50$ mA/V, $R_f = 20$ kΩ, $R_i = 2.5$ kΩ, and $R_L = 5$ kΩ.

Solution: The Y-parameters of circuit a are calculated in example 5.2.4. $Y_f = 1/R_F = 50\,\mu$S. Since $|Y_{a_{22}}| \gg |Y_f|$ and $|Y_{a_{21}}| \gg |Y_f|$ we can write:

$$\begin{aligned} A_v &\simeq -\frac{Y_{a_{21}}}{Y_{a_{22}}} \\ &= -g_m\,R_L \\ &= -250 \end{aligned}$$

The input impedance Z_{in} is given by the parallel connection of R_i with $Y_1^{-1} = [Y_f\,(1-A_v)]^{-1}$ (see also fig. 5.14 b). That is, the input impedance can be calculated as follows:

$$\begin{aligned} Z_{in} &= R_i\,\|\,[Y_f\,(1-A_v)]^{-1} \\ &= R_i\,\|\,\frac{R_f}{251} \\ &= 77.5\,\Omega \end{aligned}$$

Note that Miller's theorem indicates that R_f is reflected to the input of the circuit reduced by a factor which is (approximately) the voltage gain. This, in turn, significantly reduces the input impedance.

5.3 Computer-aided electrical analysis

In this section we describe a matrix-based method for the computation of the electrical response of linear electronic circuits. This method is based on the representation of any complex two-port circuit as an interconnection of elementary two-port circuits such as admittances, impedances and voltage- and current-controlled sources. The electrical response of such elementary sub-circuits can be characterised by one of the electrical representations discussed previously. Starting from these basic two-port circuits the electrical analysis is performed by interconnecting these elementary circuits in order to obtain the electrical characterisation of the whole circuit. As discussed in the previous section the types of connection are series, parallel and chain (or cascade). Depending on the type of connection appropriate electrical representations are adopted for the two-port sub-circuits in question.

The matrix-based analysis method can be stated as follows:

1. **Decompose the circuit to be analysed into its elementary two-port sub-circuits such as series impedances, shunt admittances, voltage and current controlled sources, etc.**

2. **Identify the type of connections between the various elementary two-port networks mentioned above (parallel, series, chain).**

3. **Characterise the electrical response of each of the elementary two-port sub-circuits according to the relevant two-port electrical representation (admittance, impedance or chain). Whenever possible use an electrical representation for the elementary two-port network taking into account the type of connection with the other elementary two-port networks:**

    (a) **If the two-port elementary circuits are interconnected in parallel use admittance representations for both elementary circuits;**

    (b) **If the two-port elementary circuits are interconnected in series use impedance representations for both elementary circuits;**

    (c) **If the two-port elementary circuits are interconnected in chain use chain representations for both elementary circuits.**

4. **According to the connections between the various elementary two-port circuits, reconstruct the overall two-port circuit. Whenever appropriate use the electrical transformation matrices shown in tables C.1 and C.2 (see appendix C) to obtain the appropriate electrical representations.**

The next example illustrates the application of these steps.

Example 5.3.1 Consider the circuit of figure 5.15 a) which is the equivalent circuit of an electronic amplifier. Determine the transimpedance, the voltage and the current gains of this circuit.

<u>Solution</u>: Figure 5.15 b) shows that the amplifier can be decomposed into a connection of elementary two-port sub-circuits. The two-port circuit composed by the voltage-controlled current source and R_i is in parallel with the series impedance R_1. The two-port circuit resulting from this connection is in series with the shunt admittance R_2. Finally, the two-port circuit resulting from this connection is in chain with the shunt admittance represented by R_3.

We start by characterising the two-port circuit constituted by the voltage-controlled current source and R_i in its admittance representation. From example 5.2.4 we know that the admittance representation for this sub-circuit can be

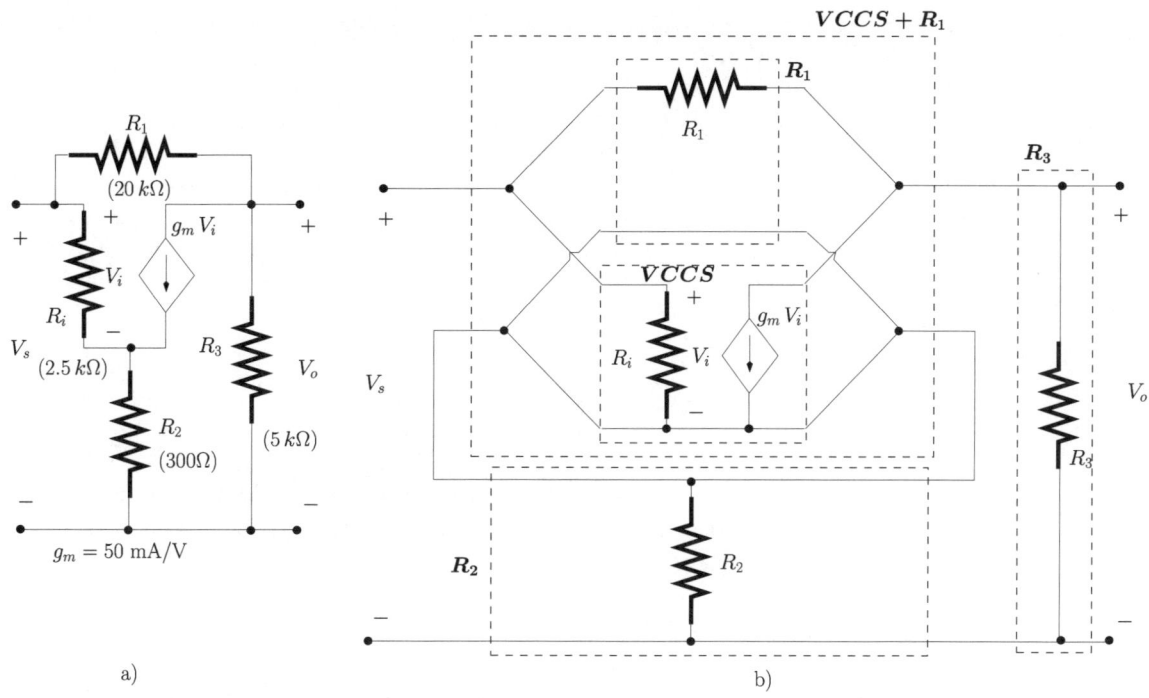

Figure 5.15: *Electronic amplifier. a) Equivalent model. b) Decomposition in terms of elementary two-port circuits.*

written as follows:

$$[Y_{VCCS}] = \begin{bmatrix} R_i^{-1} & 0 \\ g_m & 0 \end{bmatrix} \quad (5.67)$$

According to eqn 5.57 we can write the admittance representation for R_1 as follows:

$$[Y_{R_1}] = \begin{bmatrix} R_1^{-1} & -R_1^{-1} \\ -R_1^{-1} & R_1^{-1} \end{bmatrix} \quad (5.68)$$

The two-port network resulting from the parallel connection of $[Y_{VCCS}]$ with $[Y_{R_1}]$ can be written, according to eqn 5.26, as follows:

$$[Y_{VCCS+R_1}] = \begin{bmatrix} R_i^{-1} + R_1^{-1} & -R_1^{-1} \\ g_m - R_1^{-1} & R_1^{-1} \end{bmatrix} \quad (5.69)$$

Since $[Y_{VCCS+R_1}]$ is in series with the shunt admittance, R_2, it is appropriate to characterise these two sub-circuits according to impedance representations. From the example 5.2.1 we can write:

$$[Z_{R_2}] = \begin{bmatrix} R_2 & R_2 \\ R_2 & R_2 \end{bmatrix} \quad (5.70)$$

From eqn 5.45 we can write the impedance representation for $VCCS + R_1$ as follows:

$$[Z_{VCCS+R_1}] = \begin{bmatrix} \frac{R_i}{1+g_m R_i} & \frac{R_i}{1+g_m R_i} \\ \frac{R_i(1-g_m R_1)}{1+g_m R_i} & \frac{R_i+R_1}{1+g_m R_i} \end{bmatrix} \quad (5.71)$$

Using eqn 5.14, the impedance representation for the two-port sub-circuit resulting from the series connection of $VCCS + R_1$ with R_2 can be expressed as follows:

$$[Z_{VCCS+R_1+R_2}] = \begin{bmatrix} \frac{R_i}{1+g_m R_i} + R_2 & \frac{R_i}{1+g_m R_i} + R_2 \\ \frac{R_i(1-g_m R_1)}{1+g_m R_i} + R_2 & \frac{R_i+R_1}{1+g_m R_i} + R_2 \end{bmatrix} \quad (5.72)$$

Since $VCCS + R_1 + R_2$ is in chain with the shunt impedance R_3 it is appropriate to express these two sub-circuits in terms of chain representations. From eqn 5.51 we have:

$[A_{VCCS+R_1+R_2}] =$

$$= \begin{bmatrix} \frac{R_i+(1+g_m R_i) R_2}{R_i(1-g_m R_1)+R_2(1+g_m R_i)} & \frac{R_1[R_i+(1+g_m R_i) R_2]}{R_i(1-g_m R_1)+R_2(1+g_m R_i)} \\ \frac{1+g_m R_i}{R_i(1-g_m R_1)+R_2(1+g_m R_i)} & \frac{R_i+R_1+(1+g_m R_i) R_2}{R_i(1-g_m R_1)+R_2(1+g_m R_i)} \end{bmatrix} \quad (5.73)$$

From eqn 5.52, we can write:

$$[A_{R_3}] = \begin{bmatrix} 1 & 0 \\ R_3^{-1} & 1 \end{bmatrix} \quad (5.74)$$

According to eqn 5.42 the overall circuit can be characterised by an equivalent chain representation $[A_{eq}]$ given by:

$$[A_{eq}] = [A_{VCCS+R_1+R_2}] \times [A_{R_3}] \quad (5.75)$$

It is left to the reader to show that the transimpedance gain, R_m, the voltage gain, A_v, and the current gain, A_i, are

$$R_m = (A_{eq_{21}})^{-1} \quad (5.76)$$
$$= -17.8 \text{ k}\Omega$$
$$A_v = (A_{eq_{11}})^{-1} \quad (5.77)$$
$$= -12.2$$
$$A_i = (A_{eq_{22}})^{-1} \quad (5.78)$$
$$= -40.8$$

5.4 Bibliography

1. W.H. Hayt, J.E. Kemmerly, *Engineering Circuit Analysis*, 2001 (McGraw-Hill) 6th edition.

2. L. Moura, *Error Analysis in Miller's Theorems*, IEEE Trans. on Circuits and Systems-I: Fundamental Theory and Applications, Vol. 48, No. 2, Feb. 2001, pp. 241–249.

5.5 Problems

5.1 For the two-port networks of figure 5.16 determine the Z-parameters.

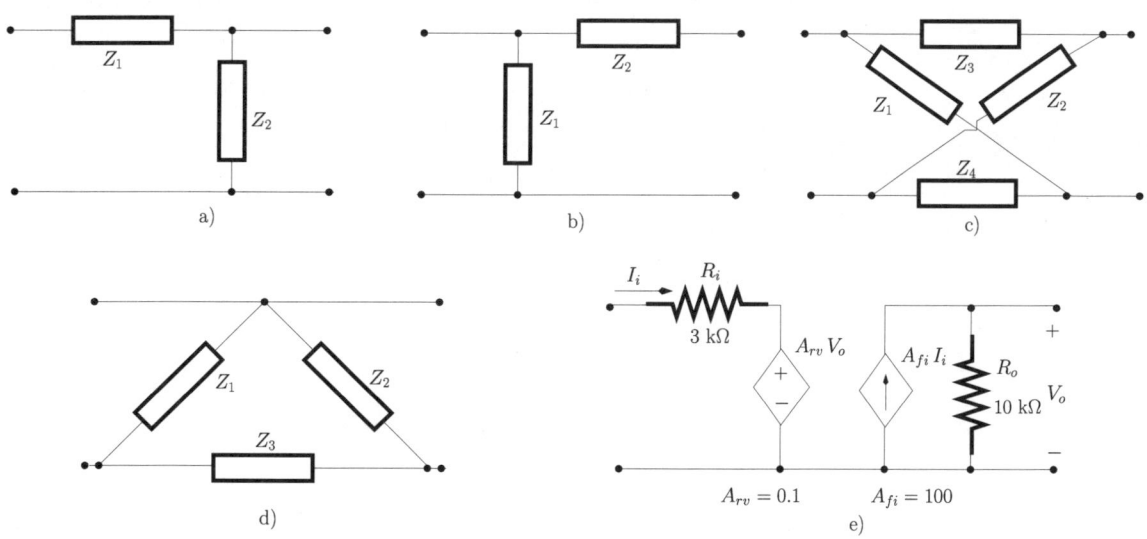

Figure 5.16: *Two-port networks.*

5.2 Show that the series electrical representation of a two-port network which results from the series connection of two two-port networks characterised by impedance representations $[\mathbf{Z}']$ and $[\mathbf{Z}'']$ can be expressed as the sum of $[\mathbf{Z}']$ with $[\mathbf{Z}'']$ (refer to fig. 5.5).

5.3 For the two-port networks of figure 5.16 determine the Y-parameters.

5.4 Show that the admittance electrical representation of a two-port network resulting from the parallel connection of two two-port networks characterised by admittance representations $[\mathbf{Y}']$ and $[\mathbf{Y}'']$ can be expressed as the sum of $[\mathbf{Y}']$ with $[\mathbf{Y}'']$ (refer to fig. 5.9).

5.5 For the two-port networks of figure 5.16 determine the chain parameters.

5.6 Show that the chain electrical representation of a two-port network which results from the chain connection of two two-port networks characterised by chain representations $[\mathbf{A}']$ and $[\mathbf{A}'']$ can be expressed as the product of $[\mathbf{A}']$ with $[\mathbf{A}'']$ (refer to fig. 5.12).

5.7 Show that the admittance representation, $[Y]$, can be obtained from the impedance representation, $[Z]$, according to the following expression: $[Y] = [Z]^{-1}$. $[Y]$ and $[Z]$ are given by eqns 5.18 and 5.6, respectively.

5.8 Show that the series admittance Y_f of figure 5.13 a) can be characterised by an admittance representation expressed by eqn 5.57.

5.9 Show that the two-port network of figure 5.13 b) can be characterised by an admittance representation expressed by eqn 5.62.

5.10 Derive expressions which allow for the conversion of chain to admittance parameters.

5.11 Consider the two-port networks of figure 5.17 with $R_i = 2.5$ kΩ, $R_o = 10$ kΩ and $g_m = 40$ mS. For each circuit determine

1. the input impedance ($I_o = 0$)
2. the output impedance ($V_s = 0$)
3. the voltage gain V_o/V_s ($I_o = 0$)
4. the current gain I_o/I_s ($V_o = 0$)
5. the transimpedance gain V_o/I_s ($I_o = 0$)
6. the transconductance gain I_o/V_s ($V_o = 0$)

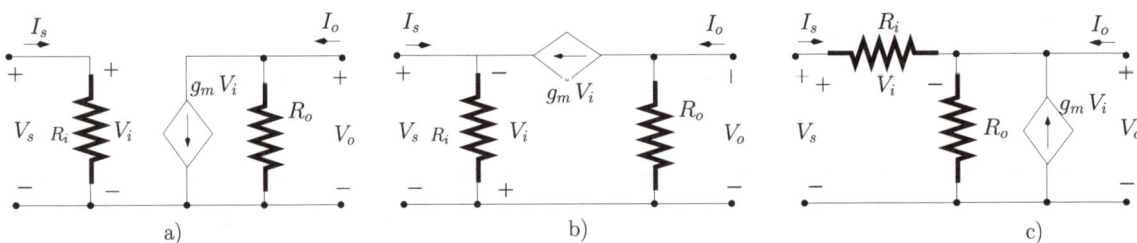

Figure 5.17: *Two-port networks.*

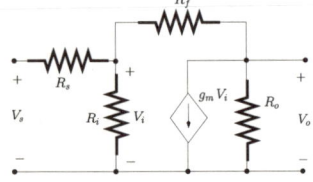

Figure 5.18: *Two-port network.*

5.12 Consider the two-port network of figure 5.18. Apply Miller's theorem to the resistance R_f and then obtain an estimate for the circuit input impedance and voltage gain V_o/V_s. Assume $R_i = 2$ kΩ, $R_o = 5$ kΩ and $g_m = 40$ mS, $R_s = 100$ Ω and $R_f = 47$ kΩ.

5.13 Consider again the two-port network of figure 5.18 (see also problem 5.12). Apply the computer-aided electrical analysis method described in section 5.3 to solve this circuit. Then determine the input impedance and voltage gain V_o/V_s and compare these values with those obtained applying Miller's theorem. Draw conclusions.

6 Basic electronic amplifier building blocks

6.1 Introduction

Linear electronic amplification is one of the most important and fundamental operations applied to electrical signals. By electrical signals we mean time varying voltage, current or power signals which represent some information such as an audio signal. In this chapter we present an introduction to various basic electronic amplifier structures. The next section addresses important amplifier 'figures of merit' such as gain, bandwidth and how to model them. Section 6.3 deals with operational amplifiers, an important integrated-circuit general purpose amplifier, which can be used for a large variety of applications ranging from signal (audio and video) amplification to analogue signal processing such as filtering. In section 6.4 we present other active devices which provide amplification. Some amplifier circuit topologies are also analysed in detail in terms of gain and bandwidth.

6.2 Modelling the amplification process

The basic role of an electronic amplifier is intuitive: it magnifies the amplitude of the electrical signals mentioned above. Hence we can identify the three main types of electronic amplification: voltage amplification, current amplification and power (voltage and current) amplification. An amplifier is characterised in terms of figures of merit. Examples of figures of merit are the gain, the bandwidth and noise figure. The first two of these are defined below while the noise figure is discussed in Chapter 8.

Amplifier gain. The gain of an amplifier is the amount of amplification provided by the amplifier. An amplifier can provide voltage gain, current gain or both, that is, power gain. If the gain is less than one we refer to it as loss or attenuation.

Amplifier bandwidth. The bandwidth of an amplifier is the range of frequencies over which the amplifier is operated so as to provide uniform gain. As discussed in Chapter 3, a signal can be expressed in the frequency domain by means of a sum of weighted phasors using the Fourier series and transform. Hence, in order to have distortion-free amplification all the frequency components of the signal must be amplified by the same amount. Also, the amplifier must provide a linear phase shift to all the frequency components of the signal, as discussed in section 3.3.4.

From the above it is clear that frequency domain tools are well suited to

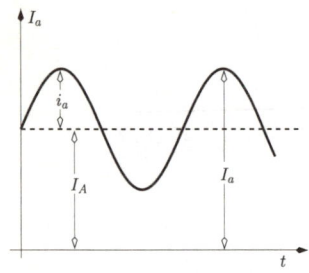

Figure 6.1: *Notation employed to describe DC and AC signals.*

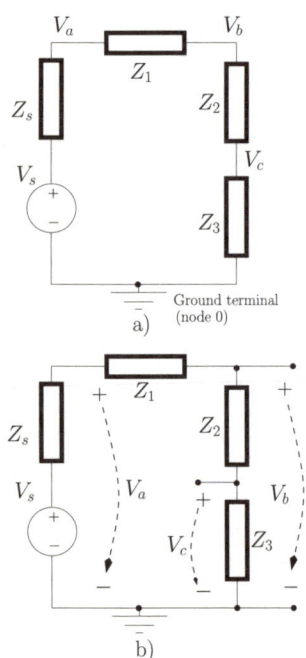

Figure 6.2: *Representation of voltages referenced to ground.*

deal with the analysis of amplifiers. Therefore, and unless stated otherwise, the analysis of amplifiers presented here is carried out using phasor analysis.

Signal notation

The notation used here is as shown in figure 6.1. The time varying component, that is, the phasor component, is expressed by a lower case with a lower case subscript, e.g. i_a, v_b. The DC component is expressed by an uppercase symbol with uppercase subscript, e.g I_A, V_B. The total instantaneous quantity, that is, the signal plus DC component, is expressed by an uppercase letter with a lowercase subscript, e.g. I_a, V_b.

When a voltage is indicated at a given node, as illustrated in figure 6.2 a), this means that the voltage is referenced to the ground terminal (or node 0) as indicated in figure 6.2 b) (see also section 1.4.3). The symbol which represents the ground terminal is shown in figure 6.2 a).

Typical amplifier transfer functions

Usually, the transfer function of an amplifier, that is the gain of an amplifier versus frequency, can be characterised as in figure 6.3 a) or figure 6.3 b). The transfer function depicted in figure 6.3 a) can be decomposed into three main frequency ranges of operation; the low-frequency, the mid-frequency (also known as mid-band) and the high-frequency ranges of operation. In many practical amplifiers, where the low-frequency range and the high frequency range are sufficiently separated, such a transfer function can be approximated by the following expression:

$$A(\omega) = \frac{j\omega/\omega_L}{1+j\omega/\omega_L} \times A_M \times \frac{1}{1+j\omega/\omega_H} \tag{6.1}$$

with $\omega_L = 2\pi f_L$ and $\omega_H = 2\pi f_H$ where f_H and f_L are the 3 dB high cut-off and low cut-off frequencies, respectively. A_M is the mid-frequency range gain. This equation can be rewritten as follows:

$$A = \begin{cases} A_M \frac{j\omega/\omega_L}{1+j\omega/\omega_L} & \omega \leq \omega_L \quad \text{(low-frequency range)} \\ A_M & \omega_L < \omega < \omega_H \quad \text{(mid-frequency range)} \\ A_M \frac{1}{1+j\omega/\omega_H} & \omega \geq \omega_H \quad \text{(high-frequency range)} \end{cases} \tag{6.2}$$

where we have dropped the explicit dependency of A with ω for the sake of simpler representation.

Low-frequency response. The fall-off of gain at low frequencies is caused by two categories of coupling capacitors. The first category, known as AC-coupling or DC-blocking capacitors, occurs at the input and output of the amplifier. The purpose of these capacitors is to block the DC level and to enable the connection of different amplifier stages with different DC bias levels. The second category, known as by-pass capacitors, is used in specific amplifier

6. Basic electronic amplifier building blocks

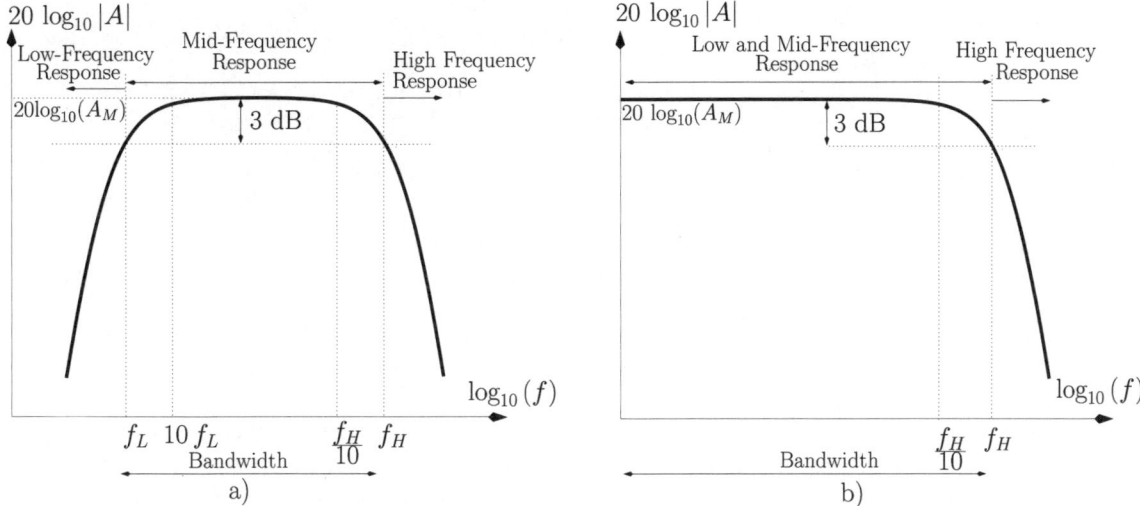

Figure 6.3: *Typical transfer functions of electronic amplifiers. a) With AC coupling capacitors. b) Without AC coupling capacitors.*

topologies where gain needs to be stabilised. Both categories of coupling capacitors normally have large values, typically tens or hundreds of micro-farads. When an amplifier does not have such coupling capacitors (as in integrated circuits) then its gain response is typically like that presented in figure 6.3 b).

High-frequency response. The gain fall-off at high frequencies is caused by internal capacitances which are an intrinsic feature of active devices (the transistors) which implement the amplifier. Basically, these capacitances result in effective short-circuiting of voltage signals to ground at high-frequencies. The values for these capacitances range, typically, from fractions of a pico-farad to a few tens of pico-farads.

Mid-frequency response. The useful range of operation of an amplifier is the mid-frequency range which defines the bandwidth of the amplifier. Over this range the effect of the coupling capacitors and the effect of the parasitic capacitances can be neglected. This means that the blocking and by-pass capacitors can be considered as short-circuits whilst the parasitic capacitances are open-circuits (see also example 6.2.1). It can be observed, from figure 6.3, that over this frequency range all frequency components are amplified by the same amount, A_M, with the exceptions of those frequencies near f_L and f_H. From the above it is clear that a given signal which is to be amplified must have a bandwidth less than or equal to the amplifier bandwidth so that all signal frequency components are equally amplified. If the bandwidth of the signal is larger than the bandwidth of the amplifier then linear distortion will occur (see section 3.3.4).

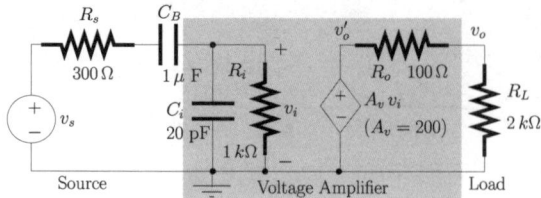

Figure 6.4: *Voltage amplifier modelled by a voltage-controlled voltage source, an input resistance R_i, a parasitic capacitance C_i and an output resistance R_o.*

Example 6.2.1 Consider a voltage amplifier modelled by a voltage-controlled voltage source, an input resistance R_i, a parasitic capacitance C_i and an output resistance R_o, as shown in figure 6.4. This amplifier is driven by a voltage signal source v_s with an output resistance R_s. C_B is a DC blocking capacitor. The amplifier drives a load R_L.

1. Determine the transfer function of the circuit, $A_{vs} = v_o/v_s$.

2. Show that for the frequency range $10\,f_L \leq f \leq f_H/10$ (that is, the centre of the mid-band) the voltage gain A_{v_M} can be determined assuming that the blocking capacitor C_B is a short-circuit and that the capacitor C_i is an open-circuit.

Solution:

1. The transfer function for the amplifier $A_{vs} = v_o/v_s$ can be written in terms of the product of partial voltage gains as follows:

$$A_{vs} = \frac{v_o}{v'_o} \times \frac{v'_o}{v_i} \times \frac{v_i}{v_s} \quad (6.3)$$

From figure 6.4 it can be seen that v_o is related to v'_o by a resistive voltage divider expression, such that;

$$v_o = \frac{R_L}{R_i + R_L} v'_o \quad (6.4)$$

v_i is, in turn, related to v_s according to the following equation:

$$v_i = \frac{R_i}{R_i + R_s}$$
$$\times \frac{j\omega C_B(R_i + R_s)}{1 + j\omega C_B(R_i + R_s) + j\omega C_i R_i - \omega^2 C_i C_B R_i R_s} v_s \quad (6.5)$$

In practical circuits where C_B is orders of magnitude larger than C_i (as in this example) v_i can be expressed by the eqn below:

$$v_i = \frac{R_i}{R_i + R_s} \times \frac{j\omega C_B(R_i + R_s)}{1 + j\omega C_B(R_i + R_s)} \times \frac{1}{1 + j\omega C_i \frac{R_i R_s}{R_i + R_s}} v_s \quad (6.6)$$

6. Basic electronic amplifier building blocks

This equation shows a wide separation between the low and the high frequency poles.

High-frequency pole: $\dfrac{1}{C_i \frac{R_i R_s}{R_i + R_s}}$

Low-frequency pole: $\dfrac{1}{C_B(R_i + R_s)}$

From the above, the voltage transfer function A_{vs} can be expressed in the format of eqn 6.1, that is,

$$A_{vs} = \underbrace{\frac{j\omega C_B(R_i + R_s)}{1 + j\omega C_B(R_i + R_s)}}_{\text{Low-frequency response}} \times \underbrace{\frac{R_i}{R_i + R_s} A_v \frac{R_L}{R_L + R_o}}_{\text{Mid-frequency range gain}}$$

$$\times \underbrace{\frac{1}{1 + j\omega C_i \frac{R_i R_s}{R_i + R_s}}}_{\text{High-frequency response}} \qquad (6.7)$$

From this eqn we can identify f_L and f_H as:

$$f_L = \frac{1}{2\pi C_B(R_i + R_s)} \qquad (6.8)$$
$$= 122.4 \text{ Hz}$$
$$f_H = \frac{1}{2\pi C_i (R_i \| R_s)} \qquad (6.9)$$
$$= 34.5 \text{ MHz}$$

and the mid-frequency range voltage gain is given by:

$$A_{vM} = \frac{R_i}{R_i + R_s} A_v \frac{R_L}{R_L + R_o} \qquad (6.10)$$
$$= 146.5$$

Figure 6.5 a) shows the three constituent parts of the voltage transfer function of eqn 6.7 where *Low f*, *Mid f* and *High f* refer to the low-frequency, mid-frequency and high-frequency constituent parts of the voltage transfer function, respectively. The product of these three parts is the overall transfer function as shown in figure 6.5 b).

2. In order to show that in the centre of the mid-band ($10 f_L \leq f \leq f_H/10$) we can consider the blocking capacitor C_B as short-circuit and the capacitor C_i as an open-circuit we consider the impedance voltage dividers shown in figure 6.6. For the impedance voltage divider of figure 6.6 a) we can write:

$$\frac{v_i}{v_s} = \frac{R_i Z_{C_i}}{R_i Z_{C_i} + R_s Z_{C_i} + R_i R_s} \qquad (6.11)$$
$$Z_{C_i} = \frac{1}{j\omega C_i}$$

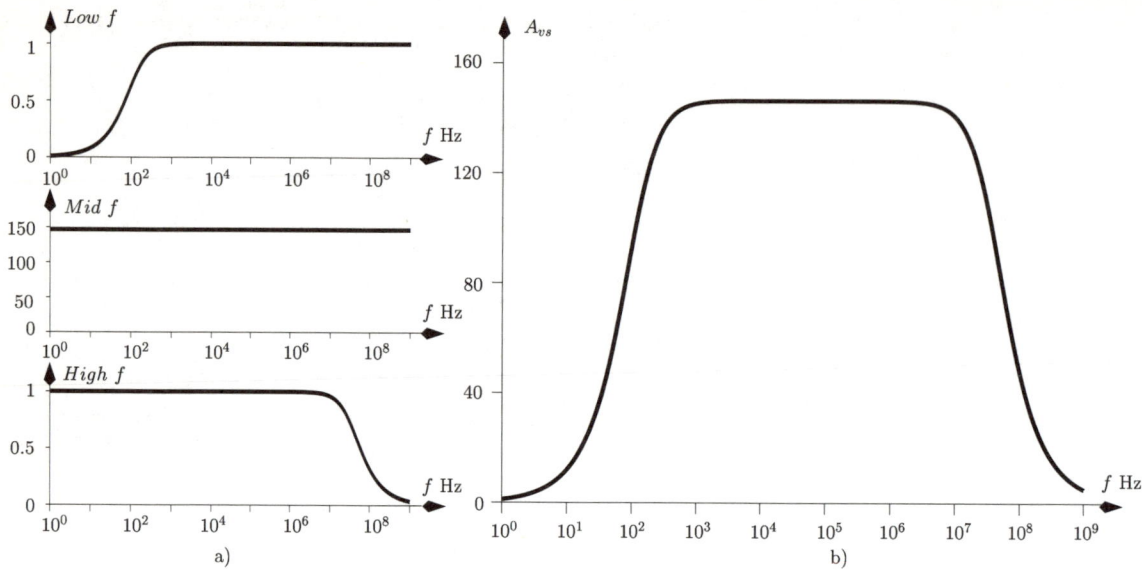

Figure 6.5: *a) The three constituent parts of the voltage transfer function of the amplifier of figure 6.4. b) Voltage transfer function.*

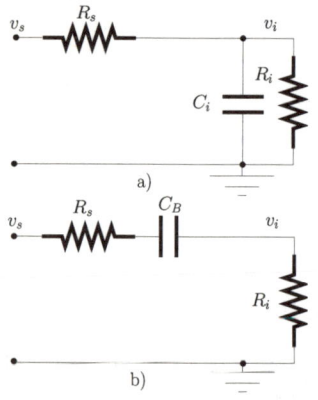

Figure 6.6: *Impedance voltage dividers.*

Equation 6.11 can be written as follows:

$$\frac{v_i}{v_s} = \underbrace{\frac{R_i}{R_s + R_i}}_{\text{Resistive loss}} \times \underbrace{\frac{1}{1 + \frac{R_s \| R_i}{Z_{C_i}}}}_{\text{Additional loss}(Z_{C_i})} \quad (6.12)$$

$$R_s \| R_i = \frac{R_s R_i}{R_s + R_i}$$

For the impedance voltage divider of figure 6.6 b) we can write:

$$\frac{v_i}{v_s} = \frac{R_i}{R_i + R_s + Z_{C_B}} \quad (6.13)$$

$$Z_{C_B} = \frac{1}{j\omega C_B} \quad (6.14)$$

Equation 6.13 can be written as follows:

$$\frac{v_i}{v_s} = \underbrace{\frac{R_i}{R_s + R_i}}_{\text{Resistive loss}} \times \underbrace{\frac{1}{1 + \frac{Z_{C_B}}{R_s + R_i}}}_{\text{Additional loss}(Z_{C_B})} \quad (6.15)$$

$$(6.16)$$

The plots of figure 6.7 a) and b) show the additional loss caused by Z_{C_i} and by Z_{C_B}, respectively, for the two impedance voltage dividers illustrated in figure 6.6. From figure 6.7 a) it is clear that for $|Z_{C_i}| \geq$

$10\,(R_i||R_s)$ the additional loss caused by the capacitance is very small, that is, for $|Z_{C_i}| \geq 10\,(R_i||R_s)$ the electrical response of the impedance voltage divider of figure 6.6 a) is approximately the same as that obtained if the capacitor C_i was replaced by an open-circuit. On the other hand, from figure 6.7 b) it can be observed that for $|Z_{C_B}| \leq 10\,(R_i + R_s)$ the electrical response of the impedance voltage divider of figure 6.6 b) is approximately the same as that obtained if the capacitor C_B was replaced by a short-circuit. For the frequency range located between $10\,f_L$

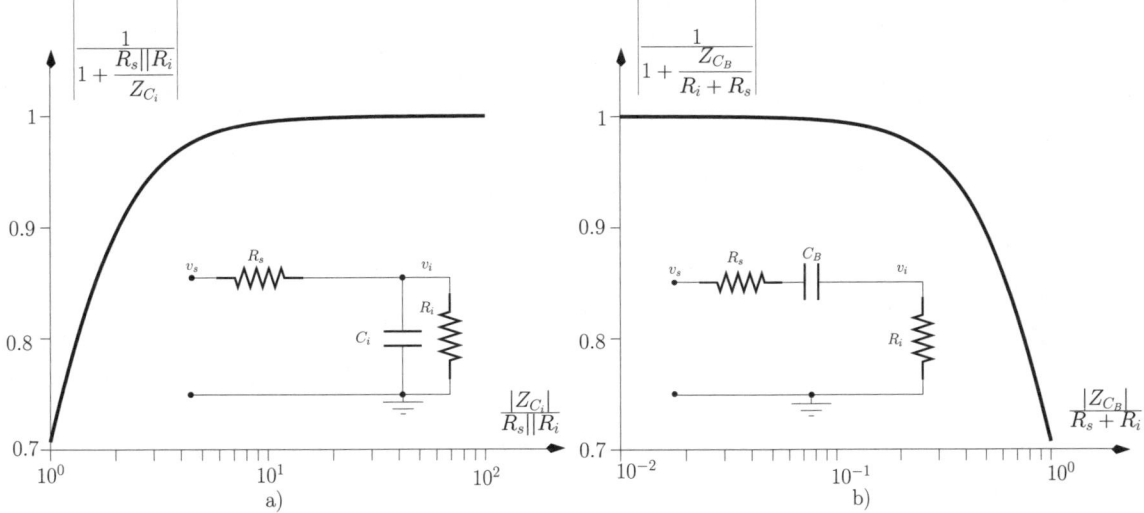

Figure 6.7: *a) Additional loss caused by Z_{C_i}. b) Additional loss caused by Z_{C_B}.*

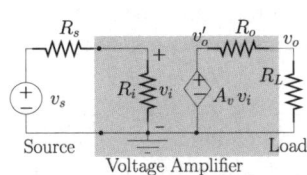

Figure 6.8: *Equivalent circuit of the amplifier of figure 6.4 valid for the mid-frequency range.*

and $f_H/10$, that is, between 1.2 kHz and 3.5 MHz we can calculate the following:

$$2.3\,\text{k}\Omega \leq |Z_{C_i}| \leq 6.5\,\text{M}\Omega \quad , \quad 1.2\,\text{kHz} \leq f \leq 3.5\,\text{MHz}$$
$$46\,\text{m}\Omega \leq |Z_{C_B}| \leq 130\,\Omega \quad , \quad 1.2\,\text{kHz} \leq f \leq 3.5\,\text{MHz}$$

Since $(R_i||R_s) = 230\,\Omega$ and $(R_i + R_s) = 1300\,\Omega$ then, from the last two eqns, we conclude that, for this frequency range, we have

$$|Z_{C_i}| \geq 10\,(R_i||R_s) \quad \text{and} \quad |Z_{C_B}| \leq 10\,(R_i + R_s) \quad (6.17)$$

Therefore, the effects of C_i and C_B can be neglected for the calculation of A_{v_M}, that is, we can consider the DC blocking capacitor C_B as a short-circuit and the capacitor C_i as an open-circuit. Figure 6.8 shows the equivalent circuit of the amplifier in this frequency range.

Small-signal amplifier models for the mid-frequency range

Although electronic amplifiers can be very complex circuits it is possible to characterise them, in the mid-frequency range, by a gain, a resistive input

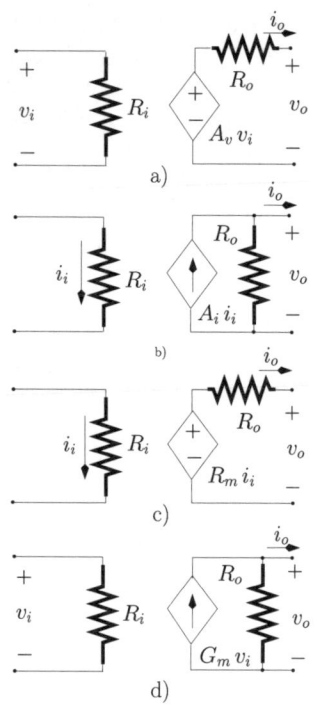

Figure 6.9: *Small-signal models for the mid-frequency range. a) Voltage amplifier. b) Current amplifier. c) Transimpedance amplifier. d) Transconductance amplifier.*

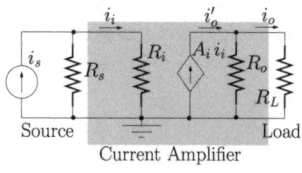

Figure 6.10: *Equivalent circuit for a current amplifier.*

impedance and a resistive output impedance. The gain is modelled by a voltage- or current-controlled current or voltage source depending on the type of amplifier. This results in the four different configurations shown in figure 6.9.

For the voltage amplifier it is desirable for the input impedance, R_i, to be as high as possible while the output impedance, R_o, should be as low as possible. In order to understand why this is so, we consider again the equivalent circuit of the voltage amplifier shown in figure 6.8 which is driven by a voltage source, v_s, with an associated output resistance, R_s. The amplifier drives a resistive load R_L. The voltage gain $A_{vs} = v_o/v_s$ can be calculated as follows:

$$A_{vs} = \frac{v_o}{v_s} \tag{6.18}$$

$$= \frac{v_o}{v_o'} \times \frac{v_o'}{v_i} \times \frac{v_i}{v_s} \tag{6.19}$$

From figure 6.8 it can be seen that v_o is related to v_o' by a voltage divider as

$$v_o = \frac{R_L}{R_i + R_L} v_o' \tag{6.20}$$

Also, v_i is related to v_s by a similar voltage divider,

$$v_i = \frac{R_i}{R_i + R_s} v_s \tag{6.21}$$

Therefore, the overall voltage gain A_{vs} can be expressed as:

$$A_{vs} = \frac{R_i}{R_i + R_s} \times A_v \times \frac{R_L}{R_i + R_L} \tag{6.22}$$

We see that R_s and R_L cause resistive loading effects at the input and output of the amplifier which, in turn, cause a decrease of the overall gain of the amplifier. For example, let us assume that the amplifier has an intrinsic voltage gain $A_v = 100$, an input impedance $R_i = 5$ kΩ and an output impedance $R_o = 1$ kΩ. If $R_s = R_i$ and if $R_L = R_o$ the voltage gain A_{vs} is 25, that is, there is a 75 % reduction of the overall gain due to these loading effects! On the other hand if $R_i = 5$ MΩ, $R_o = 50$ Ω, $R_s = 5$ kΩ and $R_L = 1$ kΩ, the overall gain is $A_{vs} \simeq A_v = 100$. It follows that an ideal voltage amplifier has a gain that does not depend on R_s or R_L. This means that this (ideal) amplifier has an input impedance which behaves as an open-circuit, $R_i \to \infty$, and a zero output impedance, $R_o = 0$.

Example 6.2.2 Show that an ideal current amplifier has a zero input impedance, $R_i = 0$, and an output impedance, R_o that is infinite.

Solution: Let us consider the equivalent circuit for a current amplifier shown in figure 6.10. The amplifier is driven by a current source with output impedance, R_s. The amplifier drives a resistive load R_L. The overall current gain, A_{is} can be written as follows:

$$A_{is} = \frac{i_o}{i_s}$$

6. Basic electronic amplifier building blocks

$$\begin{aligned} &= \frac{i_o}{i'_o} \times \frac{i'_o}{i_i} \times \frac{i_i}{i_s} \\ &= \frac{R_s}{R_s + R_i} \times A_i \times \frac{R_o}{R_o + R_L} \end{aligned} \qquad (6.23)$$

If we let $R_i \to 0$ and $R_o \to \infty$ we obtain

$$\begin{aligned} \lim_{\substack{R_i \to 0 \\ R_o \to \infty}} A_{is} &= \lim_{\substack{R_i \to 0 \\ R_o \to \infty}} \frac{R_s}{R_s + R_i} \times A_i \times \frac{R_o}{R_o + R_L} \\ &= A_i \end{aligned} \qquad (6.24)$$

and the overall current gain A_{is} is maximised at a value of A_i, when $R_i = 0$ and $R_o \to \infty$.

6.3 Operational amplifiers

Operational amplifiers (op-amps) are electronic, integrated-circuit amplifiers which are important in the implementation of a large variety of analogue circuits and systems such as audio and video amplifiers, analogue filters, instrumentation amplifiers, etc. Figure 6.11 a) shows the op-amp circuit symbol. Terminal 1 is called the non-inverting terminal while terminal 2 is called the inverting terminal. Terminal 3 indicates the output of the amplifier. The op-amp is a voltage amplifier whose input is the voltage across the input terminals and the output is referred to ground, which is usually the potential midway between the power rails. These are shown in figure 6.11 a) connected to terminals 4 and 5. Frequently, the circuit symbol for the op-amp omits the terminals for the connection of the power rails. The differential voltage gain is the ratio of the output to input voltages and is called the differential open-loop voltage gain.

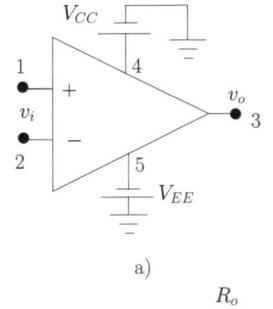

The ideal op-amp

Although the op-amp is an electronic circuit of some complexity its electrical behaviour can be modelled according to the circuit model shown in figure 6.11 b). The input impedance of an ideal op-amp is infinity, that is the op-amp does not draw any current from the input source by its input terminals. On the other hand the output impedance of the amplifier is zero. The differential voltage gain is infinity. Note that the ideal op-amp amplifies only the voltage difference between the input terminals. The last ideal characteristic considered for the ideal op-amp is infinite bandwidth. As will be shown shortly, these four ideal characteristics make the analysis (and the design) of circuits with op-amps quite simple.

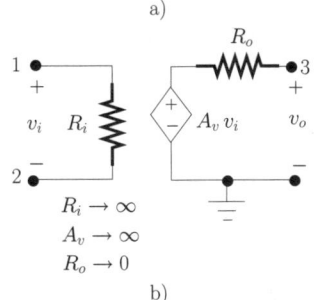

Figure 6.11: *Operational amplifier. a) Circuit symbol. b) Equivalent circuit.*

6.3.1 Open-loop and feedback concepts

The practical op-amp is rarely used as an open-loop amplifier. One of the main reasons for this being its extremely high voltage gain (typically 10^5–10^6) that easily results in output voltage saturation. Rather the op-amp is employed using feedback. It is the use of feedback which allows the implementation of a broad variety of circuits using op-amps and some of them are presented in section 6.3.2.

The concept of virtual short-circuit between input terminals

Let us consider the circuit of figure 6.12 constituted by an op-amp and a resistive voltage divider (R_1 and R_2) which implements *feedback*. This circuit is commonly called a non-inverting amplifier. For this circuit let us assume that $R_i \to \infty$ and $R_o = 0$. We can write the following equations:

$$v_s = v_i + v_f \qquad (6.25)$$
$$v_o = A_v v_i \qquad (6.26)$$

where v_f represents the feedback voltage and A_v is the open-loop gain. Since

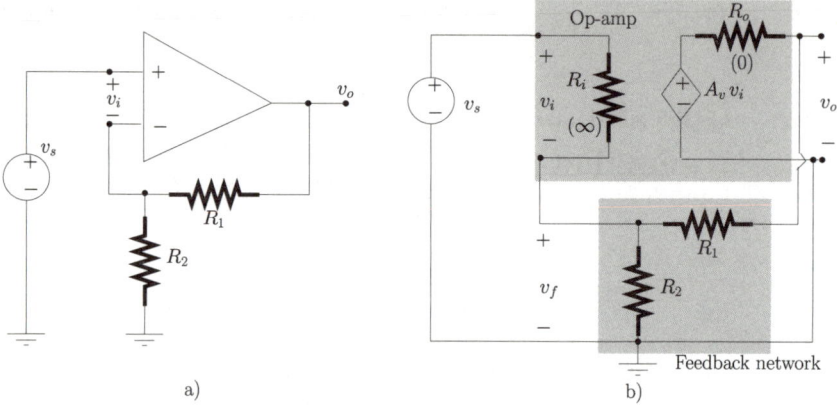

Figure 6.12: *Non-inverting amplifier. a) Block diagram. b) Equivalent electrical model.*

there is no current flowing through the op-amp input terminals (due to its infinite input impedance), v_f can be related to v_o using the resistive voltage divider expression:

$$v_f = \frac{R_2}{R_2 + R_1} v_o \qquad (6.27)$$
$$\beta = \frac{R_2}{R_2 + R_1} \qquad (6.28)$$

β represents the fraction of the output voltage which is fed back to the input of the circuit via the voltage divider. Equation 6.25 can be written as:

$$v_s = \frac{v_o}{A_v} + \beta v_o \qquad (6.29)$$

Solving this last equation to obtain v_o we get:

$$v_o = \frac{A_v}{1 + \beta A_v} v_s \qquad (6.30)$$

and eqn 6.26 can be solved to obtain v_i as follows;

$$v_i = \frac{1}{1 + \beta A_v} v_s \qquad (6.31)$$

6. Basic electronic amplifier building blocks

If we now let A_v tend towards infinity in eqns 6.30 and 6.31 then we obtain

$$v_o = \lim_{A_v \to \infty} \frac{A_v}{1 + \beta A_v} v_s$$
$$= \frac{1}{\beta} v_s$$
$$= \frac{R_2 + R_1}{R_1} v_s \quad (6.32)$$

and

$$v_i = \lim_{A_v \to \infty} \frac{1}{1 + \beta A_v} v_s = 0 \quad (6.33)$$

From eqn 6.32 we can write the closed-loop gain A_{vf} as follows:

$$A_{vf} = \frac{v_o}{v_s} = \frac{R_2 + R_1}{R_1} = 1 + \frac{R_2}{R_1} \quad (6.34)$$

Equation 6.33 reveals that when the op-amp gain is very large then the voltage difference between the two input terminals of the op-amp tends to zero. This gives rise to a 'virtual short-circuit' between the two input terminals of the op-amp. This virtual short-circuit is always valid when the feedback applied to the op-amp is negative, and is a valuable concept which contributes to the simplicity of analysis and design of circuits with op-amps.

Another very important result is the one expressed by eqn 6.34 which states that the voltage gain depends *only* on the values of the external resistances used to implement the feedback network. Again, this is a consequence of the large differential open-loop voltage gain.

6.3.2 Other examples and applications

The inverting amplifier

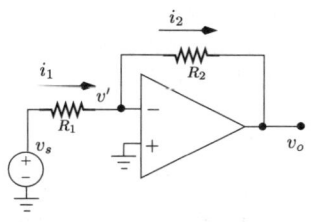

Figure 6.13: *Inverter amplifier*

Figure 6.13 shows another important voltage amplifier topology; the inverting amplifier. Since there is no current flowing through the input terminals we have:

$$i_1 = i_2 \quad (6.35)$$

Using the concept of virtual short-circuit between the op-amp input terminals we can write $v' = 0$. Therefore, the voltage v_s is applied across R_1 and v_o occurs across R_2. Equation 6.35 can be written as follows:

$$\frac{v_s}{R_1} = -\frac{v_o}{R_2} \quad (6.36)$$

and the closed-loop voltage gain is

$$A_{vf} = \frac{v_o}{v_s}$$
$$= -\frac{R_2}{R_1} \quad (6.37)$$

Example 6.3.1 Design two voltage amplifiers with voltage gains of $+15$ and -15.

Solution: Considering the non-inverting amplifier and using eqn 6.34 we have $R_2 = 14\,R_1$. Choosing $R_1 = 1$ kΩ we have $R_2 = 14$ kΩ.

Now for the gain of -15 we choose the inverting amplifier. Using eqn 6.37 we have $R_2 = 15\,R_1$. Choosing $R_1 = 1$ kΩ we have $R_2 = 15$ kΩ.

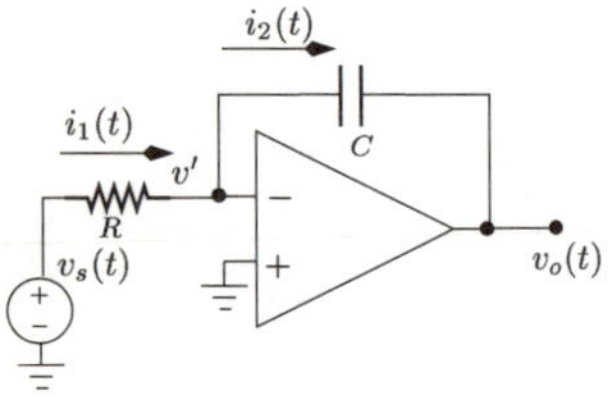

The integrator amplifier

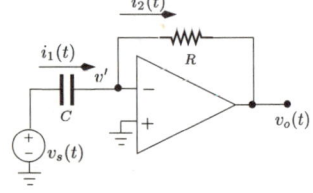

Figure 6.14: *Integrator amplifier.*

Figure 6.14 shows an integrator amplifier. It should be noted that this amplifier is a structure that is quite similar to that of the inverting amplifier discussed previously. The main difference is that R_2 is replaced by a capacitor, C. We use now a time domain analysis.

Using the concept of virtual short-circuit we have $v' = 0$. Also $i_1(t) = i_2(t)$. Therefore, we can write;

$$\frac{v_s(t)}{R} = -C\,\frac{d\,v_o(t)}{d\,t} \tag{6.38}$$

Solving the last equation in order to obtain $v_o(t)$ we get;

$$v_o(t) = -\frac{1}{RC}\int_0^t v_s(t)\,d\,t + V_{co} \tag{6.39}$$

where V_{co} is the initial voltage across the capacitor terminals (at $t = 0$). Equation 6.39 shows that the output voltage is proportional to the time integral of the input voltage. Note that the DC gain of such an amplifier is equal to the open-loop gain and is, therefore, very high. In practice a large resistor is placed across the capacitor to define a finite DC gain which prevents the saturation of the output voltage.

Example 6.3.2 Show that if we interchange the positions of the capacitor and the resistor in figure 6.14 we obtain an amplifier that produces an output that is proportional to the time differential of the input.

Solution: If we swap the position of the resistor and the capacitor in figure 6.14 we obtain the circuit shown in figure 6.15. Once again we can write $v' = 0$ and $i_1(t) = i_2(t)$. Hence we have that

$$C\frac{d\,v_s(t)}{d\,t} = -\frac{v_o(t)}{R} \tag{6.40}$$

that is

$$v_o(t) = -RC\,\frac{d\,v_s(t)}{d\,t} \tag{6.41}$$

Figure 6.15: *Differentiator amplifier.*

The last equation shows that the output voltage of the amplifier is proportional to the derivative of $v_s(t)$.

The adder amplifier

Figure 6.16 shows a circuit which allows the addition of voltage signals. Since $v' = 0$ and each voltage v_{sk} is applied across each resistor R_k we can write

$$i_k = \frac{v_{sk}}{R_k}, \quad k = 1, 2, \ldots, n \tag{6.42}$$

The output current, i_o, is the sum of each input current i_k:

$$i_o = i_1 + i_2 + \ldots + i_n \tag{6.43}$$

that is:

$$-\frac{v_o}{R_o} = \frac{v_{s1}}{R_1} + \frac{v_{s2}}{R_2} + \ldots + \frac{v_{sn}}{R_n} \tag{6.44}$$

The last equation can be written as:

$$v_o = -R_o \sum_{k=1}^{n} \frac{v_{sk}}{R_k} \tag{6.45}$$

If all resistances R_k are equal to R then we can write;

$$v_o = -\frac{R_o}{R} \sum_{k=1}^{n} v_{sk} \tag{6.46}$$

and the output voltage is proportional to the sum of the input voltages.

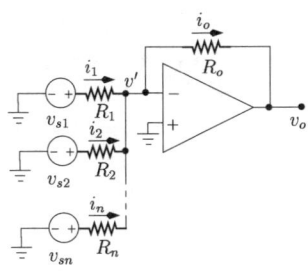

Figure 6.16: *The adder amplifier.*

The difference amplifier

The circuit shown in figure 6.17 amplifies the difference between the two input voltages v_{sa} and v_{sb}. v' can be calculated as the voltage given by a voltage divider:

$$v' = \frac{R_2}{R_1 + R_2} v_{sb} \tag{6.47}$$

Since $i_2 = i_1$ we can write;

$$\frac{v_{sa} - v'}{R_1} = \frac{v' - v_o}{R_2} \tag{6.48}$$

Solving the last eqn in order to obtain v_o we get:

$$v_o = \frac{R_2 + R_1}{R_1} v' - \frac{R_2}{R_1} v_{sa} \tag{6.49}$$

and using eqn 6.47 we can write v_o as follows:

$$v_o = \frac{R_2}{R_1}(v_{sb} - v_{sa}) \tag{6.50}$$

showing that the output voltage is proportional to the difference of the input voltages.

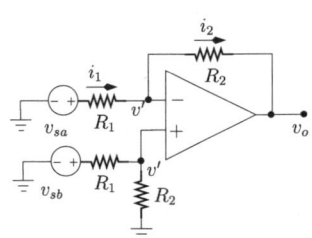

Figure 6.17: *The difference amplifier.*

The instrumentation amplifier

The instrumentation amplifier (IA) is a difference amplifier with a very high input impedance. Since the difference amplifier of figure 6.17 presents a relatively low input impedance equal to $2\,R_1$, it is not suitable for this use. Figure 6.18 shows the basic structure of an instrumentation amplifier. From this figure

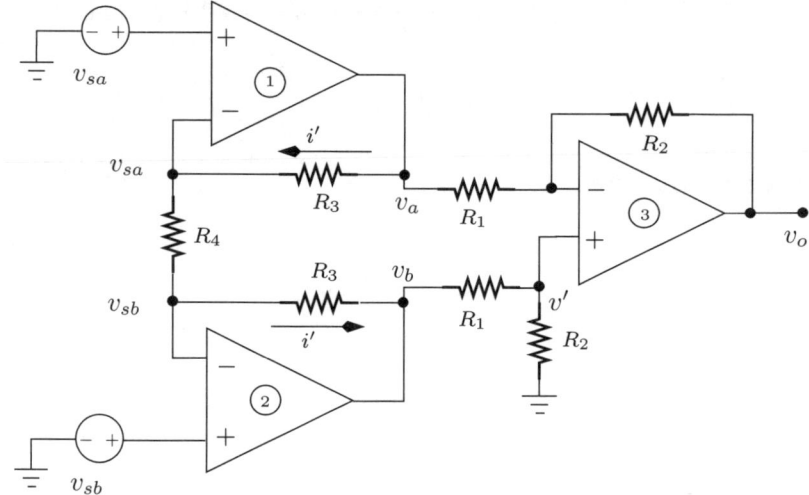

Figure 6.18: *Instrumentation amplifier.*

we observe that resistances marked R_1 and R_2 and the op-amp 3 constitute a difference amplifier. Hence, from eqn 6.50 we can write:

$$v_o = \frac{R_2}{R_1}(v_b - v_a) \tag{6.51}$$

Also we can write:

$$\frac{v_a - v_{sa}}{R_3} = \frac{v_{sa} - v_{sb}}{R_4} \tag{6.52}$$

$$\frac{v_{sa} - v_{sb}}{R_4} = \frac{v_{sb} - v_b}{R_3} \tag{6.53}$$

Solving these eqns to obtain $v_b - v_a$ we get:

$$v_b - v_a = \frac{2\,R_3 + R_4}{R_4}(v_{sb} - v_{sa}) \tag{6.54}$$

Hence, the output voltage can be written as follows:

$$v_o = \frac{R_2}{R_1}\frac{2\,R_3 + R_4}{R_4}(v_{sb} - v_{sa}) \tag{6.55}$$

Note that, in theory the input impedance of this amplifier is infinite when op-amps 1 and 2 are assumed to be ideal.

6.4 Active devices

We now present the main active electronic devices[1] which are fundamental in the amplification process: the Bipolar Junction Transistor (BJT) and the Field-Effect Transistor (FET). We also present and analyse some important amplifier topologies such as the common-emitter amplifier and the differential-pair amplifier.

6.4.1 The junction or p–n diode

The diode is a two terminal device usually manufactured using a silicon (Si) semiconductor. Figure 6.19 shows the geometry of a diode where we observe the existence of two regions which form a p–n junction. The p region is Si doped with an acceptor such as boron while the n region is Si doped with a donor such as phosphorus.

The electrical model for the diode is expressed as follows:

$$I_D = I_{SD}\left(e^{V_D/V_T} - 1\right) \qquad (6.56)$$

where I_{SD} is the 'reverse saturation current' of the diode. I_{SD} is of the order of 10^{-15} ampere and depends on the area of the diode and the temperature. The voltage V_T is called the thermal voltage given by:

$$V_T = \frac{\mathcal{K}T}{q} \qquad (6.57)$$

with $\mathcal{K} = 1.38 \times 10^{-23}$ joule/kelvin, representing Boltzmann's constant, T is the temperature in kelvin and $q = 1.6 \times 10^{-19}$ coulomb is the electronic charge. At room temperature $V_T \simeq 25$ mV. The DC electrical characteristics for a typical small area diode are illustrated in figure 6.20. From this figure we observe that for voltages V_D less than 0.7 volt the diode does not conduct a significant current. Note that for $V_D < 0$, I_D is nearly constant and equal to $-I_{DS}$, a very small value and not visible in figure 6.20. However, for a voltage $V_D \simeq 0.7$ the diode starts conducting. From this figure it is clear that diodes are a non-linear electronic devices.

The p–n junction forms the basis of the bipolar junction transistor which is discussed in the next section.

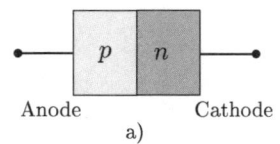
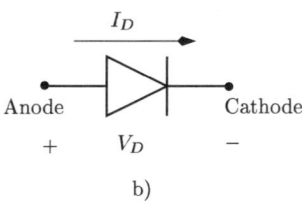

Figure 6.19: *The diode. a) Geometry (p–n junction). b) Symbol.*

6.4.2 The bipolar junction transistor

The bipolar junction transistor (BJT) is a three terminal active device usually manufactured using silicon (Si). Figure 6.21 shows the physical structure, symbol and direction of current flow for the NPN transistor while figure 6.22 represents the PNP transistor. Both transistors are implemented with two p–n junctions. For the NPN transistor both the emitter and the collector areas are n type silicon. However, these two regions are not interchangeable as the emitter is usually significantly more heavily doped (n^+) than the collector region.

---

[1]Our discussion of active devices is limited to their characteristics and behaviour as circuit elements. Explanations based on the physical nature of these devices has been limited and is only used so as to aid understanding of the circuit characteristics and models. Readers are referred to references [1, 2] for detailed descriptions of the physics of such devices.

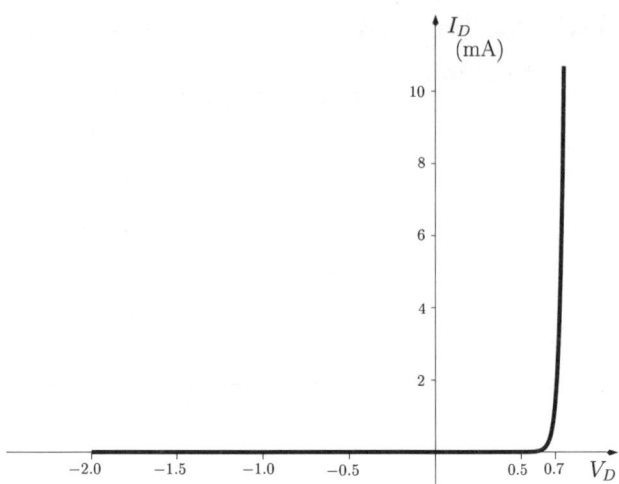

Figure 6.20: *Typical DC electrical characteristic for the junction diode.*

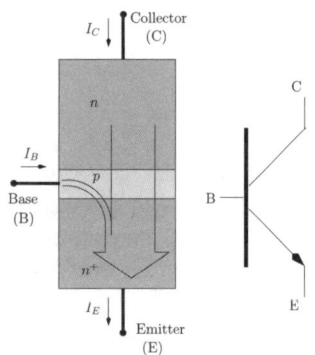

Figure 6.21: *Geometry and symbol for an NPN bipolar transistor.*

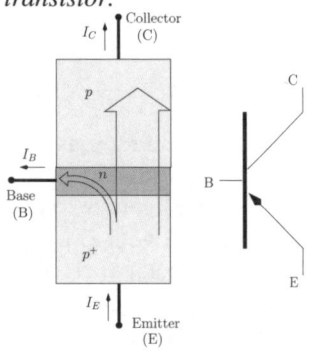

Figure 6.22: *Geometry and symbol for a PNP bipolar transistor.*

Also, the collector area is usually greater than the emitter area. The base region is very thin when compared to the emitter and collector regions. Conversely, for the PNP transistor both the emitter and the collector areas are p type silicon and the base region is n type silicon.

Figures 6.21 and 6.22 also indicate the directions of the current flowing through each terminal of each transistor. It can be seen that

$$I_E = I_C + I_B \tag{6.58}$$

where I_E I_C and I_B are the emitter current, the collector current and the base current, respectively. From this figure it can be seen that base current is usually significantly smaller than either the emitter current or the collector current. The base terminal is usually the input terminal for both devices while either the collector or the emitter is the output terminal.

The Ebers-Moll model for the BJT

The Ebers-Moll model, shown in figure 6.23, describes the relationships between the various currents and voltages of the bipolar transistor. The model consists of two diodes and of two current-controlled current sources. We discuss the application of this model to the NPN transistor. The PNP model is similar. For the NPN transistor the diode currents I_{D_E} and I_{D_C} are given by the diode equation (see eqn 6.56);

$$I_{D_E} = I_{S_E}\left(e^{V_{BE}/V_T} - 1\right) \tag{6.59}$$

$$I_{D_C} = I_{S_C}\left(e^{V_{BC}/V_T} - 1\right) \tag{6.60}$$

where I_{S_E} and I_{S_C} are the saturation currents of the two diodes. Since the collector region is usually larger than the emitter region, I_{S_C} is usually larger than I_{S_E} (by a factor up to 50).

6. Basic electronic amplifier building blocks

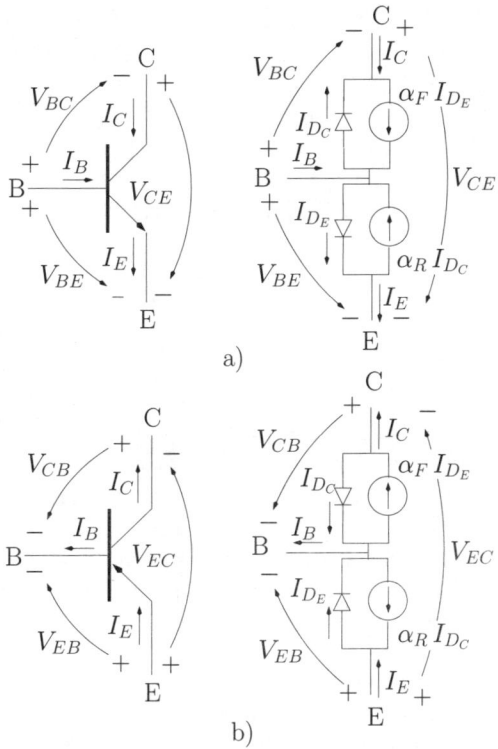

Figure 6.23: *Ebers-Moll model. a) NPN bipolar transistor. b) PNP bipolar transistor.*

The relationship between the terminal currents and the junction voltages can be written, from figure 6.23 a), as follows:

$$I_E = I_{D_E} - \alpha_R I_{D_C} \qquad (6.61)$$
$$I_C = -I_{D_C} + \alpha_F I_{D_E} \qquad (6.62)$$
$$I_B = (1 - \alpha_F) I_{D_E} + (1 - \alpha_R) I_{D_C} \qquad (6.63)$$

α_F is the 'forward gain' of the transistor and is typically near unity (0.98 to 0.998). α_R is the 'reverse gain' of the transistor and is typically near zero (0.02 to 0.1). The last equations can be written, using eqns 6.59 and 6.60, as follows

$$I_E = \underbrace{I_{S_E} \left(e^{V_{BE}/V_T} - 1 \right)}_{\text{Diode effect}} - \underbrace{\alpha_R I_{S_C} \left(e^{V_{BC}/V_T} - 1 \right)}_{\text{Reverse transistor effect}} \qquad (6.64)$$

$$I_C = \underbrace{\alpha_F I_{S_E} \left(e^{V_{BE}/V_T} - 1 \right)}_{\text{Forward transistor effect}} - \underbrace{I_{S_C} \left(e^{V_{BC}/V_T} - 1 \right)}_{\text{Diode effect}} \qquad (6.65)$$

$$I_B = I_E - I_C \qquad (6.66)$$

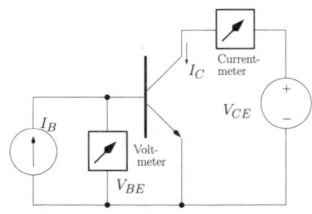

Figure 6.24: *Measurement of the DC characteristic for an NPN bipolar transistor.*

Figure 6.24 shows how we can obtain the variation of the collector current I_C with the collector-emitter voltage V_{CE} for various values of I_B (V_{BE}). These curves are the I–V curves for the transistor and an example is shown in figure 6.25. From this figure we identify three distinct regions of operation for the transistor; the active region, the cut-off region and the saturation region.

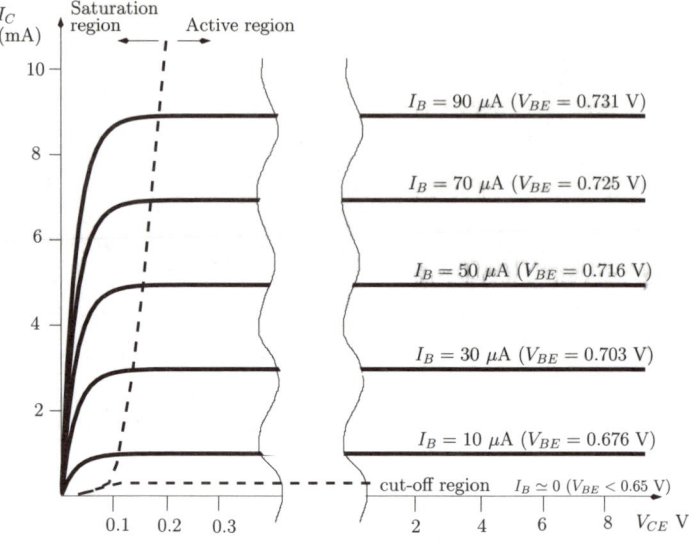

Figure 6.25: *DC characteristic for a small NPN bipolar transistor.*

Active region

The active region is the relevant region of operation when the transistor is used as an amplification device. In this region of operation the transistor can be characterised by its V_{BE} and by its V_{BC} as follows:

$$0.65 \lesssim V_{BE} \lesssim 0.75 \text{ volt} \qquad (6.67)$$
$$V_{BC} \leq 0 \qquad (6.68)$$

Under the conditions expressed by eqns 6.67 and 6.68 we can write:

$$\begin{aligned} I_E &\simeq I_{S_E}\left(e^{V_{BE}/V_T} - 1\right) \\ &\simeq I_{S_E} e^{V_{BE}/V_T} \text{ for } V_{BE} > 4V_T \qquad (6.69) \\ I_C &\simeq \alpha_F I_{S_E} e^{V_{BE}/V_T} \qquad (6.70) \\ &= \alpha_F I_E \\ I_B &= I_E - I_C \\ &= \frac{I_C}{\beta_F} \qquad (6.71) \end{aligned}$$

with

$$\beta_F \triangleq \frac{\alpha_F}{1 - \alpha_F}$$

6. Basic electronic amplifier building blocks

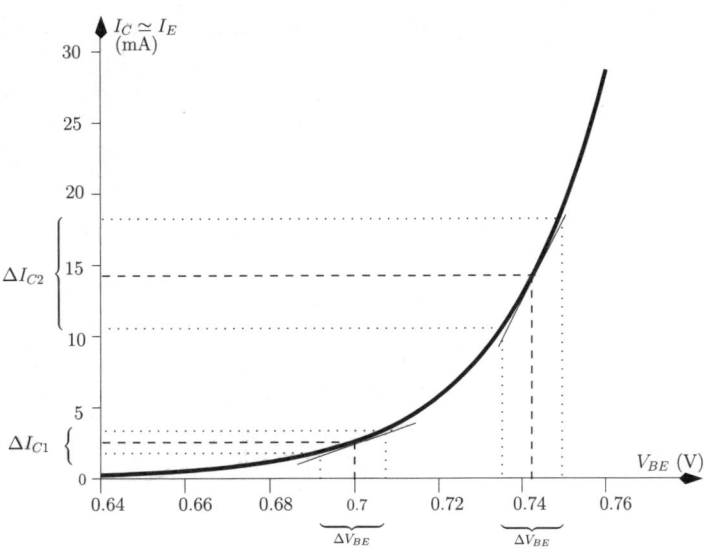

Figure 6.26: *Variation of I_C with V_{BE} when the transistor is in the linear region. Example ($I_{S_E} = 1.8 \times 10^{-15}$ A).*

where β_F is called the 'forward current gain'. β_F is often written as β and simply called the 'current gain'. For α_F ranging from 0.98 to 0.998, β_F varies from 49 to 499. From eqns 6.69 to 6.71 and from figure 6.25 a) we observe that the behaviour of the transistor in the linear region is similar to a current-controlled current source since the output current does not depend on V_{CE}. It should be noted that the variation of I_E or I_C with V_{BE} is highly non-linear as illustrated in figure 6.26 which presents the variation of I_C with V_{BE}. We observe that for a V_{BE} variation from 0.65 V to 0.75 V the collector current varies (in an exponential manner) from about 0.5 mA to 20 mA. Fortunately, for small V_{BE} variations the collector current varies in an approximately linear manner.

It is important to realise that although the I_C–V_{BE} relation is non-linear, the relation between I_C and I_B is linear. This is a direct consequence of the forward transistor effect described in eqn 6.65.

The bipolar transistor as an amplifier

Figure 6.27 a) shows a very simple transistor circuit which illustrates the role of the BJT as an amplification device. An alternative representation for this circuit is presented in 6.27 b) where the DC voltage source, V_{CC}, is described by a small horizontal trace with its value. This value is assumed to be referenced to ground. The input signal source symbol is often omitted for simplicity.

From this figure we observe that the input voltage signal is between the base and emitter while the output is between the collector and emitter. Since the emitter is the common terminal, in terms of input and output ports, the circuit configuration is known as the 'common-emitter amplifier'. We assume

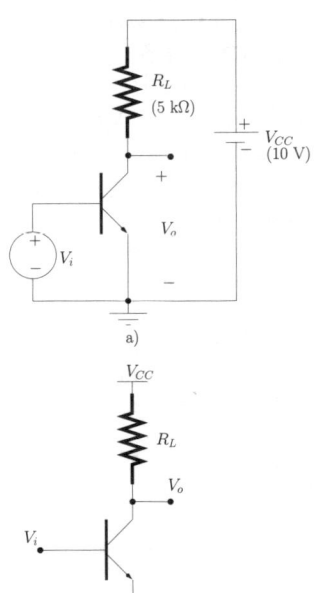

Figure 6.27: *a) Common-emitter amplifier. b) Alternative representation.*

the following characteristics for the BJT: $\alpha_F = 0.99$, $\alpha_R = 0.02$, $I_{S_E} = 1.8 \times 10^{-15}$ A, and $I_{S_C} = 18 \times 10^{-15}$ A.

We can write the following equations:

$$V_{be} = V_i \tag{6.72}$$
$$V_{ce} = V_o \tag{6.73}$$
$$V_{CC} = V_{ce} + R_L I_c \tag{6.74}$$

From the above we can easily derive a linear relation describing the variation of the collector current with the collector-emitter voltage for a given load R_L and supply voltage V_{CC}. This is known as the *load-line* equation and, in this case, is given by:

$$\begin{aligned} I_c &= \frac{V_{CC}}{R_L} - \frac{V_{ce}}{R_L} \tag{6.75} \\ &= 2 - 0.2 V_{ce} \text{ mA} \tag{6.76} \end{aligned}$$

In figure 6.28 we present a graphical explanation of the amplification process provided by the common-emitter circuit. Figure 6.28 a) illustrates the Ebers-Moll equations, discussed above, which relate the collector current, I_c, to the collector-emitter voltage V_{ce} for various base-emitter voltages V_{be}. This figure also shows the load-line given by eqn. 6.75. Figure 6.28 b) shows the input voltage signal, V_i, while figure 6.28 c) shows the output voltage signal, V_o. It should be noted that the input voltage sits on top of a DC voltage; the base-emitter bias voltage, V_{BE_Q}. This voltage intersects the load-line defining the transistor operating point Q (Quiescent) which, in turn, defines the collector bias current, I_{C_Q}, and the corresponding collector-emitter bias voltage, V_{CE_Q}. From figure 6.28 a) we observe that $I_{C_Q} = 1$ mA and that $V_{CE_Q} = 5$ V. It is important to note that the ideal value for V_{CE_Q} is $V_{CC}/2$ in order to have a maximum and symmetrical output voltage swing without distortion. Note that this distortion is non-linear and known as 'clipping'.

The voltage gain, A_v is defined as the ratio between the output voltage and the input voltage:

$$\begin{aligned} A_v &= \frac{v_o}{v_i} \\ &= -\frac{3 \text{ V}}{15 \text{ mV}} \\ &= -200 \end{aligned}$$

Note that, in order to have distortionless amplification, the maximum input voltage swing must not exceed 22.5 mV (see next example) otherwise the output voltage waveform will suffer distortion, resulting from clipping as illustrated in figure 6.29.

Example 6.4.1 Show that if the input voltage v_i exceeds about 22.5 mA distortion will occur in the output voltage.

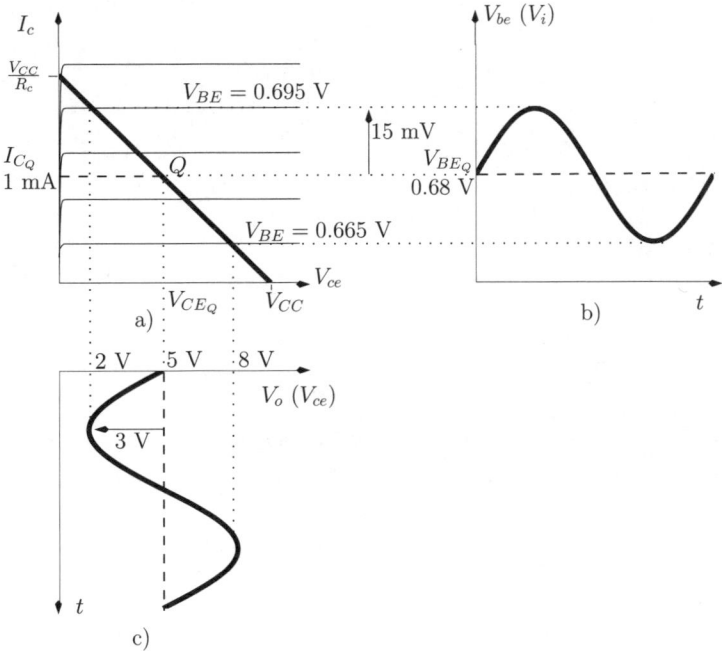

Figure 6.28: *The amplification process provided by the common-emitter.*

Solution: According to figure 6.28 the maximum amplitude for the output signal voltage is about ±5 volts. Taking 0.5 volt as a safety margin we establish the maximum amplitude for the output signal voltage as ±4.5 volts. Dividing this value by the voltage gain, $A_v = 200$, we obtain the maximum input voltage of 22.5 mV.

The analysis of the circuit presented in figure 6.27 considers the bias (DC) and the (AC) signals simultaneously. However, it is possible, and often desirable, to separate these two analyses. In fact from eqns 6.72 – 6.74 we can separate the signal component from the DC components as follows:

$$\begin{aligned} V_{be} &= V_{BE_Q} + v_{be} \\ &= V_{BE_Q} + v_i \end{aligned} \quad (6.77)$$

$$\begin{aligned} V_{ce} &= V_{CE_Q} + v_{ce} \\ &= V_{CE_Q} + v_o \end{aligned} \quad (6.78)$$

Hence, eqn 6.74 can be written as follows:

$$\begin{aligned} V_{CC} &= V_{CE_Q} + v_{ce} + R_L \left(I_{C_Q} + i_c \right) \\ &= V_{CE_Q} + v_o + R_L \left(I_{C_Q} + i_c \right) \end{aligned} \quad (6.79)$$

Equation 6.79 can be separated into two equations: one relates to the signal

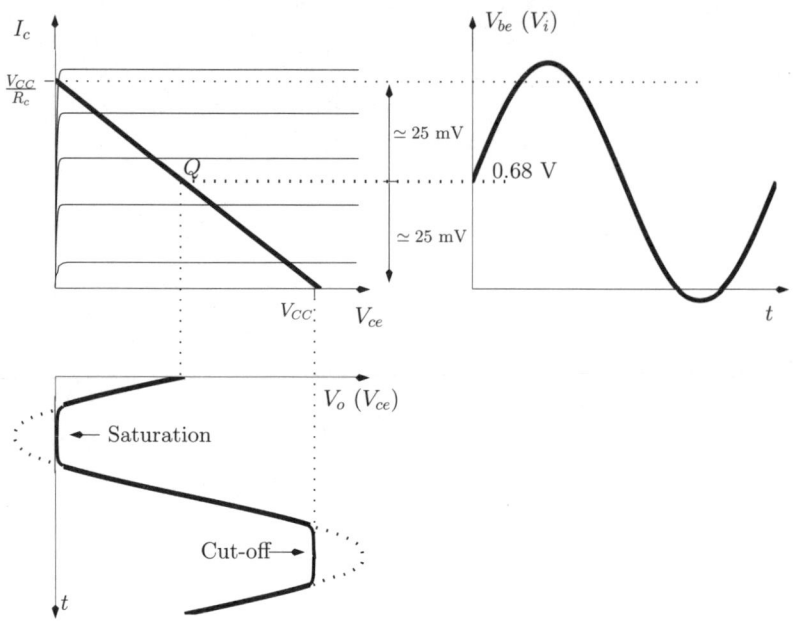

Figure 6.29: *Clipping in the amplification process.*

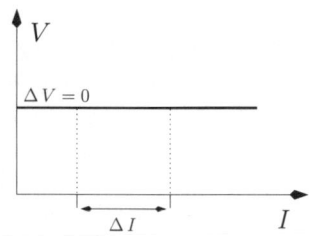

Figure 6.30: *Ideal DC voltage source.*

component while the other equation relates to the DC bias component:

$$\begin{cases} V_{CC} = V_{CE_Q} + R_L I_{C_Q} & \Leftarrow \text{DC} \\ 0 = v_o + R_L i_c & \Leftarrow \text{signal} \end{cases} \quad (6.80)$$

Equation 6.80 reveals that, **in terms of signal analysis the DC voltage source is equivalent to a ground terminal**. This is reasonable and expected since an ideal voltage source has a zero output impedance which implies that (see also figure 6.30):

$$\frac{\Delta V}{\Delta I} = 0 \quad (6.81)$$

The hybrid-π linear model for the transistor

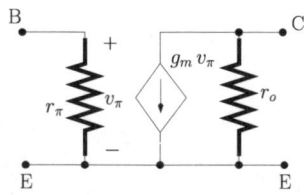

Figure 6.31: *The hybrid-π model for a BJT.*

Figure 6.31 shows the hybrid-π model which is a small-signal linear model for the transistor valid for the mid-frequency range. This model is partially derived from the Ebers-Moll large-signal model and it considerably simplifies the AC analysis of bipolar transistor amplifiers.

The hybrid-π model assumes a given operation point in the linear operation region, defined by a collector current I_{C_Q} and by a V_{BE_Q}. As mentioned above $V_{BE} = V_{BE_{Q1}} \simeq 0.7$ volt. For small V_{be} variations the corresponding variations of I_c are approximately linear. From Taylor's series (see appendix A) we can write the following expression for the collector current (DC plus signal):

$$I_c = I_{CQ} + v_{be} \left. \frac{dI_c}{dV_{be}} \right|_{I_C=I_{CQ}} \qquad (6.82)$$

The transconductance gain, g_m, is defined as:

$$g_m \triangleq \left. \frac{dI_c}{dV_{be}} \right|_{I_C=I_{CQ}} \qquad (6.83)$$

From eqn 6.70 we can write

$$\left. \frac{dI_c}{dV_{be}} \right|_{I_C=I_{CQ}} = \left. \frac{\alpha_F I_{S_E}}{V_T} e^{V_{BE}/V_T} \right|_{I_C=I_{CQ}} \qquad (6.84)$$

giving

$$g_m = \frac{I_{CQ}}{V_T} \qquad (6.85)$$

The dynamic resistance between the base and the emitter terminals, r_π, defines the small-signal v_{be} variations with small-signal i_b variations according to the following expression:

$$r_\pi \triangleq \left. \frac{dV_{be}}{dI_b} \right|_{I_b=I_{BQ}} \qquad (6.86)$$

where $I_B = I_{BQ}$ is the bias base current equal to I_{CQ}/β. Hence, from eqns 6.70 and 6.71 we can write:

$$V_{be} = V_T \ln \left(\frac{\alpha_F \beta I_b}{I_{S_E}} \right) \qquad (6.87)$$

and r_π can be determined as follows:

$$\left. \frac{dV_{be}}{dI_b} \right|_{I_B=I_{BQ}} = V_T \left. \frac{\frac{\alpha_F \beta}{I_{S_E}}}{\frac{\alpha_F \beta I_b}{I_{S_E}}} \right|_{I_b=I_{BQ}}$$

$$= \frac{V_T}{I_{CQ}} \beta \qquad (6.88)$$

Therefore,

$$r_\pi = \frac{\beta}{g_m} \qquad (6.89)$$

According to the Ebers-Moll model when the transistor is in the active region the collector current does not vary with V_{CE}. In practice this is not true since there is a slight increase of the collector current as V_{CE} increases. This phenomena is the Early effect. In terms of an electrical model this effect

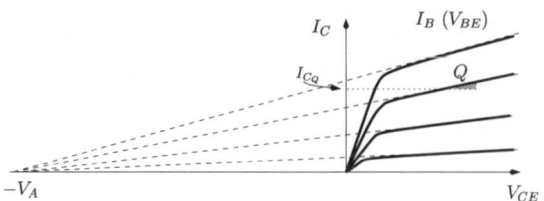

Figure 6.32: *The Early effect.*

is described by a dynamic resistance located between the collector and the emitter, r_o, and its value is given by

$$r_o = \frac{|V_A|}{I_{C_Q}} \qquad (6.90)$$

V_A is a theoretical voltage known as the Early voltage and can be obtained from the convergence point of the output characteristics as shown in figure 6.32. For a BJT r_o is of the order of hundreds of kΩ.

Example 6.4.2 Apply the hybrid-π model to the common-emitter amplifier depicted in figure 6.27 and determine the small-signal voltage gain v_s/v_i for the following values: $I_C = 1$ mA, $\beta = 100$ and $V_A = 120$ V.

Solution: Since for AC analysis V_{CC} is replaced by a ground terminal, the equivalent AC circuit for the amplifier of figure 6.27 is as shown in figure 6.33 a). Substituting the transistor symbol by its hybrid-π equivalent model we obtain the circuit shown in figure 6.33 b). From this circuit it is straightforward to determine the small-signal voltage gain, $A_v = v_o/v_i$:

$$\begin{aligned} A_v &= \frac{v_o}{v_i} \\ &= \frac{-g_m(R_L||r_o)\,v_\pi}{v_\pi} \end{aligned} \qquad (6.91)$$

According to eqn 6.85 $g_m = 40$ mA/V. From eqn 6.90 we have $r_o = 120$ kΩ. Since $r_o \gg R_L$

$$\begin{aligned} A_v &\simeq -g_m R_L \\ &= -200 \end{aligned} \qquad (6.92)$$

Note that this result is the same as that obtained using the large signal analysis.

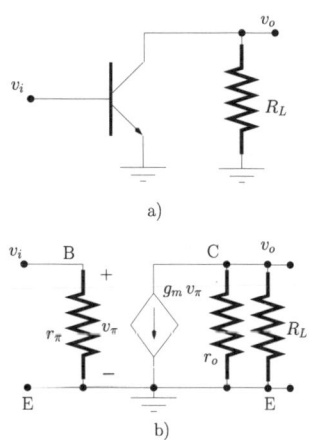

Figure 6.33: *Common-emitter amplifier. a) AC equivalent circuit. b) Small-signal equivalent circuit using the hybrid-π model.*

6.4.3 The insulated gate field-effect transistor

Like the BJT the field-effect transistor (FET) is also a three terminal device. Figure 6.34 shows the symbol and the physical structure of a popular mem-

6. Basic electronic amplifier building blocks

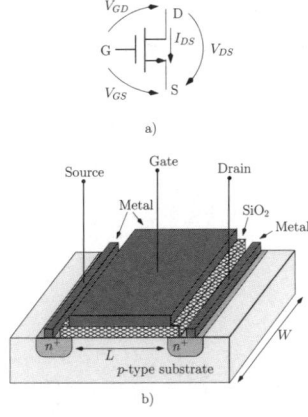

Figure 6.34: *n-channel FET.* a) Symbol. b) Geometry.

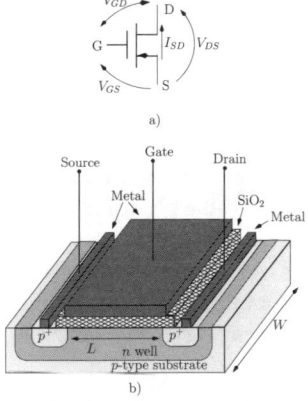

Figure 6.35: *p-channel FET.* a) Symbol. b) Geometry.

ber of the FET family known as the enhancement n-channel FET[2]. Both the drain and the source are heavily doped n regions. The gate is isolated from the source, the drain and the substrate by a layer of silicon dioxide (SiO$_2$). Hence, this FET is often called Insulated Gate FET (IGFET) or Metal-Oxide Semiconductor FET (MOSFET). Unlike the BJT transistor the FET is a symmetrical device, that is, the source and the drain of the device can be interchanged.

Figure 6.35 shows the symbol and the physical structure of an enhancement p-channel FET. Note that the p-channel FET can also be constructed in an n substrate. However, with the integrated technology it is common to implement an n well in the p substrate and this n well serves the purpose of a substrate for the implementation of the p-channel transistor.

DC large signal model

A popular DC large signal model used to characterise n-channel FETs is defined by the following set of equations:

$$I_{DS} = \begin{cases} k_n \frac{W}{L}\left[(V_{GS}-V_{Th})V_{DS} - \frac{1}{2}V_{DS}^2\right], & V_{DS} \leq V_{GS}-V_{Th} \\ & \text{and } V_{GS} > V_{Th} \\ & \text{(Triode region)} \\ \frac{1}{2}k_n\frac{W}{L}\left[V_{GS}-V_{Th}\right]^2, & V_{DS} > V_{GS}-V_{Th} \\ & \text{and } V_{GS} > V_{Th} \\ & \text{(Saturation region)} \\ 0, & V_{GS} \leq V_{Th} \\ & \text{(Cut-off region)} \end{cases} \quad (6.93)$$

where I_{DS} is the current which flows from the drain to the source of the transistor. V_{GS} and V_{DS} are the gate-source and the drain-source voltages, respectively, as indicated in figure 6.34. L is the channel length and W is the width of the channel. V_{Th} is the threshold voltage which is often between 1 and 3 volts. k_n is defined as follows:

$$k_n = \mu_n C_{ox} \quad \text{A/V}^2 \quad (6.94)$$

Here μ_n is the mobility of the electrons in the channel measured in m^2/Vs. C_{ox} is the capacitance per unit of area of the capacitor produced by the insulator (SiO$_2$) which acts as a dielectric material between the two plates formed by the gate and the channel. As mentioned above, silicon dioxide is an insulator and, therefore, the gate current is extremely small. Figure 6.36 shows the DC characteristic of an n-channel FET provided by the large signal model as described by eqn 6.93. From this figure we observe the three regions of operation predicted by the model of eqn 6.93; the triode region, the saturation region and the cut-off region. It can be seen that in the triode region the FET behaves

---

[2]The channel is the charge layer under the gate of the device. n-channel indicates that the charges are negative (electrons). Such charges can be attracted towards the top of the device when a positive voltage is applied to the gate, thus creating the channel between the source and drain. For detailed discussion of the physics of FET devices the reader is referred to [1, 2].

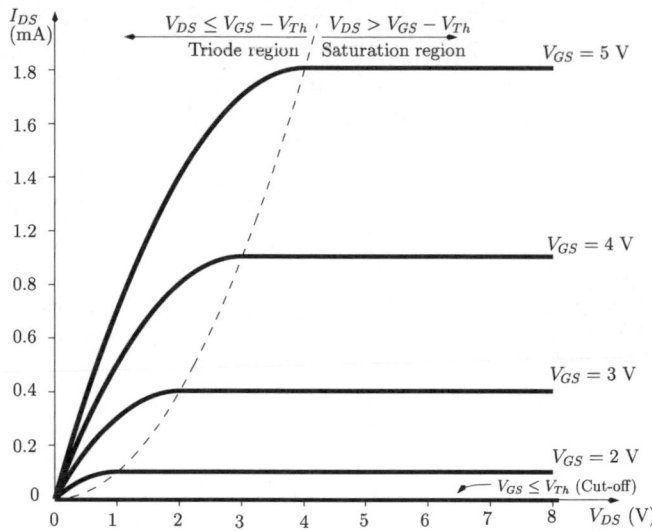

Figure 6.36: *DC curves of n-channel FET according to the large signal model (eqn 6.93).* $V_{th} = 1$ V *and* $k_n W/L = 0.2$ mA/V^2.

nearly as a linear resistance, controlled by V_{GS}. In the saturation region the FET behaves as a current source controlled by V_{GS}. This is the region employed when the FET is used as an amplification device. Note that, according to this model, I_{DS} does not vary with V_{DS}. Hence, we can represent the variations of I_{DS} with V_{GS} as shown in figure 6.37. From this figure we observe that for V_{GS} above the threshold voltage the drain current varies in a quadratic manner with V_{GS} (see also eqn 6.93).

The DC large signal model for p channel FETs can be obtained from eqn 6.93 after changing the voltages V_{GS} and V_{DS} by V_{SG} and V_{SD}, respectively, as indicated in figure 6.35.

Low frequency small-signal for the IGFET

Figure 6.38 shows the low frequency small-signal for the IGFET when the device is operating in the saturation region. Assuming an operating point defined by V_{GS_Q} and I_{D_Q} (see also fig 6.37), the transconductance for the FET is defined as follows:

$$\begin{aligned} g_m &\triangleq \left. \frac{dI_{DS}}{dV_{GS}} \right|_{V_{GS}=V_{GS_Q}} \\ &= k_n \frac{W}{L} \left[V_{GS_Q} - V_{Th} \right] \end{aligned} \quad (6.95)$$

Since we have $I_{DS_Q} = 1/2\, k_n \frac{W}{L} \left[V_{GS_Q} - V_{Th} \right]^2$ we can write g_m as follows:

$$g_m = \frac{2\, I_{DS_Q}}{V_{GS_Q} - V_{Th}} \quad (6.96)$$

6. Basic electronic amplifier building blocks

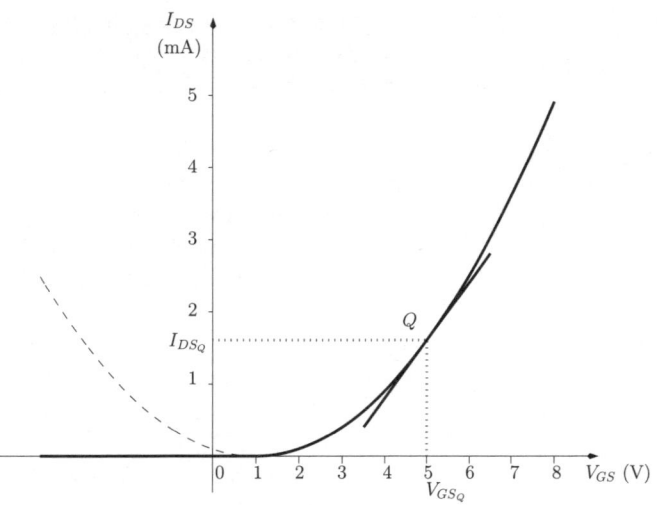

Figure 6.37: I_{DS} versus V_{GS} in the saturation region. $V_{Th} = 1\ V$ and $k_n\ W/L = 0.2\ mA/V^2$.

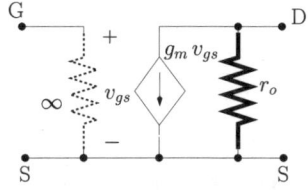

Figure 6.38: *FET small-signal equivalent circuit.*

According to the DC large signal model described by eqn 6.93 the drain current of the FET does not vary with V_{DS} when the device is operating in the saturation region. However, in practice we observe a slight increase of this current as V_{DS} increases, similar to the Early effect in BJTs. In FETs this phenomenon is known as channel-length modulation and is described, in terms of small signal model as a resistance r_o which can be calculated as follows:

$$r_o = \frac{V_A}{I_{DS_Q}} \qquad (6.97)$$

where V_A is defined by the convergence of the output characteristics in a similar way to the Early voltage of the BJT. For many FETs V_A is normally between 40 to 100 volts.

High-frequency models for active devices

Bipolar transistors and field-effect transistors have potential barriers in the various p–n junctions which induce charge storage effects. Such effects can be modelled as equivalent capacitances between the terminals of these devices. Figure 6.39 a) shows the high-frequency hybrid-π model for the BJT including the capacitive effects mentioned above. There is an equivalent capacitance between the base and the emitter terminal; C_π (1–15 pF), and a collector-base capacitance C_μ (0.1–5 pF). The model also includes a resistance r_x accounting for a small voltage drop within the base region. Values for this resistance vary typically between 2 and 10 Ω. The effect of r_x is usually negligible in the mid-band frequency range since r_π is much larger than r_x. However, at high frequencies its effect can be significant specially if the signal source is a voltage source with an output impedance of the order of r_x.

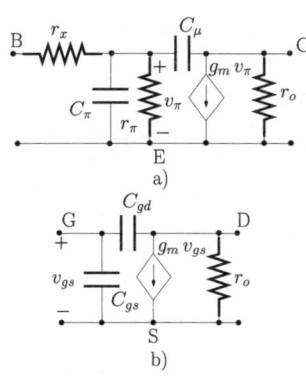

Figure 6.39: *High-frequency small signal models. a) BJT. b) FET.*

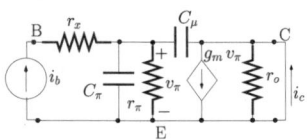

Figure 6.40: *Calculation of f_T for a BJT.*

Figure 6.39 b) shows the high-frequency small-signal model for the FET which includes two capacitance: C_{gs} and C_{gd}. These two capacitors vary typically between fractions of pico-farad to 10 pF and C_{gs} is usually larger than C_{gd}.

A very important figure of merit for both BJTs and FETs is the unity-gain bandwidth, f_T, which indicates the frequency at which the short-circuit current gain drops to unity (0 dB). For BJTs this frequency can be calculated from the circuit shown in figure 6.40. Since the collector is short-circuited to the emitter v_π is also applied to C_μ. Hence we can write:

$$i_b = \frac{v_\pi(1 + j\omega r_\pi C_\pi)}{r_\pi} + v_\pi j\omega C_\mu \qquad (6.98)$$

$$i_c = (g_m - j\omega C_\mu) v_\pi \qquad (6.99)$$

For the frequency range for which this model is valid, ωC_μ is much less than gm and, therefore, we can calculate the short-circuit current gain as follows;

$$\frac{i_c}{i_b} \simeq \frac{g_m r_\pi}{1 + j\omega (C_\mu + C_\pi) r_\pi} \qquad (6.100)$$

The unity-gain bandwidth, f_T, can be calculated from the last eqn (see problem 6.8) as follows:

$$f_T = \frac{g_m}{2\pi (C_\mu + C_\pi)} \qquad (6.101)$$

Similarly, for FET devices it can be shown (see problem 6.9) that f_T is given by:

$$f_T = \frac{g_m}{2\pi (C_{gd} + C_{gs})} \qquad (6.102)$$

For either type of transistor f_T can vary from hundreds of MHz to a few tens of GHz depending on the technology and size of the devices.

The high-frequency models presented in figure 6.39 are valid for frequencies up to about $f_T/3$. For frequencies higher than $f_T/3$ other parasitic elements, like the parasitic inductances and resistances associated with terminal connectors, must be taken into account in order to obtain an accurate characterisation of the devices.

6.4.4 The common-emitter amplifier

Figure 6.41 a) shows a common-emitter amplifier with an input bias circuit consisting of resistors R_1 and R_2. C_B and C_L are the DC blocking (or AC coupling) capacitors while C_E is a bypass capacitor which, at mid-frequency range, short-circuits R_E allowing for a greater voltage gain for this frequency range. We take a BJT characterised by $\beta = 200$, $C_\pi = 8$ pF, $C_\mu = 3$ pF. We also assume $r_x \simeq 0$ and that the Early effect can be neglected, that is, $r_o \to \infty$.

6. Basic electronic amplifier building blocks

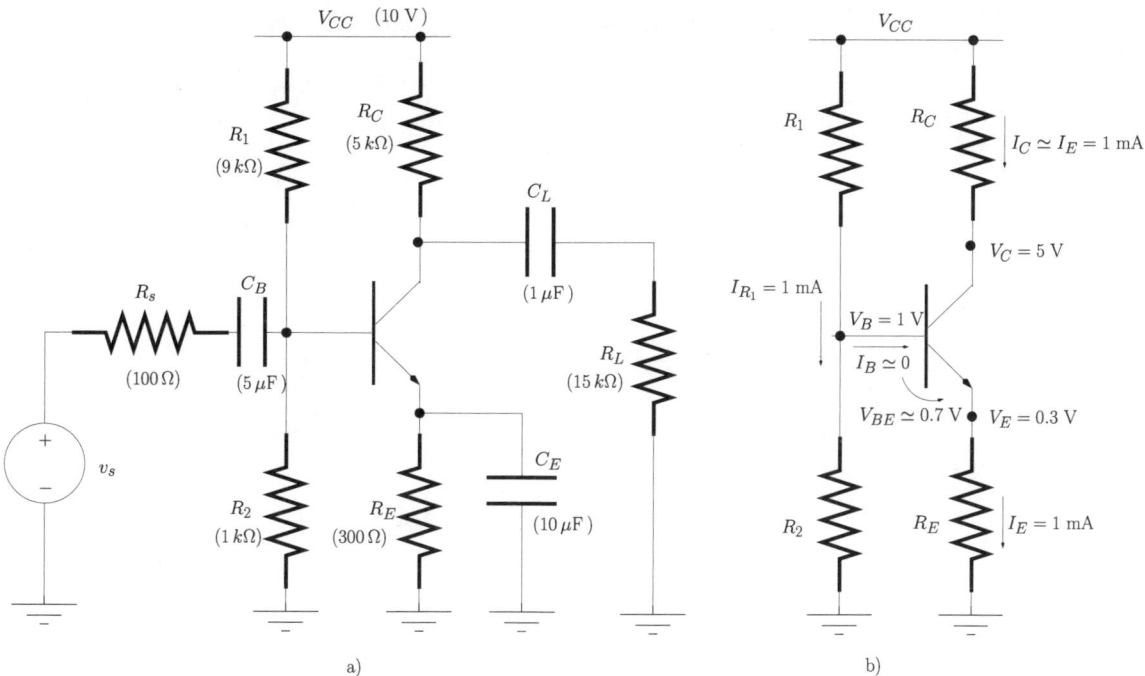

Figure 6.41: *a) Common-emitter amplifier. b) Equivalent circuit for the DC analysis.*

DC analysis

Figure 6.41 b) shows the equivalent circuit used in the DC analysis. Recall that at DC each capacitor is an open-circuit. We assume that the BJT is operating within the linear range of operation and $V_{BE} \simeq 0.7$ volt.

Since the bias currents passing through R_1 and R_2 (I_{R_1} and I_{R_2}) are much larger than I_B we can assume, *for DC analysis*, that $I_B \simeq 0$. These two approximations[3] ($I_B \simeq 0$ and $V_{BE} \simeq 0.7$ V) simplify considerably this type of analysis.

The current that flows through R_1 is nearly the same as the current that flows through R_2, that is $I_{R_1} = I_{R_2}$. Since V_{CC} is applied across these two resistors we can write:

$$I_{R_1} = I_{R_2} = \frac{V_{CC}}{R_1 + R_2} \quad (6.103)$$
$$= 1\,\text{mA}$$

The voltage at the base of the transistor is given by:

$$V_B = I_{R_2} R_2 \quad (6.104)$$
$$= 1\,\text{V}$$

---

[3]It should be clear that I_B in reality is not zero! However, the statements $I_B \ll I_{R_1}$ and $I_B \ll I_{R_2}$ are equivalent to saying that $I_B = 0$ for the purpose of DC analysis.

The voltage across R_E can be calculated as follows:

$$V_E = V_B - V_{BE} \quad (6.105)$$
$$= 0.3\,\text{V}$$

Hence the emitter current is given by:

$$I_E = \frac{V_E}{R_E} \quad (6.106)$$
$$= 1\,\text{mA}$$

Since $I_C \simeq I_E$ the voltage at the collector terminal is:

$$V_C = V_{CC} - R_C\,I_C \quad (6.107)$$
$$= 5\,\text{V}$$

Note that $I_B = I_C/\beta = 10\,\mu\text{A}$ and, therefore $I_B \ll I_{R_1}$ as assumed initially.

Low-frequency analysis

Now that the DC analysis is complete we are able to calculate g_m and r_π as follows:

$$g_m = \frac{I_C}{V_T} \quad (6.108)$$
$$= 40\,\text{mA/V}$$
$$r_\pi = \frac{\beta}{g_m} \quad (6.109)$$
$$= 5\,\text{k}\Omega$$

Figure 6.42 shows the equivalent low-frequency small-signal of the common-emitter amplifier. We start by calculating the voltage gain, $A_v = v_o/v_s$ which

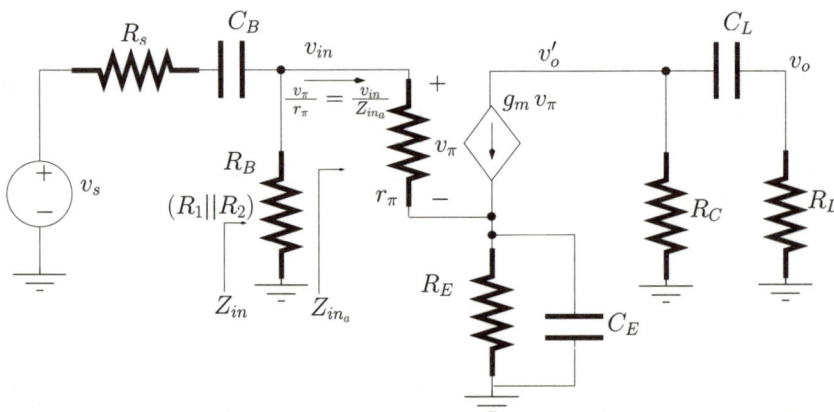

Figure 6.42: *Equivalent small-signal (low-frequency) circuit of the common-emitter amplifier.*

can be written as the product of three partial gains:

$$A_v = \frac{v_o}{v'_o} \times \frac{v'_o}{v_{in}} \times \frac{v_{in}}{v_s} \quad (6.110)$$

The partial gain v_o/v'_o is given by the impedance voltage divider consisting of the resistance R_L and the impedance of the capacitor $(j\,\omega\,C_L)^{-1}$:

$$\frac{v_o}{v'_o} = \frac{j\,\omega\,C_L\,R_L}{j\,\omega\,C_L\,R_L + 1} \quad (6.111)$$

The small signal base current, i_b, can be written as follows:

$$i_b = \frac{v_{in}}{Z_{in}} = \frac{v_\pi}{r_\pi} \quad (6.112)$$

and the partial gain v'_o/v_{in} can be written as:

$$\frac{v'_o}{v_{in}} = \frac{-g_m\,v_\pi\,(R_C||Z_L)}{Z_{in_a}\frac{v_\pi}{r_\pi}}$$
$$= \frac{-g_m\,r_\pi\,(R_C||Z_L)}{Z_{in_a}} \quad (6.113)$$

where

$$Z_L = R_L + \frac{1}{j\,\omega\,C_L} \quad (6.114)$$

and Z_{in_a} is the impedance seen looking into the base of the BJT. By applying a test voltage, v_t to the base, as shown in figure 6.43, Z_{in_a} can be simply found as shown below.

$$Z_{in_a} = \frac{v_t}{i_t}$$
$$= \frac{v_\pi + Z_E\left(\frac{v_\pi}{r_\pi} + g_m\,v_\pi\right)}{\frac{v_\pi}{r_\pi}}$$
$$= r_\pi + (\beta + 1)\,Z_E \quad (6.115)$$

with Z_E given by

$$Z_E = R_E || \frac{1}{j\,\omega\,C_E}$$
$$= \frac{R_E}{1 + j\,\omega\,C_E\,R_E} \quad (6.116)$$

The partial gain v_{in}/v_s is given by the impedance voltage divider consisting of R_B in parallel with Z_{in_a} and the impedance consisting of the series of R_s with the impedance of the capacitor $(j\,\omega\,C_B)^{-1}$:

$$\frac{v_{in}}{v_s} = \frac{(R_B||Z_{in_a})}{(R_B||Z_{in_a}) + R_s + (j\,\omega\,C_B)^{-1}} \quad (6.117)$$

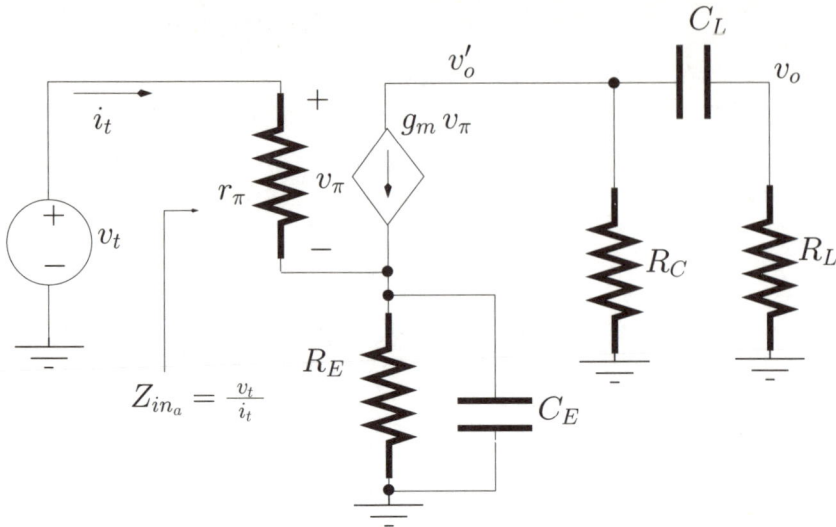

Figure 6.43: *Equivalent circuit for the calculation of Z_{in_a} (see also fig 6.42).*

The total gain is obtained by multiplying the three partial gains derived above. It is obvious that the expression for this gain is very lengthy. Therefore, it is difficult to extract the relevant information directly, namely the 3 dB cut-off frequency and the knowledge of which capacitor (or capacitors) mostly influence this frequency. Nevertheless, it is possible to plot this gain and extract quantitative information. Figure 6.44 shows $|A_v|$ versus the frequency. From this figure it is possible to see that the mid-frequency gain tends to $|A_{v_m}| = 132.6$ and that the low cut-off frequency is $f_L = 680$ Hz.

Short-circuit time constants method

The short-circuit time constants method (see [4] for a detailed study) is a very straightforward means of determining an estimate for f_L without the need for a graphical representation of the exact low-frequency transfer function of the voltage gain. This method also provides useful insight into the contribution of each DC-blocking and bypass capacitor to this cut-off frequency.

The short-circuit time constants method is applied according to the following steps:

1. **The estimate for f_L is obtained by the following expression:**

$$f_L \simeq \frac{1}{2\pi} \sum_{k=1}^{N} \frac{1}{\tau_k} \qquad (6.118)$$

where N is the number of time constants given by the number of DC blocking capacitors plus the number of by-pass capacitors which are present in the low-frequency equivalent circuit of the amplifier.

For the amplifier of figure 6.42 there are two DC blocking capacitors and one by-pass capacitor. Hence, $N = 3$.

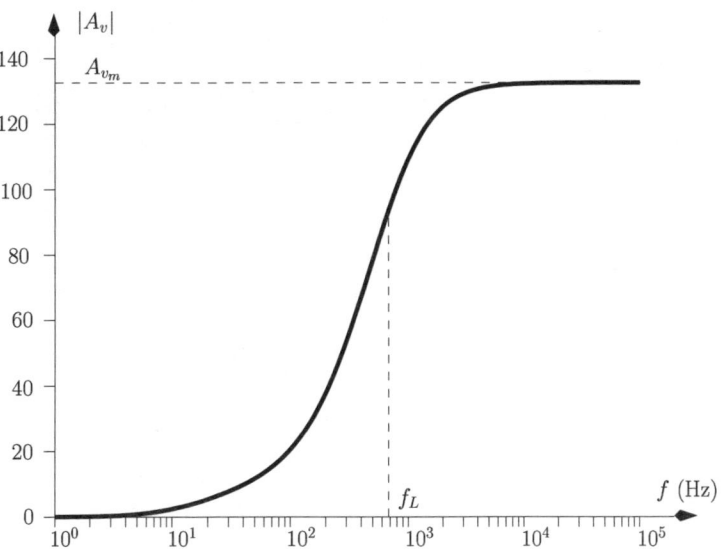

Figure 6.44: *Voltage gain for the low-frequency range.*

2. **Each time constant, τ_k, is calculated as the product of each DC blocking or bypass capacitor and the resistance, R_{eq_k} 'seen' by the capacitor when all the remaining capacitors are substituted by short-circuits and the input signal source is replaced by its output resistance. Hence, an ideal voltage source is replaced by a short-circuit and an ideal current source is replaced by an open-circuit.**

The first time constant of the circuit of figure 6.42, τ_1, is that associated with C_B. Figure 6.45 a) shows the equivalent circuit to determine R_{eq_1}. It can be seen that all other capacitances and the voltage source v_s are replaced by short-circuits and a test voltage source, v_t is located in the place of C_B in order to determine the resistance seen by this capacitor as follows:

$$R_{eq_1} = \frac{v_t}{i_t} \tag{6.119}$$

From this figure it can be seen that v_t is applied to the series connection of R_s with the parallel combination of R_B with r_π. Hence

$$\begin{aligned} R_{eq_1} &= R_s + (R_B || r_\pi) \\ &= R_s + \frac{R_B\, r_\pi}{R_B + r_\pi} \end{aligned} \tag{6.120}$$

Therefore τ_1 is given by:

$$\begin{aligned} \tau_1 &= C_B \left(R_s + \frac{R_B\, r_\pi}{R_B + r_\pi} \right) \\ &= 4.3\,\text{ms} \end{aligned} \tag{6.121}$$

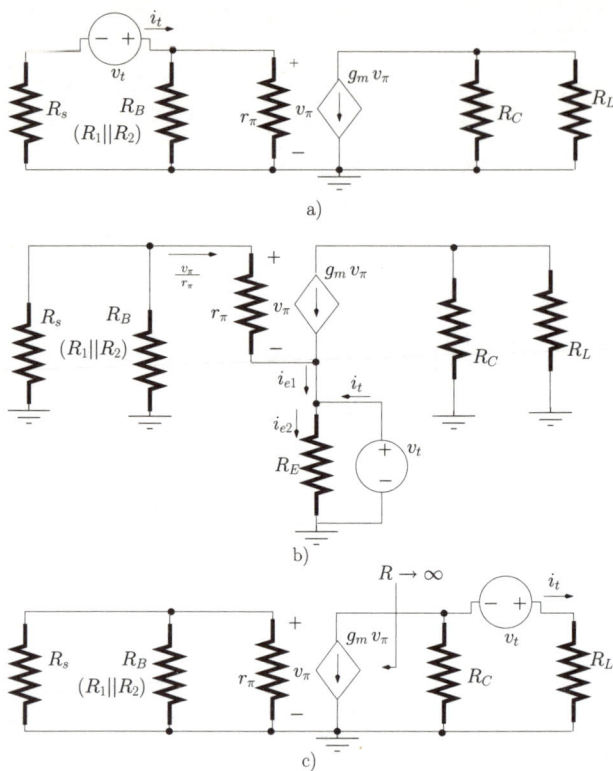

Figure 6.45: *Application of the short-circuit time constants method. a) Calculation of the resistance seen by C_B. b) Calculation of the resistance seen by C_E. c) Calculation of the resistance seen by C_L.*

The time constant associated with C_E is τ_2. Figure 6.45 b) shows the equivalent circuit to determine R_{eq_2}. It can be seen that a test voltage source, v_t, is located in the place of C_E in order to determine the resistance seen by this capacitor as follows:

$$R_{eq_2} = \frac{v_t}{i_t} \tag{6.122}$$

We can write the following equation:

$$i_t = i_{e2} - i_{e1} \tag{6.123}$$

with

$$i_{e2} = \frac{v_t}{R_E} \tag{6.124}$$

and

$$i_{e1} = g_m v_\pi + \frac{v_\pi}{r_\pi} \tag{6.125}$$

Also we can write

$$v_t = -v_\pi - \frac{v_\pi}{r_\pi} \times (R_s \| R_B) \tag{6.126}$$

or

$$v_\pi = \frac{v_t}{1 + \frac{(R_s \| R_B)}{r_\pi}} \tag{6.127}$$

Using eqns 6.123–6.127 we can write i_t as follows:

$$\begin{aligned} i_t &= \frac{v_t}{R_E} - g_m v_\pi - \frac{v_\pi}{r_\pi} \\ &= \frac{v_t}{R_E} - \frac{g_m r_\pi + 1}{r_\pi} v_\pi \\ &= \frac{v_t}{R_E} + \frac{\beta + 1}{r_\pi} \times \frac{v_t}{1 + \frac{(R_s \| R_B)}{r_\pi}} \end{aligned} \tag{6.128}$$

From the last eqn we can write $v_t/i_t = R_{eq_2}$ as follows

$$R_{eq_2} = R_E \| \left[\frac{r_\pi + (R_s \| R_B)}{\beta + 1} \right] \tag{6.129}$$

Therefore, τ_2 is given by:

$$\begin{aligned} \tau_2 &= C_E \, R_{eq_2} \\ &= 0.2 \, \text{ms} \end{aligned} \tag{6.130}$$

τ_3 is the time constant associated with C_L. Figure 6.45 c) shows the equivalent circuit to determine R_{eq_3}. We see that $v_\pi = 0$. Hence, the output impedance of the voltage-controlled current source tends to infinity and v_t is effectively applied to the series of R_L with R_C;

$$R_{eq_3} = R_L + R_C \tag{6.131}$$

and

$$\begin{aligned} \tau_3 &= C_L \, R_{eq_3} \\ &= 20 \, \text{ms} \end{aligned} \tag{6.132}$$

This analysis clearly indicates that the time constant which dominates and determines f_L is that associated with C_E. In many practical circuits, the resistance seen by this capacitor tends to be relatively small when compared to R_{eq_1} and R_{eq_3} and so this dominates f_L.

The estimate for f_L given by this method is:

$$f_L = 726 \, \text{Hz} \tag{6.133}$$

Comparing this value with that obtained from figure 6.44 we observe that the error in the estimate of f_L is only 7%.

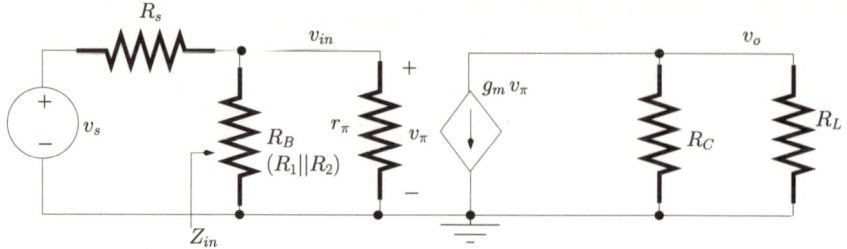

Figure 6.46: *Common-emitter equivalent circuit for the mid-frequency range.*

Mid-frequency range analysis

Figure 6.46 shows the equivalent small-signal circuit of the common-emitter amplifier at medium frequencies. The gain, $A_{vm} = v_o/v_s$, can be written as follows:

$$\begin{aligned} A_{vm} &= \frac{v_o}{v_{in}} \times \frac{v_{in}}{v_s} \\ &= -g_m \left(R_C \| R_L \right) \times \frac{Z_{in}}{Z_{in} + R_s} \end{aligned} \quad (6.134)$$

where Z_{in} is given by R_B in parallel with r_π:

$$\begin{aligned} Z_{in} &= R_B \| r_\pi \\ &= 763 \ \Omega \end{aligned}$$

Hence, $A_{vm} = -132.6$. Note that this value agrees with that obtained earlier from figure 6.44.

High-frequency analysis

Figure 6.47 shows the equivalent circuit at high frequencies. It can be shown

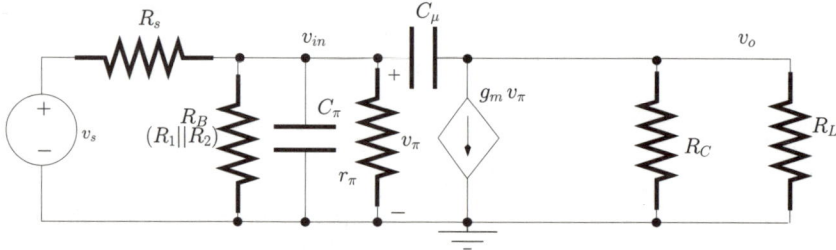

Figure 6.47: *Common-emitter equivalent circuit for the high-frequency range.*

(see problem 6.6) that the voltage transfer function, v_o/v_s, for this circuit can be written as follows:

$$A_v = \frac{Y_s \, R'_L \, (j\omega C_\mu - g_m)}{Y'_\pi (1 + j\omega R'_L C_\mu) + j\omega [C_\pi + C_\mu(1 + g_m R'_L)] - \omega^2 C_\mu C_\pi R'_L} \quad (6.135)$$

with

$$Y'_\pi = \frac{1}{R_s} + \frac{1}{R_B} + \frac{1}{r_\pi} \qquad (6.136)$$

$$Y_s = \frac{1}{R_s} \qquad (6.137)$$

and

$$R'_L = R_L \| R_C \qquad (6.138)$$

Equation 6.135 is lengthy and it is not straightforward to extract the 3-dB cut-off frequency, f_H. Figure 6.48 plots the magnitude of the gain at high-frequencies (eqn 6.135) from which f_H can be determined as 4.6 MHz.

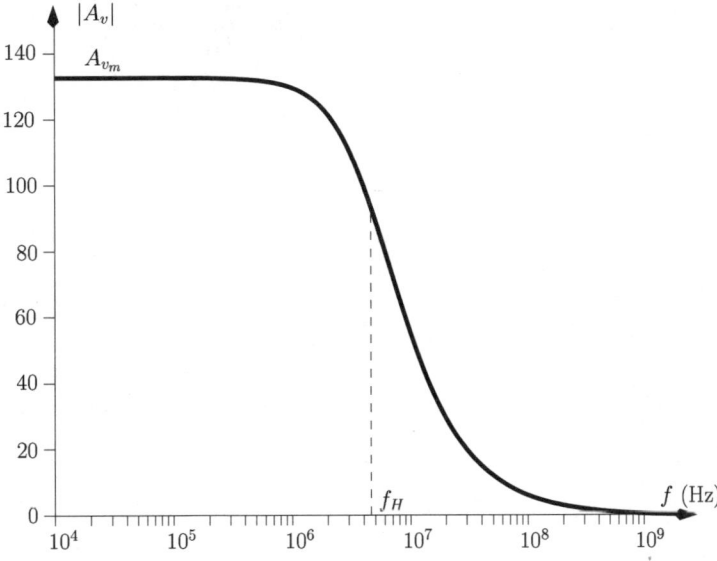

Figure 6.48: *Voltage gain for the high-frequency range.*

The use of Miller's theorem, discussed in the last chapter (section 5.2.5), provides insight into the high frequency response of the common-emitter amplifier. In fact, by applying Miller's theorem to the impedance associated with C_μ, $Z_\mu = (j\omega C_\mu)^{-1}$, in figure 6.47 we obtain the circuit of figure 6.49 a). These two circuits are equivalent in terms of input impedance and voltage gain. For this circuit we can write the following eqns:

$$A_{v_{BC}} = \frac{v_o}{v_{in}}$$

$$Z_1 = \frac{Z_\mu}{1 - A_{v_{BC}}}$$

$$= \frac{1}{j\omega C_\mu (1 - A_{v_{BC}})} \qquad (6.139)$$

$$Z_2 = \frac{Z_\mu A_{v_{BC}}}{A_{v_{BC}} - 1}$$

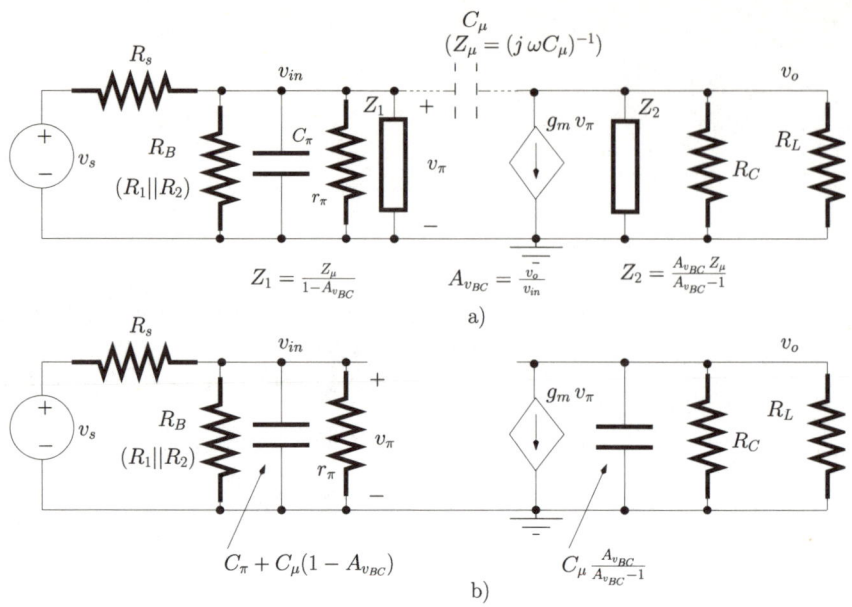

Figure 6.49: *Application of Miller's theorem to C_μ.*

$$= \frac{1}{j\omega C_\mu \frac{A_{v_{BC}}-1}{A_{v_{BC}}}} \qquad (6.140)$$

$A_{v_{BC}}$ is the voltage gain between the base and the collector and can be calculated as follows:

$$A_{v_{BC}} = -\frac{g_m - j\omega C_\mu}{\frac{1}{R'_L} + j\omega C_\mu} \qquad (6.141)$$

with $R'_L = R_C || R_L$. At the 3 dB cut-off frequency, f_H we can write:

$$g_m > |j\omega C_\mu|_{\omega=2\pi f_H} \quad \text{and} \quad \frac{1}{R'_L} > |j\omega C_\mu|_{\omega=2\pi f_H}$$

$$\left(4 \times 10^{-2} > 5.8 \times 10^{-5}\right) \quad \text{and} \quad \left(2.7 \times 10^{-4} > 5.8 \times 10^{-5}\right)$$

Therefore, we can approximate $A_{v_{BC}}$ as follows:

$$\begin{aligned} A_{v_{BC}} &\simeq -g_m R'_L \qquad (6.142) \\ &= -132.6 \end{aligned}$$

Equation 6.139 indicates that Z_1 is the impedance associated with an equivalent capacitance with value $C_\mu (1 - A_{v_{BC}})$. Since this capacitance is in parallel with C_π we can add them in order to obtain an input equivalent capacitance of $C_\pi + C_\mu (1 - A_{v_{BC}})$, as illustrated in figure 6.49 b). Similarly, Z_2 is the impedance associated with a capacitance with value $C_\mu \frac{A_{v_{BC}}-1}{A_{v_{BC}}}$ which is also represented in figure 6.49 b). Note that, according to Miller's theorem, the capacitance C_μ is reflected to the input of the amplifier after being multiplied by

6. Basic electronic amplifier building blocks

a factor of 133.6! As is shown next, this has a dramatic impact on the amplifier bandwidth.

Open-circuit time constants method

In a similar fashion to the low-frequency analysis, the open-circuit time constants method (see [4] for a detailed study) is a straightforward means of obtaining an estimate for f_H.

The open-circuit time constants method is applied according to the following steps:

1. **The estimate for f_H is obtained by the following expression:**

$$f_H \simeq \frac{1}{2\pi \sum_{n=1}^{N} \tau_n} \quad (6.143)$$

   where N is the number of time constants produced by the number of the capacitors that are present in the high-frequency equivalent circuit of the amplifier.

   For the circuit of figure 6.49 b) we identify two (equivalent) capacitances, hence, $N = 2$.

2. **Each time constant, τ_n, is given by the product of each capacitor by the resistance, R_{eq_n}, seen by that capacitor when all the remaining capacitors are substituted by open-circuits and the input signal source is replaced by its output resistance. Hence, an ideal voltage source is replaced by a short-circuit and an ideal current source is replaced by an open-circuit.**

   The first time constant of the circuit of figure 6.49, τ_1, is that associated with $C_\pi + C_\mu (1 + g_m R'_L)$. It can be shown (see problem 6.7) that the resistance R_{eq_1} seen by this capacitance is:

$$\begin{aligned} R_{eq_1} &= R_s \| R_B \| r_\pi \\ &= 88 \ \Omega \end{aligned} \quad (6.144)$$

   Hence τ_1 can be written as:

$$\begin{aligned} \tau_1 &= [C_\pi + C_\mu (1 + g_m R'_L)] \ R_{eq_1} \\ &= 27.7 \text{ ns} \end{aligned} \quad (6.145)$$

   The second time constant of the circuit of figure 6.49, τ_2, is that associated with $C_\mu (1 + g_m R'_L)/(g_m R'_L) \simeq C_\mu$. It can be shown (see problem 6.7) that the resistance R_{eq_2} seen by this capacitance is:

$$\begin{aligned} R_{eq_2} &= R'_L \\ &= 3.75 \text{ k}\Omega \end{aligned} \quad (6.146)$$

Hence τ_2 can be written as:

$$\begin{aligned}\tau_2 &= C_\mu R'_L \quad (6.147)\\ &= 7.5 \text{ ns}\end{aligned}$$

f_H can be estimated as follows:

$$\begin{aligned}f_H &= \frac{1}{2\pi(\tau_1+\tau_2)} \quad (6.148)\\ &= 4.5 \text{ MHz}\end{aligned}$$

It can be seen that the cut-off frequency provided by the exact analysis is virtually the same as that provided by applying Miller's theorem together with the open-circuit time constants method. Also, from this discussion it is clear that $\tau_1 > \tau_2$ and that τ_1 dominates the high frequency response, that is:

$$f_H \simeq \frac{1}{2\pi[C_\pi + C_\mu(1+g_m R'_L)]R_{eq_1}} \quad (6.149)$$

This equation indicates one of the most important trends for this type of amplifier: the larger the mid-frequency voltage gain, $-g_m R'_L$, the smaller the bandwidth of the amplifier becomes; a direct result of the Miller effect.

Example 6.4.3 The amplifier of figure 6.50 a) is known as the common-base amplifier. Determine:

1. The voltage gain in the mid-frequency range;
2. The current gain in the mid-frequency range;
3. The input impedance in the mid-frequency range;
4. An estimate for f_H.

<u>Solution</u>:

1. In order to determine the hybrid-π parameters, namely r_π and g_m, we need to determine the bias collector current of the BJT. Fortunately, the equivalent circuit for the DC calculations is the same as that presented in figure 6.41 b). Therefore $g_m = 40$ mA/V and $r_\pi = 5$ kΩ. Since the DC blocking capacitors are short-circuits in the mid-frequency range and since, for AC signal analysis purposes, V_{CC} is modelled by a short-circuit to ground, the equivalent circuit for AC analysis is as shown in figure 6.51 a). Replacing the transistor, in figure 6.51 a) by its hybrid-π equivalent circuit we obtain the small-signal equivalent circuit for the common-base amplifier shown in figure 6.51 b).

6. Basic electronic amplifier building blocks

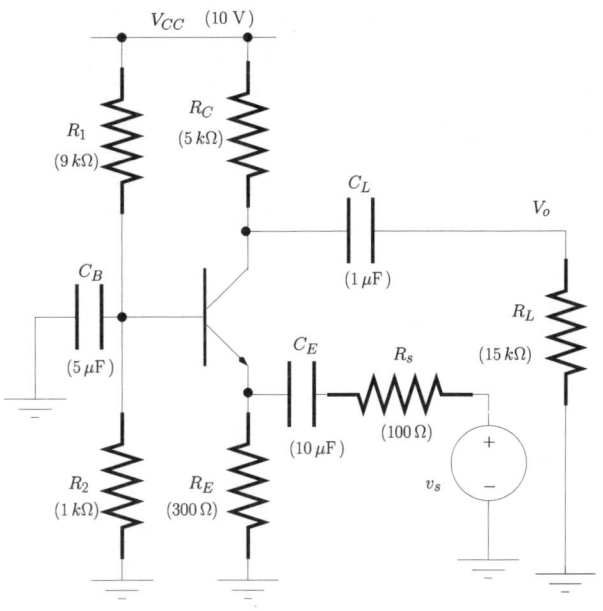

Figure 6.50: *Common-base amplifier.*

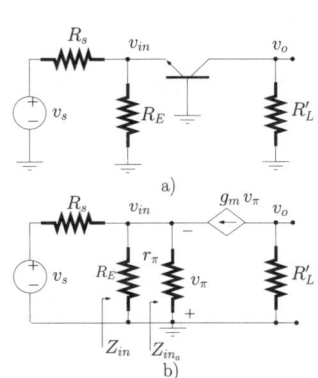

Figure 6.51: *Common-base amplifier.* a) *Equivalent circuit for AC analysis.* b) *Equivalent circuit for small-signal AC analysis.*

The voltage gain $A_v = v_o/v_s$ is given by

$$\begin{aligned}
A_v &= \frac{v_o}{v_{in}} \times \frac{v_{in}}{v_s} \\
&= \frac{-g_m\, v_\pi\, R'_L}{-v_\pi} \times \frac{Z_{in}}{Z_{in} + R_s} \\
&= g_m\, R'_L \times \frac{Z_{in}}{Z_{in} + R_s}
\end{aligned} \qquad (6.150)$$

where

$$\begin{aligned}
R'_L &= R_L || R_C \qquad (6.151) \\
&= 3.75 \text{ k}\Omega
\end{aligned}$$

and where Z_{in} is the input impedance of the amplifier.

From figure 6.51 b) it is clear that Z_{in} results from the parallel connection of R_E with Z_{in_a}. Z_{in_a} is the impedance looking into the emitter of the BJT with the base connected to ground. Figure 6.52 shows the equivalent circuit for the calculation of Z_{in_a}. From this circuit we can write:

$$\begin{aligned}
Z_{in_a} &= \frac{v_t}{i_t} \\
&= \frac{-v_\pi}{-\frac{v_\pi}{r_\pi} - g_m\, r_\pi} \\
&= \frac{r_\pi}{1 + g_m\, r_\pi}
\end{aligned}$$

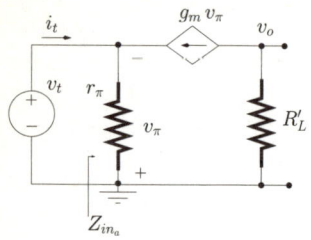

Figure 6.52: *Equivalent circuit for the calculation of Z_{in_a}.*

$$\simeq \frac{1}{g_m} \quad \text{since } g_m\, r_\pi = \beta \gg 1 \quad (6.152)$$
$$= 25\ \Omega$$

Now eqn 6.150 gives $A_V = 28$. Note that, this gain is significantly smaller than that obtained for the common-emitter, a result of the low input impedance of the amplifier in comparison with the source output impedance.

2. Figure 6.53 shows the equivalent circuit for the calculation of the current

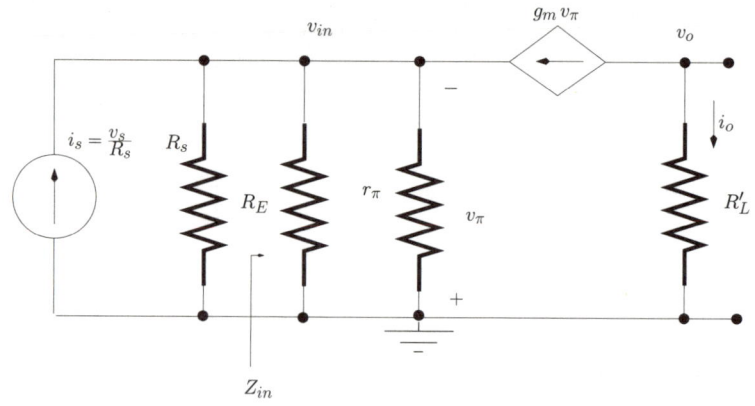

Figure 6.53: *Equivalent circuit for the calculation of the current gain A_i.*

gain $A_i = i_o/i_s$, where the voltage signal source has been replaced by its equivalent Norton model. The gain can be calculated as follows:

$$\begin{aligned}
A_i &= \frac{i_o}{v_{in}} \times \frac{v_{in}}{i_s} \\
&= \frac{-g_m\, v_\pi}{-v_\pi} \times \frac{i_s(Z_{in}\|R_s)}{i_s} \\
&= g_m\,(Z_{in}\|R_s) \qquad (6.153)\\
&= 0.75
\end{aligned}$$

For the common-base amplifier, the current gain is always less than unity. In fact, if R_s and R_E are much larger than Z_{in_a}, then we have $(R_s\|Z_{in}) \simeq Z_{in_a}$ and, under this condition we can write the current gain as:

$$\begin{aligned}
A_{i_{max}} &= g_m\, Z_{in_a} \\
&= \frac{g_m\, r_\pi}{g_m\, r_\pi + 1} \\
&= \frac{\beta}{\beta + 1} \qquad (6.154)
\end{aligned}$$

which can approach but never achieve a value of unity.

6. Basic electronic amplifier building blocks

3. The amplifier input impedance is $Z_{in} = Z_{in_a} || R_E = 23 \ \Omega$. Note that this impedance is very low.

4. Figure 6.54 shows the equivalent circuit of the common-base amplifier

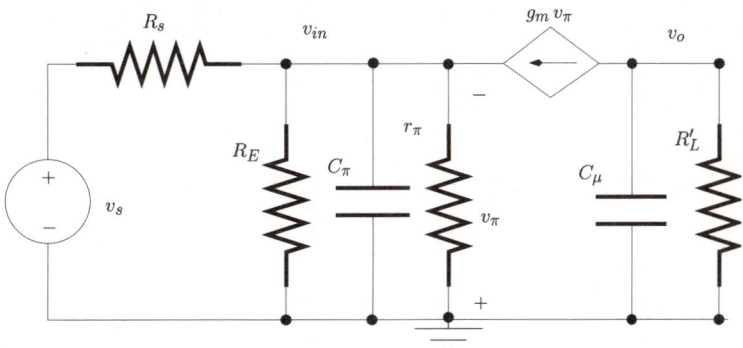

Figure 6.54: *Equivalent circuit at high frequencies.*

at high frequencies where we can identify two capacitors. We assume $r_x \simeq 0$. The first time constant is that associated with C_π. The resistance seen by this capacitor is:

$$R_{eq_1} = Z_{in} || R_s = 19 \ \Omega$$

and $\tau_1 = (Z_{in} || R_s) \, C_\pi = 0.2$ ns.

The second time constant is associated with C_μ. The resistance seen by this capacitor is:

$$R_{eq_2} = R_L' = 3.75 \text{ k}\Omega$$

and $\tau_2 = R_L' \, C_\mu = 7.5$ ns. The estimate for f_H is given by:

$$\begin{aligned} f_H &= \frac{1}{2\pi(\tau_1 + \tau_2)} \\ &= 20.6 \text{ MHz} \end{aligned} \qquad (6.155)$$

This is much greater than the 4.6 MHz value of the common-emitter amplifier. This is because the Miller effect does not operate on C_μ.

Example 6.4.4 The amplifier of figure 6.55 a) is called a common-collector (or emitter follower) amplifier. Determine;

1. The DC bias voltages and currents of the circuit;

2. The voltage gain in the mid-frequency range;

3. The input impedance in the mid-frequency range;

4. The output impedance in the mid-frequency range;

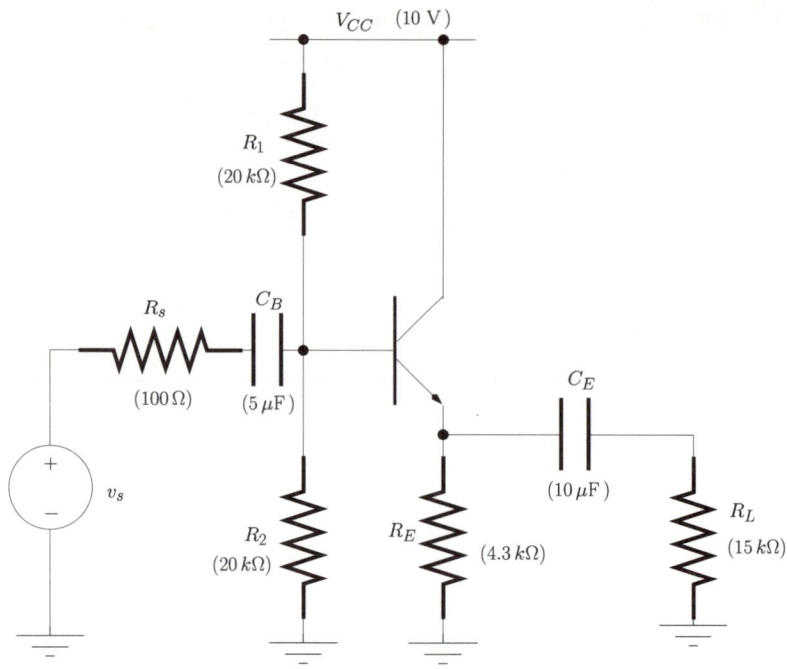

Figure 6.55: *Common-collector amplifier.*

5. The current gain in the mid-frequency range. For this calculation assume that $R_s = 40$ kΩ.

Solution:

1. Figure 6.56 shows the equivalent circuit used in the DC analysis. Assuming that $V_{BE} \simeq 0.7$ V and that $I_B \simeq 0$, we can write:

$$I_1 = \frac{V_{CC}}{R_1 + R_2} \quad (6.156)$$
$$= 0.25 \text{ mA}$$

The voltage at the base of the transistor is given by:

$$V_B = I_1 R_1 \quad (6.157)$$
$$= 5 \text{ V}$$

The voltage across R_E can be calculated as follows:

$$V_E = V_B - V_{BE} \quad (6.158)$$
$$= 4.3 \text{ V}$$

Hence the emitter current is given by:

$$I_E = \frac{V_E}{R_E} \quad (6.159)$$
$$= 1 \text{ mA}$$

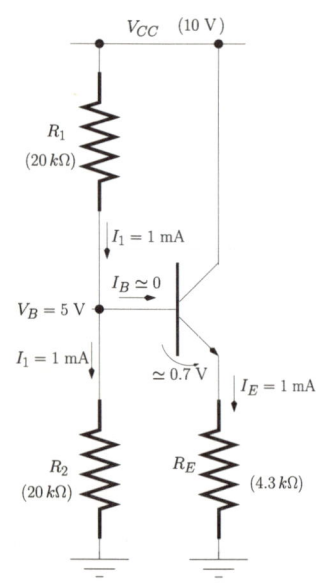

Figure 6.56: *Common-collector amplifier. Equivalent circuit for DC analysis.*

6. Basic electronic amplifier building blocks

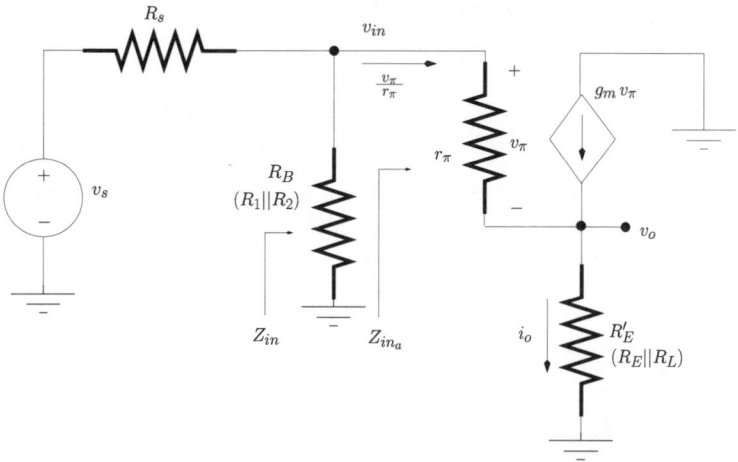

Figure 6.57: *Common-collector amplifier. Equivalent circuit for small-signal AC analysis.*

The collector current I_C is approximately the same as I_E and is equal to $= 1$ mA.

2. Figure 6.57 shows the small-signal equivalent circuit for the mid-frequency range AC analysis. The voltage gain, $A_v = v_o/V_s$ can be calculated as follows:

$$
\begin{aligned}
A_v &= \frac{v_o}{v_{in}} \times \frac{v_{in}}{v_s} \\
&= \frac{R'_E \, i_o}{R'_E \, i_o + v_\pi} \times \frac{R_B || Z_{in_a}}{R_s + (Z_{in_a} || R_B)} \\
&= \frac{R'_E \left(\frac{v_\pi}{r_\pi} + g_m v_\pi\right)}{R'_E \left(\frac{v_\pi}{r_\pi} + g_m v_\pi\right) + v_\pi} \times \frac{R_B || Z_{in_a}}{R_s + (Z_{in_a} || R_B)} \\
&= \frac{R'_E (\beta + 1)}{r_\pi + R'_E (\beta + 1)} \times \frac{R_B || Z_{in_a}}{R_s + (Z_{in_a} || R_B)} \quad (6.160)
\end{aligned}
$$

$$R_B = R_1 || R_2 \quad (6.161)$$
$$= 10 \text{ k}\Omega$$
$$R'_E = R_E || R_L \quad (6.162)$$
$$= 3.3 \text{ k}\Omega$$

Z_{in_a} is the impedance looking into the base of the amplifier as indicated in figure 6.57. Figure 6.58 shows the circuit for the calculation of Z_{in_a}:

$$
\begin{aligned}
Z_{in_a} &= \frac{v_t}{i_t} \\
&= \frac{v_\pi + R'_E \, i_o}{\frac{v_\pi}{r_\pi}}
\end{aligned}
$$

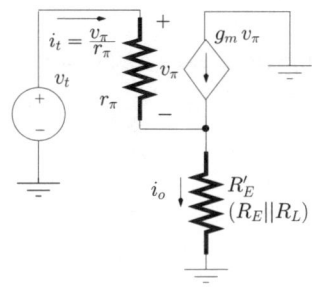

Figure 6.58: *Equivalent circuit for the calculation of Z_{in_a}.*

$$= \frac{v_\pi + R'_E\left(\frac{v_\pi}{r_\pi} + g_m v_\pi\right)}{v_\pi} r_\pi$$

$$= r_\pi + R'_E(\beta + 1) \quad (6.163)$$

$$= 676 \text{ k}\Omega$$

From eqn 6.160 the voltage gain A_v is found to be 0.95. In the common-collector amplifier the voltage gain can approach (but is always less than) one.

3. The input impedance of the amplifier, Z_{in} as indicated in figure 6.57 is:

$$Z_{in} = R_B \| Z_{in_a}$$
$$\simeq R_B \quad (6.164)$$
$$= 10 \text{ k}\Omega$$

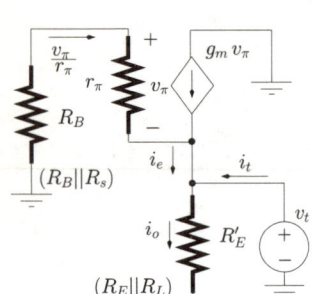

Figure 6.59: Circuit for the calculation of the output impedance.

4. The circuit of figure 6.59 shows the equivalent circuit for the calculation of the output impedance.

$$Z_o = \frac{v_t}{i_t} \quad (6.165)$$

$$v_t = -v_\pi - R'_B \frac{v_\pi}{r_\pi} \quad (6.166)$$

$$i_t = \frac{v_t}{R'_E} - i_e$$

$$= \frac{v_t}{R'_E} - \frac{v_\pi}{r_\pi} - g_m v_\pi \quad (6.167)$$

$$R'_B = R_B \| R_s \quad (6.168)$$
$$= 96.2 \, \Omega$$

Solving eqns 6.166 and 6.167 in order to obtain Z_o as given by eqn 6.165 we obtain

$$Z_o = R'_E \| \left(\frac{r_\pi + R'_B}{\beta + 1}\right) \quad (6.169)$$
$$= 25.1 \, \Omega$$

5. Figure 6.60 shows the equivalent circuit for the calculation of the current gain, A_i:

$$A_i = \frac{i_o}{i_s}$$

$$= \frac{i_o}{v_{in}} \times \frac{v_{in}}{i_s}$$

$$= \frac{g_m v_\pi + \frac{v_\pi}{r_\pi}}{\frac{Z_{in_a} v_\pi}{r_\pi}} \times (R_s \| Z_{in})$$

$$= \frac{\beta + 1}{Z_{in_a}} \times (R_s \| Z_{in}) \quad (6.170)$$

$$= 2.3$$

6. Basic electronic amplifier building blocks

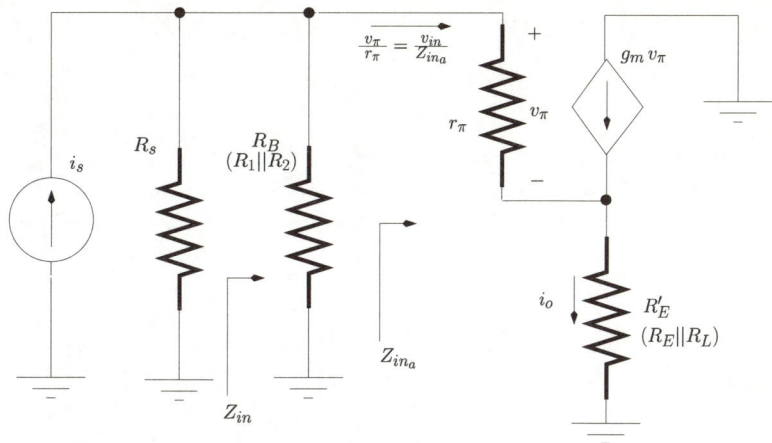

Figure 6.60: *Equivalent circuit for the calculation of the current gain.*

The common-collector amplifier is characterised by a high input impedance, a very low output impedance and a voltage gain close to unity. The current gain is, however, larger than unity. In fact, from eqn 6.170 we can observe that if $R_s||Z_{in} \simeq Z_{in_a}$ then the current gain can be as large as $\beta + 1$. This circuit is often used as a buffer or as the final stage of a circuit composed of a cascade or chain of amplifiers.

6.4.5 The differential pair amplifier

All the configurations discussed above can be implemented with IGFETs instead of BJTs. The analysis of such configurations follows the same principles discussed previously. However, the IGFET is used almost exclusively in integrated circuits where large DC blocking capacitors are difficult to fabricate because of their large physical areas. A circuit which overcomes this problem is the differential pair, a very important circuit in electronics which forms the basis of the differential inputs of op-amps, for example. The circuit configuration with IGFETs is illustrated in figure 6.61 a).

It can be seen that this configuration comprises two transistors with both sources connected to a current source I_Q. This current source biases these two devices and it is usually connected to a negative voltage source, $-V_{SS}$. The drain of each FET is connected to an identical resistor represented by R_D.

For this circuit we can observe that there are two input terminals, V_{in_1} and V_{in_2}, and two outputs represented by V_{o1} and V_{o2}. Figure 6.61 b) illustrates this circuit in **common-mode** operation. The term common-mode arises from the fact that both inputs are driven by the same voltage, V_c. Let us assume first that V_c is zero. From symmetry it is reasonable to conclude that the current I_Q is equally divided between the two branches of the differential pair and that the drain current of each transistor is equal to $I_Q/2$. Assuming that both transistors are operating in the saturation region this means that there is a bias

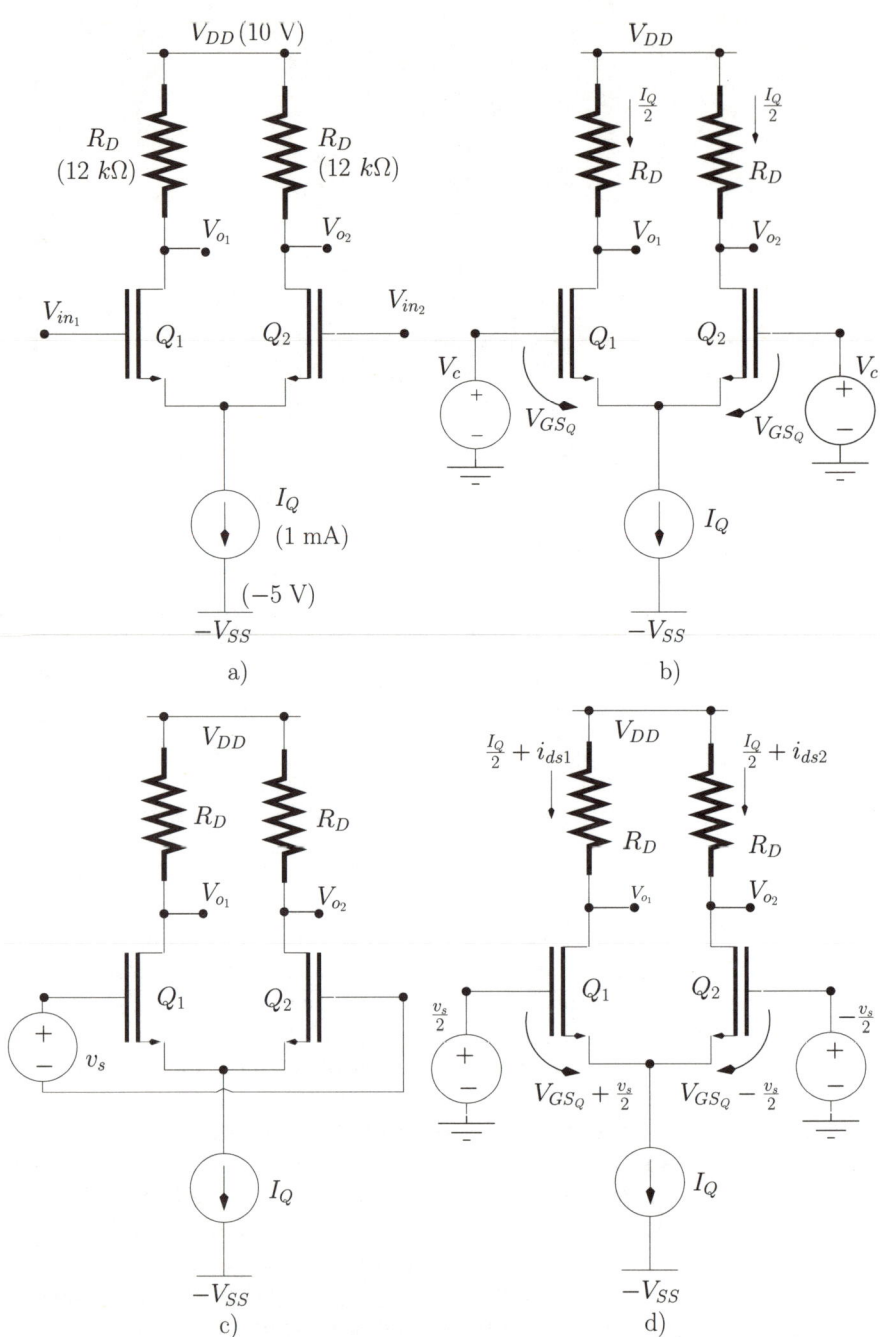

Figure 6.61: *Differential pair.   a) Implementation with IGFETs.   b) Common-mode operation.   c) Differential-mode operation. d) Equivalent differential-mode operation.*

voltage V_{GS_Q} for each transistor which satisfies the following equation:

$$I_{DS_Q} = \frac{I_Q}{2} = \frac{1}{2} k_n \frac{W}{L} \left(V_{GS_Q} - V_{Th}\right)^2 \qquad (6.171)$$

where I_{DS_Q} is the drain bias current of each transistor. Assuming $V_{Th} = 1$ V and $k_n \frac{W}{L} = 0.2$ mA/V^2, we can calculate the bias voltage V_{GS_Q} as,

$$\begin{aligned} V_{GS_Q} &= V_{Th} + \sqrt{\frac{I_Q}{k_n \frac{W}{L}}} \qquad (6.172) \\ &= 2.6 \text{ V} \end{aligned}$$

For each transistor the drain voltage is:

$$\begin{aligned} V_{o1} = V_{o2} &= V_{DD} - R_D \frac{I_Q}{2} \qquad (6.173) \\ &= 4 \text{ V} \end{aligned}$$

If now we increase V_c to 0.5 V, the symmetry of operation of the circuit is maintained. Hence, the current I_Q is still equally divided between the two branches of the circuit which means that the gate-source voltage of both transistors is equal to V_{GS_Q} (as given by eqn 6.172) and the output voltages V_{o1} and V_{o2} maintain their values (as given by eqn 6.173). This means that the output voltages of the differential pair do not vary with common-mode voltages, that is, **the differential pair does not respond to common-mode input signals**.

Let us consider now the situation described in figure 6.61 c) where an input voltage signal is applied across the two input terminals. This situation is equivalent, from an electrical point-of-view, to that shown in figure 6.61 d) where the voltage signal v_s is replaced by two signals sources with symmetrical voltage values: $\pm v_s/2$. In this situation the differential pair is said to be operating in the **differential mode** i.e:

$$V_{gs1} = V_{GS_Q} + \frac{v_s}{2} \qquad (6.174)$$

$$V_{gs2} = V_{GS_Q} - \frac{v_s}{2} \qquad (6.175)$$

This alters the balance of currents in the two branches of the circuit. The drain current of Q_1 increases while the current of Q_2 decreases by the same amount. The currents I_{ds1} and I_{ds2} can be quantified as follows:

$$I_{ds1} = \frac{1}{2} k_n \frac{W}{L} \left(V_{GS_Q} + \frac{v_s}{2} - V_{Th}\right)^2 \qquad (6.176)$$

$$I_{ds2} = \frac{1}{2} k_n \frac{W}{L} \left(V_{GS_Q} - \frac{v_s}{2} - V_{Th}\right)^2 \qquad (6.177)$$

Subtracting the square root of these two drain currents we get:

$$\sqrt{I_{ds1}} - \sqrt{I_{ds2}} = \sqrt{\frac{1}{2} k_n \frac{W}{L}} v_s \qquad (6.178)$$

but the sum of the two drain currents is equal to I_Q:

$$I_{ds_1} + I_{ds_2} = I_Q \tag{6.179}$$

Solving eqns 6.178 and 6.179 in order to obtain I_{ds1} and I_{ds2} it can be shown (see problem 6.12) that:

$$I_{ds1} = \frac{I_Q}{2} + \sqrt{I_Q\, k_n \frac{W}{L}}\, \frac{v_s}{2} \sqrt{1 - \frac{\frac{v_s^2}{4} k_n \frac{W}{L}}{I_Q}} \tag{6.180}$$

$$I_{ds2} = \frac{I_Q}{2} - \sqrt{I_Q\, k_n \frac{W}{L}}\, \frac{v_s}{2} \sqrt{1 - \frac{\frac{v_s^2}{4} k_n \frac{W}{L}}{I_Q}} \tag{6.181}$$

From eqn 6.171 we find that;

$$k_n \frac{W}{L} = \frac{I_Q}{\left(V_{GS_Q} - V_{Th}\right)^2} \tag{6.182}$$

Using this result in eqns 6.180 and 6.181, we obtain:

$$I_{ds1} = \frac{I_Q}{2} + \frac{I_Q}{V_{GS_Q} - V_{Th}} \frac{v_s}{2} \sqrt{1 - \left(\frac{v_s/2}{V_{GS_Q} - V_{Th}}\right)^2} \tag{6.183}$$

$$I_{ds2} = \frac{I_Q}{2} - \frac{I_Q}{V_{GS_Q} - V_{Th}} \frac{v_s}{2} \sqrt{1 - \left(\frac{v_s/2}{V_{GS_Q} - V_{Th}}\right)^2} \tag{6.184}$$

Figure 6.62 shows the variation of the drain current of each transistor, normalised to I_Q, with the variation of v_s normalised to $V_{GS_Q} - V_{Th}$. From this figure we observe that for $v_s = 0$ V, I_Q is divided equally between I_{ds1} and

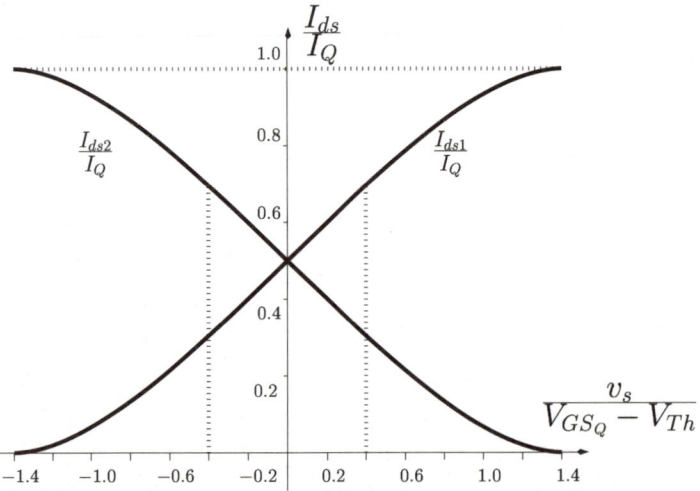

Figure 6.62: *Large signal operation of the differential pair for differential input voltage v_s.*

6. Basic electronic amplifier building blocks

I_{ds2} as mentioned previously. Also, for $|v_s| \simeq 1.4\,(V_{GS_Q} - V_{Th})$ the current I_Q can be fully switched to either of the branches of the differential pair. For this situation one of the transistors conducts I_Q while the other is cut-off. From this figure we also observe that for values of $|v_s| \leq 0.4\,(V_{GS_Q} - V_{Th})$ the operation of the differential pair is approximately linear. This conclusion can also be inferred from eqns 6.183 and 6.184 where, for $|v_s| \ll 2\,(V_{GS_Q} - V_{Th})$, they can be approximated as follows:

$$I_{ds1} = \underbrace{\frac{I_Q}{2}}_{\text{Bias}} + \underbrace{\frac{I_Q}{V_{GS_Q} - V_{Th}}\frac{v_s}{2}}_{\text{AC signal}} \qquad (6.185)$$

$$I_{ds2} = \underbrace{\frac{I_Q}{2}}_{\text{Bias}} - \underbrace{\frac{I_Q}{V_{GS_Q} - V_{Th}}\frac{v_s}{2}}_{\text{AC signal}} \qquad (6.186)$$

From the definition of the FET transconductance (see eqn 6.96) we can write the last two equations as:

$$I_{ds1} = \frac{I_Q}{2} + i_{ds1} \qquad (6.187)$$

$$I_{ds2} = \frac{I_Q}{2} + i_{ds2} \qquad (6.188)$$

with

$$i_{ds1} = +g_m\frac{v_s}{2} \qquad (6.189)$$

$$i_{ds2} = -g_m\frac{v_s}{2} \qquad (6.190)$$

The output voltages V_{o1} and V_{o2} can be determined to be:

$$\begin{aligned} V_{o1} &= V_{DD} - R_D\, I_{ds1} \\ &= \underbrace{V_{DD} - R_D\frac{I_Q}{2}}_{\text{Bias}} - \underbrace{g_m\, R_D\frac{v_s}{2}}_{\text{AC signal}} \end{aligned} \qquad (6.191)$$

$$\begin{aligned} V_{o2} &= V_{DD} - R_D\, I_{ds2} \\ &= \underbrace{V_{DD} - R_D\frac{I_Q}{2}}_{\text{Bias}} + \underbrace{g_m\, R_D\frac{v_s}{2}}_{\text{AC signal}} \end{aligned} \qquad (6.192)$$

The differential voltage gain can be defined as follows:

$$\begin{aligned} A_{vd} &= \frac{v_{o1} - v_{o2}}{v_s} \\ &= -g_m\, R_D \\ &= -7.5 \end{aligned} \qquad (6.193)$$

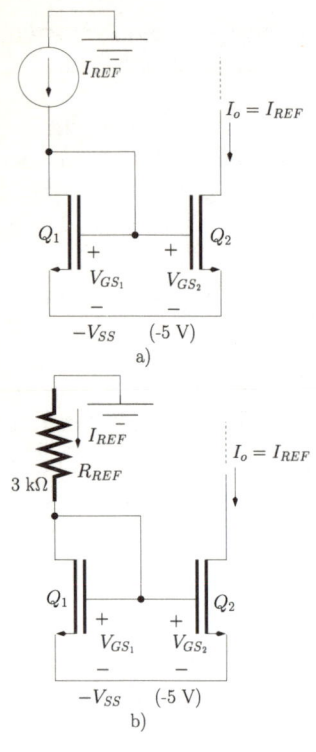

Figure 6.63: *Current mirrors.*

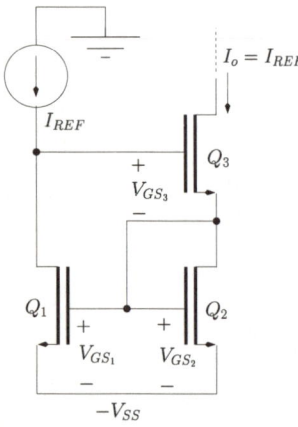

Figure 6.64: *Improved current mirror.*

It should be noted that the analysis presented above does not take into account the channel width modulation effect, that is we assume that r_o is infinite. If this effect is accounted for then the differential gain is given by:

$$A_{vd} = -g_m \left(R_D \| r_o\right) \quad (6.194)$$

Finally, it is important to note that the input impedance of the differential pair is very high and can be approximated to an open-circuit whilst the output impedance at each of the output terminals is $R_D \| r_o$.

The current mirror

In integrated circuits the current source which biases the differential pair is usually implemented with a circuit called the 'current mirror'. Figure 6.63 a) shows a basic current mirror constructed using a reference current source I_{REF} and two IGFETs, Q_1 and Q_2 which are assumed to be identical. The operation of this circuit is quite straightforward; Q_1, with its drain and gate connected together, conducts the current I_{REF} and a voltage $V_{DS_1} = V_{GS_1}$ is established. Note that Q_1 is operating in the saturation region. Since V_{GS_1} is equal to V_{GS_2}, Q_2 will conduct I_{REF} as long as it also operates in its saturation region.

Figure 6.63 b) shows a current mirror where the reference current is set by the resistance, R_{REF}, together with Q_1. For both transistors we assume $V_{Th} = 1$ V and $k_n \frac{W}{L} = 2$ mA/V². For this circuit we can write:

$$I_{REF} R_{REF} + V_{GS_1} - V_{SS} = 0 \quad (6.195)$$

$$\frac{1}{2} k_n \frac{W}{L} (V_{GS_1} - V_{Th})^2 = I_{REF} \quad (6.196)$$

Solving these two eqns we get the two following solutions for I_{REF}

$$I_{REF} = \frac{1}{k_n \frac{W}{L} R_{REF}^2} + \frac{V_{SS} - V_{Th}}{R_{REF}}$$
$$\pm \frac{\sqrt{1 + (V_{SS} - V_{Th}) 2 k_n \frac{W}{L} R_{REF}}}{k_n \frac{W}{L} R_{REF}^2} \quad (6.197)$$

that is

$$I_{REF} = 1.6 \text{ mA} \quad \text{or} \quad 1.0 \text{ mA}$$

Only one of the above solutions is valid. $I_{REF} = 1.6$ mA is not valid since, from eqn 6.196, we obtain a corresponding $V_{GS1} = 0.1$ V which is below the threshold voltage. On the other hand, for $I_{REF} = 1.0$ mA eqn 6.196 gives us a V_{GS1} equal to 1.74 volts, clearly greater than V_{Th}.

Each of the current mirrors shown in figure 6.63 has an output resistance equal to r_o of Q_2. This can be improved by using the three transistor current mirror configuration shown in figure 6.64 where all transistors are identical. In this circuit Q_2 and Q_3 have the same drain current which is equal to I_{REF} since the three transistors have equal gate-source voltages $V_{GS_1} = V_{GS_2} = V_{GS_3}$.

In order to have Q_3 operating in its saturation region, V_{DS_3} must be greater than $V_{GS_1} - V_{Th}$. Note that Q_1 and Q_2 operate in the saturation region since $V_{DS_1} = 2\,V_{GS_1}$ and $V_{DS_2} = V_{GS_1}$.

It can be shown that the output resistance of this current source is approximately equal to $g_m\, r_o^2$ (see problem 6.14).

6.5 Bibliography

1. S.M. Sze, *Physics of Semiconductor Devices*, 1981 (Wiley Interscience), 2nd edition.

2. M. Shur, *Physics of Semiconductor Devices*, 1990 (Prentice Hall).

3. G. Clayton and S. Winder, *Operational Amplifiers*, 2003 (Newnes), 5th edition.

4. P.E. Gray and C.L. Searle, *Electronic Principles*, 1969 (Wiley).

5. P.R. Gray and R.G. Meyer, *Analysis and Design of Analog Integrated Circuits*, 1985 (Wiley).

6. A.S. Sedra and K.C. Smith, *Microelectronic Circuits*, 1998 (Oxford University Press), 4th edition.

6.6 Problems

6.1 Design a voltage amplifier using one op-amp and two resistors with a gain of $+20$.

6.2 Consider the voltage amplifier of figure 6.13. Assume $R_1 = 1$ kΩ and $R_2 = 5$ kΩ. Determine the voltage gain and the input impedance.

6.3 Design a circuit such that its output voltage, v_o is related to its input voltage, v_{in}, by $v_o = -v_{in}$.

6.4 Design a circuit such that its output voltage is equal to the sum of its three input voltages, that is $v_o = v_{i1} + v_{i2} + v_{i3}$.

6.5 Consider the instrumentation amplifier of figure 6.18 with $R_1 = R_2 = R_3 = R_4 = 1$ kΩ. Determine v_o as a function of v_{sa} and v_{sb}.

6.6 Show that the voltage transfer function of the common-emitter amplifier at high frequencies can be written as in eqn 6.135.

6.7 Show that the resistances R_{eq_1} and R_{eq_2} seen by the capacitances $C_\pi + C_\mu (1 + g_m\, R'_L)$ and C_μ are as given by eqn 6.144 and by eqn 6.146, respectively.

6.8 Show that the unity-gain bandwidth, f_T, for a BJT can be expressed by eqn 6.101.

6.9 Show that the unity-gain bandwidth, f_T, for a FET can be expressed by eqn 6.102.

6.10 Consider the amplifier of figure 6.65. Determine the voltage gain, v_o/v_s, in the mid-frequency range. Also calculate the amplifier's bandwidth. Assume $V_{Th} = 1$ V, $k_n\, W/L = 4$ mA/V^2, $V_A = 80$ V, $C_{gd} = 2$ pF and $C_{gs} = 20$ pF.

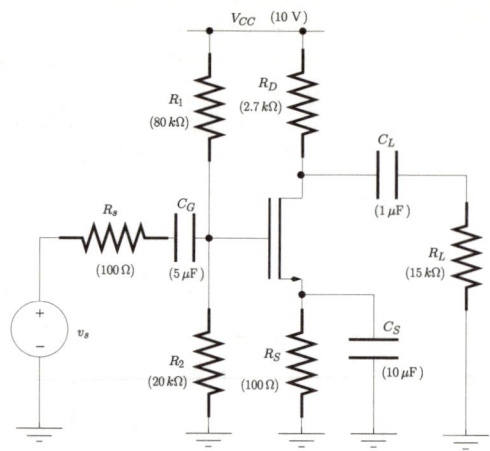

Figure 6.65: *Circuit of problem 6.10.*

6.11 Consider the amplifier of figure 6.66. Determine the current gain, i_o/i_s, in the mid-frequency range. Also calculate the amplifier's bandwidth. Assume $\beta = 200$, $V_A \to \infty$, $C_\mu = 3$ pF and $C_\pi = 18$ pF.

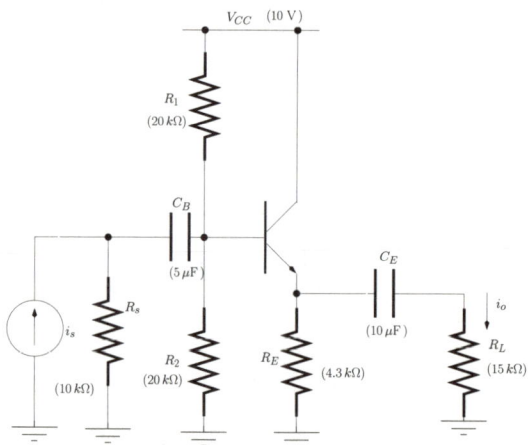

Figure 6.66: *Circuit of problem 6.11.*

6.12 Show that the large-signal drain currents of the differential pair of figure 6.61 can be written as in eqns 6.180 and 6.181.

6.13 Consider the amplifier of figure 6.67. Determine its voltage gain, v_o/v_s, in the mid-frequency range. Assume that all transistors have the same electrical characteristics; $V_{Th} = 1$ V, $k_n\,W/L = 2$ mA/V^2, and $V_A = 80$ V.

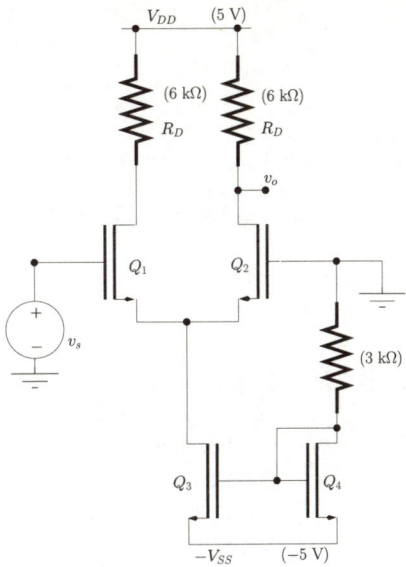

Figure 6.67: *Circuit of problem 6.13.*

6.14 Consider the current mirror of figure 6.68. Derive an expression to describe its output impedance and show that it is approximately equal to $g_m\, r_o^2$ if $R \gg r_o$.

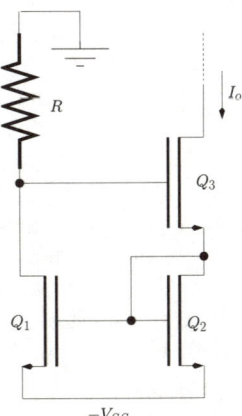

Figure 6.68: *Circuit of problem 6.14.*

7 RF circuit analysis techniques

7.1 Introduction

In this chapter we discuss the main tools used in the analysis of Radio-Frequency (RF), or microwave[1], signals and RF linear electronic circuits.

These tools are based on phasor analysis, presented in Chapter 3, but with the addition of another dimension: the space or physical length. This is because, as will be discussed in sections 7.2 and 7.3, in the microwave frequency range the corresponding wavelength (1 cm to 100 cm) is of the order of the physical size of the vast majority of the electrical components and the physical electrical connections between them. Thus, signal propagation issues must be considered in this type of analysis.

Another significant difference between RF analysis tools and the analysis methods presented previously, is that the electrical measurements and the characterisation of electronic devices using open- and short-circuit methods is difficult to achieve over the entire RF frequency range which is located between 300 MHz and 30 GHz. In fact, at these frequencies short- and open-circuits are very difficult to implement due to the existence of parasitic inductances and capacitances in a practical measurement set-up. Such a problem implies that it is difficult to characterise RF circuits in terms of voltage or current gains and input and output impedances. This problem is overcome using the *Scattering* parameters. These parameters, which are presented in section 7.4, are defined in terms of travelling waves and completely characterise the behaviour of RF and microwave electronic circuits. In addition these parameters are closely related with practical RF measurements.

Finally, in section 7.5 we present the Smith chart which is a powerful graphical method to handle the analysis, modelling and design of RF circuits. We also discuss, in detail, the problem of impedance matching using transmission lines and the uses of L-section based circuits.

7.2 Lumped versus distributed

In essence, the analysis of electronic and electrical circuits presented in the previous chapters uses the concept of the phasor, where the electrical entity considered (voltage or current) depends only on the time dimension for a certain phasor angular frequency, ω.

$$v(t) = V \, \text{Real}\left[e^{j\omega t}\right] \qquad (7.1)$$

---

[1]Traditionally, RF referred to signals with frequencies extending from 100 kHz to tens of MHz and Microwaves covered a higher frequency range extending into the tens of GHz. Today, as a result of the high frequencies used in wireless communication systems, it is common to use the two terms "RF" and "microwave" interchangeably to refer to circuits operating at frequencies beyond few hundreds of MHz.

7. RF circuit analysis techniques

$$i(t) = I\,\text{Real}\left[e^{j\omega t+\phi}\right] \tag{7.2}$$

Such a definition assumes, in an implicit way, that the amplitude of the electrical signal does not depend on the space dimension or, in other words, the amplitude of the signal varies 'simultaneously' at any physical point of the circuit as if the propagation speed, v_p, of the signal was infinite. What actually is assumed is that the wavelength, λ, of the electrical signal is so much greater than the circuit physical dimensions that this signal occurs at the same phase angle and amplitude, at any time, anywhere in the circuit. In other words the circuit physical dimension is zero. This assumption allows us to consider any circuit element, such as resistors, capacitors, and connecting cables or copper lines in circuit boards, as lumped elements.

However, for any type of propagation medium, the propagation speed for an electrical signal (voltage or current) is finite and is related to the signal frequency, f, and wavelength, λ, as follows:

$$v_p = \lambda f \tag{7.3}$$

Hence, assuming that v_p is constant, the signal wavelength decreases as its frequency increases. When the wavelength of a high frequency signal is approximately equal or less than the circuit physical length, l, then the amplitude of such a signal *also* varies significantly as it propagates along the circuit physical length.

Figure 7.1 illustrates this. Figure 7.1 a) shows the amplitude of a 28 MHz sine wave propagating along a lossless connecting cable of length 1 m. According to eqn 7.3 the wavelength is about 10 metres assuming $v_p = 2.8 \times 10^8$ m/s. This means that the wavelength is 10 times greater than the physical dimensions of the cable. From this figure it is clear that the signal amplitude is almost independent of the physical dimensions of the circuit at any time. This approximation is, in general, valid for frequency signals up to 100 MHz for which the wavelength is greater or equal than 300 cm since the physical dimensions of a typical circuit implementation rarely exceeds 20 to 30 cm in total length.

On the other hand, figure 7.1 b) shows the amplitude of a 280 MHz sine wave propagating along the same lossless connecting cable. The wavelength is about 1 metre which is equal to the cable physical dimension. From this figure it is clear that the signal amplitude depends not only on the instant of time considered but also on where the amplitude measurement is taken. Microwave signals feature wavelengths which range from 1 cm to 100 cm. For these signals the lumped concept for most circuit elements is no longer valid and there is a need to adopt distributed models which take into account the physical dimensions of the electrical elements.

An appropriate model for a signal phasor describing a voltage travelling wave, such as that illustrated in Figure 7.1 b), is given by:

$$v(t,x) = \text{Real}\left[V_A e^{j\omega t - j\beta x}\right] \tag{7.4}$$

$$v(t,x) = \text{Real}\left[e^{j\omega t} V(x)\right] \tag{7.5}$$

where

$$V(x) = V_A e^{-j\beta x} \tag{7.6}$$

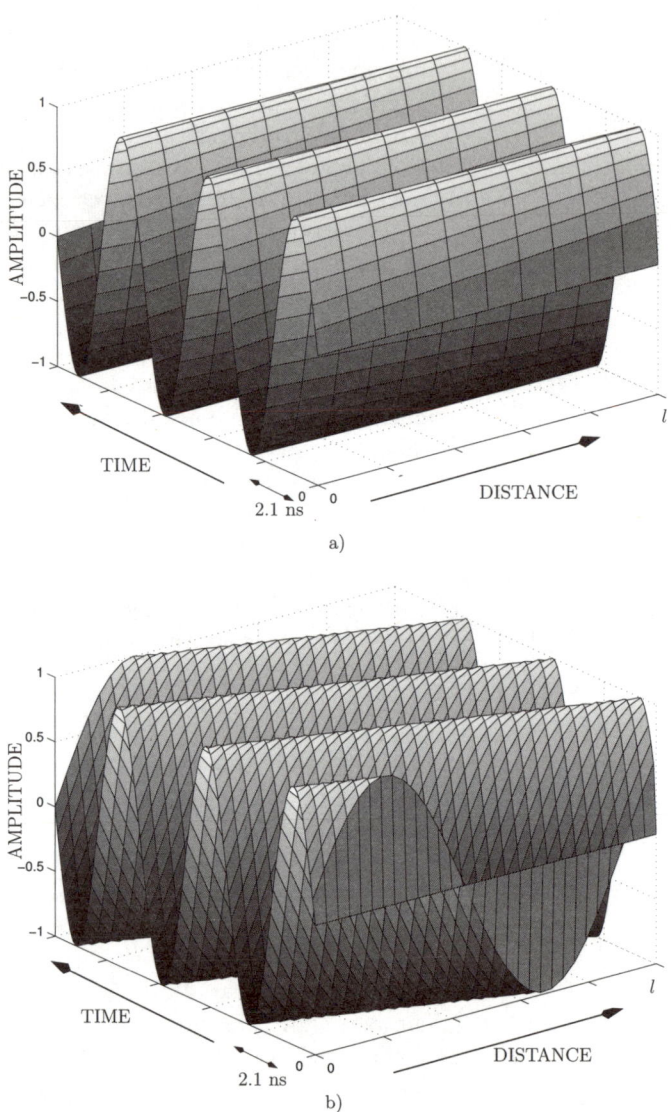

Figure 7.1: *Propagating sine wave along a lossless connecting cable with $l = 1$ m. a) 28 MHz sine wave ($\lambda = 10 \times l$). b) 280 MHz sine wave ($\lambda = l$).*

7. RF circuit analysis techniques

V_A is the amplitude, x is the physical distance and β is defined as the propagation constant. β is related to λ according to the following eqn:

$$\lambda = \frac{2\pi}{\beta} \qquad (7.7)$$

and the propagation speed can be related to β and ω as follows:

$$v_p = \frac{\omega}{2\pi}\lambda \qquad (7.8)$$

$$= \frac{\omega}{\beta} \qquad (7.9)$$

It should be noted that $V(x)$ is a static phasor (see eqn 3.57) which represents the phase dependence of the voltage signal on the physical length.

Example 7.2.1 Consider a signal with a bandwidth of 7 GHz which is processed by a filter. Give an estimate for the maximum size for this filter which allows the use of lumped models in the analysis of such a filter. The propagation speed of the signal is 2.4×10^8 m/s.

Solution: The wavelength corresponding to 7 GHz is:

$$\lambda = \frac{2.4 \times 10^8}{7 \times 10^9} = 3.4 \text{ cm}$$

Since the use of lumped models requires the signal wavelength to be about 10 times greater than the circuit size, then the maximum size must not exceed about 3.4 mm. This can be achieved using integrated circuit technology.

7.3 Electrical model for ideal transmission lines

In order to characterise an ideal (or lossless) transmission line in terms of an electrical model we consider the circuit[2] of figure 7.2 where maximum power transfer, from a source with an output impedance Z_s to a load $Z_L = Z_s^*$, is intended. The transmission line electrical characteristics are such that the impedance at its input terminals is equal to the load impedance, Z_L. In this situation the transmission line allows for maximum power transfer from Z_s to Z_L and the transfer function for the line in figure 7.2, $H(f, x)$ must impose only a time (phase) delay to the propagating voltage. This delay is a function of its length. In fact, from eqn 7.6 we can conclude that:

$$H(f, x) = \frac{V(x)}{V(x=0)}$$

$$= e^{-j\beta x} \qquad (7.10)$$

Recall that if $x(t)$ and $X(f)$ form a Fourier transform pair then we have that

$$x(t-\tau) \overset{\mathfrak{F}}{\longleftrightarrow} X(f)\, e^{-j\omega\tau} \qquad (7.11)$$

---

[2] A single line connecting two elements represents an ideal conductor with zero physical dimension.

where $x(t-\tau)$ is a delayed replica of $x(t)$ with τ representing this time delay.

Figure 7.3 shows an ideal transmission line and its equivalent circuit model. This transmission line can be sub-divided into N equal sections where each one has a length $\Delta x = l/N$. Each section has a capacity per unit of length of C and an inductance per unit of length of L. From the above the impedance which terminates the line, Z_L, is equal to the impedance seen at the input of each section. Hence, we can write:

$$Z_L = Z_1 + \frac{Z_2 Z_L}{Z_2 + Z_L} \quad (7.12)$$

$$\text{with} \quad Z_1 = j\omega L \Delta x$$

$$\text{and} \quad Z_2 = \frac{1}{j\omega C \Delta x}$$

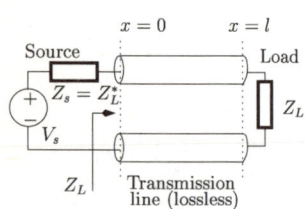

Figure 7.2: *Ideal transmission line allowing for maximum power transfer to the load Z_L.*

Solving eqn 7.12 in order to obtain Z_L we get

$$Z_L = \frac{Z_1 + \sqrt{Z_1^2 + 4Z_2 Z_1}}{2} \quad (7.13)$$

The voltage transfer function between any two adjacent sections is equal and it is given by the impedance voltage divider:

$$\frac{V(k\Delta x)}{V([k-1]\Delta x)} = \frac{\frac{Z_2 Z_L}{Z_2 + Z_L}}{\frac{Z_2 Z_L}{Z_2 + Z_L} + Z_1} \quad (7.14)$$

with $k = 1, 2, \ldots, N$. From eqn 7.12 we write:

$$\frac{Z_2 Z_L}{Z_2 + Z_L} = Z_L - Z_1 \quad (7.15)$$

Using the result of eqn 7.15 in eqn 7.14, the voltage transfer function between any two adjacent sections can be written as:

$$\frac{V(k\Delta x)}{V([k-1]\Delta x)} = \frac{Z_L - Z_1}{Z_L} \quad (7.16)$$

The voltage transfer function considered in a particular section of the line, $x = k\Delta x$, is given as:

$$\begin{aligned}\frac{V(x = k\Delta x)}{V(x = 0)} &= \frac{V(x = k\Delta x)}{V(x = [k-1]\Delta x)} \times \frac{V(x = [k-1]\Delta x)}{V(x = [k-2]\Delta x)} \times \cdots \\ &\quad \times \frac{V(x = 2\Delta x)}{V(x = \Delta x)} \times \frac{V(x = \Delta x)}{V(x = 0)} \\ &= \left(\frac{V(k\Delta x)}{V([k-1]\Delta x)}\right)^k \end{aligned} \quad (7.17)$$

We can write eqn 7.16 as follows:

$$\frac{V(k\Delta x)}{V([k-1]\Delta x)} = \frac{\sqrt{\frac{L}{C} - \omega^2 \frac{L^2 \Delta x^2}{4}} - j\omega \frac{L\Delta x}{2}}{\sqrt{\frac{L}{C} - \omega^2 \frac{L^2 \Delta x^2}{4}} + j\omega \frac{L\Delta x}{2}}$$

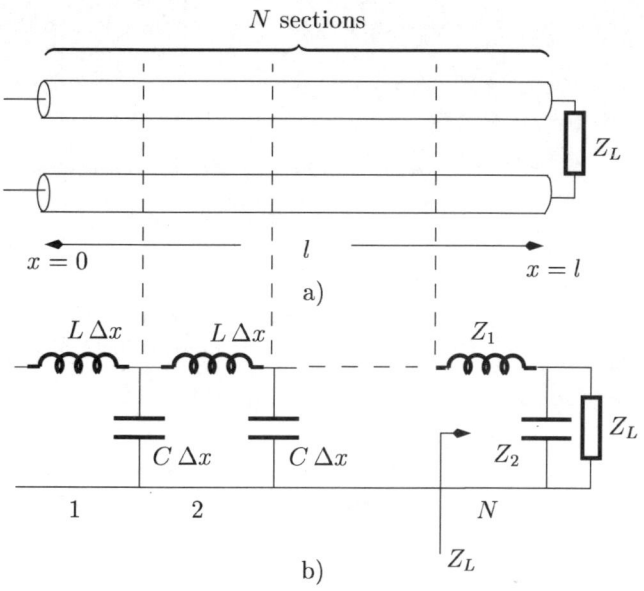

Figure 7.3: *a) Representation of a transmission line. b) Equivalent model for an ideal transmission line.*

$$= \frac{\sqrt{\left(\frac{L}{C} - \frac{\omega^2 L^2 \Delta x^2}{2}\right)^2 + \omega^2 L^2 \Delta x^2 \left(\frac{L}{C} - \frac{\omega^2 L^2 \Delta x^2}{4}\right)}}{\frac{L}{C}}$$

$$\times \quad \exp\left[-j\tan^{-1}\left(\frac{\omega L \Delta x \sqrt{\frac{L}{C} - \frac{\omega^2 L^2 \Delta x^2}{4}}}{\frac{L}{C} - \frac{\omega^2 L^2 \Delta x^2}{2}}\right)\right] \quad (7.18)$$

If we increase the number of sections, $N \to \infty$, and decrease the length of each section, $\Delta x \to 0$, in such a way that the product $l = \Delta x\, N$ is kept constant, then we have:

$$\lim_{\substack{k \to \infty \\ \Delta x \to 0}} k\, \Delta x \quad = \quad x \quad (7.19)$$

where x is now a continuous variable representing the physical length. Also, we can expand the arc-tangent function in a series as follows:

$$\tan^{-1}(x) \quad = \quad x - \frac{1}{2}x^3 - 24x^5 + \ldots \quad (7.20)$$

Using the result of eqns 7.19 and 7.20 it can be shown (see problem 7.1) that $H(f, x)$ can be written as:

$$\lim_{\substack{k \to \infty \\ \Delta x \to 0}} H(f, k\,\Delta x) \quad = \quad 1 \times \exp\left(-j\omega\sqrt{LC}\,x\right) \quad (7.21)$$

or

$$H(f, x) \quad = \quad e^{-j\omega\sqrt{LC}\,x} \quad (7.22)$$

and

$$Z_L = \sqrt{\frac{L}{C}} \qquad (7.23)$$

Comparing eqn 7.10 with eqn 7.22 we observe that the propagation constant, β, is equal to

$$\beta = \omega\sqrt{LC} \qquad (7.24)$$

From eqn 7.22 we can relate the voltage at any point of the transmission line, x, with the input voltage, $V(0)$, according to the expression below:

$$V(x) = e^{-j\omega\sqrt{LC}\,x}\, V(0) \qquad (7.25)$$

Similarly, it can be shown (see problem 7.2) that we can relate the current at any point of the transmission line, x, with the input voltage according to:

$$I(x) = e^{-j\omega\sqrt{LC}\,x}\, \frac{V(0)}{Z_o} \qquad (7.26)$$

where

$$Z_o = \sqrt{\frac{L}{C}} \qquad (7.27)$$

$Z_o = \sqrt{L/C}$ is termed the 'characteristic impedance' of the transmission line. It should be noted that Z_o is real and relates only the amplitude of the voltage propagating wave with the amplitude of the current propagating wave. Therefore, Z_o does not represent any dissipative effect along the transmission line!

The eqns derived above, for the voltage and current travelling waves, were calculated under the assumption that the transmission line was terminated by a load impedance Z_L equal to the characteristic impedance, Z_o. Under this condition there is maximum power transfer from the transmission line to the load and, therefore, the propagating signal (voltage and current) is totally absorbed by this load. In this situation the transmission line is said to be matched to the load.

Example 7.3.1 Figure 7.4 a) illustrates the transmission of a square voltage pulse, shown in figure 7.4 b), through a 10 metre transmission line which is lossless. The transmission line is matched ($Z_L = Z_o$) and is characterised by an inductance per metre of 250 nH and a capacitance per metre of 100 pF. Determine:

1. An expression for the voltage, in the time domain, at any physical point, x, of the transmission line;

2. The delay of the voltage across the load;

3. An expression for the current, in the time domain, at any physical point, x, of the transmission line.

Solution:

1. The voltage $v_s(t)$ has a Fourier transform, $V_s(f)$ given by (see appendix A)

$$V_s(f) = V_A T \operatorname{sinc}(fT) \tag{7.28}$$

Since the transmission line is terminated by a load impedance equal to its characteristic impedance, the voltage at the input of the transmission line, $V(f, x = 0)$, is given by:

$$\begin{aligned} V(x=0) &= V_A T \operatorname{sinc}(fT) \frac{Z_o}{Z_o + Z_o} \\ &= \frac{V_A}{2} T \operatorname{sinc}(fT) \end{aligned} \tag{7.29}$$

where we drop the explicit dependency of V on f for simplicity. The transmission line imposes only a time delay to the propagating voltage $V(x)$ and, from eqn 7.10, we can write:

$$\begin{aligned} V(x) &= V(x=0)\, e^{-j\beta x} \\ &= \frac{V_A}{2} T \operatorname{sinc}(fT)\, e^{-j\beta x} \end{aligned} \tag{7.30}$$

From eqn 7.9 we can write β:

$$\beta = \frac{2\pi f}{v_p} \tag{7.31}$$

where v_p is the propagation speed (see also eqn 7.24):

$$\begin{aligned} v_p &= \frac{1}{\sqrt{LC}} \\ &= 2 \times 10^8 \text{ m/s} \end{aligned} \tag{7.32}$$

From eqns 7.31 and 7.32 we can write eqn 7.30 as follows:

$$V(x) = \frac{V_A}{2} T \operatorname{sinc}(fT)\, e^{-j\, 2\pi f \sqrt{LC}\, x} \tag{7.33}$$

and, using eqn 7.11 we can write the time domain voltage on the transmission line as

$$v_o(t, x) = \frac{V_A}{2} \operatorname{rect}\left(\frac{t - \sqrt{LC}\, x}{T}\right) \tag{7.34}$$

2. From eqn 7.34 we can write the voltage across the load as

$$v_o(t, l) = \frac{V_A}{2} \operatorname{rect}\left(\frac{t - \sqrt{LC}\, l}{T}\right) \tag{7.35}$$

where the delay $\sqrt{LC}\, l$ is equal to 50 ns.

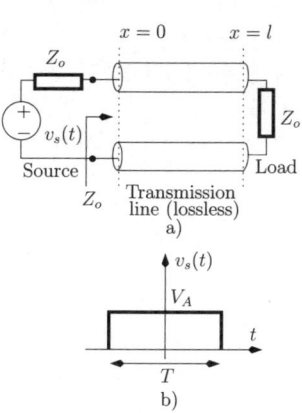

Figure 7.4: a) Transmission of a square voltage pulse through an ideal transmission line. b) Square voltage pulse.

3. Using eqn 7.26 we can determine the current $I(x)$ as follows:

$$\begin{aligned} I(x) &= \frac{V(x=0)}{Z_o} e^{-j\beta x} \\ &= \frac{V_A}{2 Z_o} T \operatorname{sinc}(fT) e^{-j 2\pi f \sqrt{LC} x} \end{aligned} \quad (7.36)$$

Using eqn 7.11 to return to the time domain, we can write the current in the transmission line as

$$i(t,x) = \frac{V_A}{2 Z_o} \operatorname{rect}\left(\frac{t - \sqrt{LC}\, x}{T}\right) \quad (7.37)$$

From this example it is clear that an ideal (lossless) transmission which is matched does not introduce any distortion to the voltage and current propagating signals.

When the load which terminates the transmission line is different from its characteristic impedance ($Z_L \neq Z_o$) the condition of maximum power transfer is not satisfied and parts of the voltage and current wave signals are reflected back to the signal source. In fact, the general solution for the voltage and current wave signals, at a given point of the transmission line, d, terminated with a general load impedance Z_L is given by the sum of two propagating waves; an incident wave, travelling towards the load, and a reflected wave travelling back towards the signal source. Such a situation can be expressed as follows:

$$V(d) = \underbrace{V_A\, e^{j\beta d}}_{\text{incident wave}} + \underbrace{V_B\, e^{-j\beta d}}_{\text{reflected wave}} \quad (7.38)$$

$$\begin{aligned} I(d) &= I_A\, e^{j\beta d} - I_B\, e^{-j\beta d} \\ &= \frac{V_A}{Z_o} e^{j\beta d} - \frac{V_B}{Z_o} e^{-j\beta d} \end{aligned} \quad (7.39)$$

It should be noted that there is a change in the distance reference: $d = 0$ is now the load reference plane while $d = l$ is the distance to the signal source from Z_L as shown in figure 7.5.

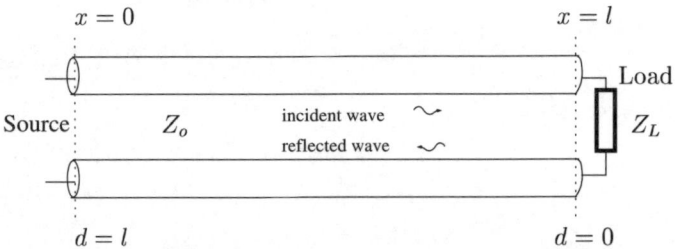

Figure 7.5: *Transmission line with characteristic impedance Z_o and unmatched load Z_L.*

7. RF circuit analysis techniques

The reflection coefficient, $\Gamma(d)$, is defined as the ratio between the voltage of the reflected wave and the voltage of the incident wave at any location on the transmission line, d:

$$\Gamma(d) = \frac{V_B\, e^{-j\beta d}}{V_A\, e^{j\beta d}} = \frac{V_B}{V_A} e^{-2j\beta d} \quad (7.40)$$

The voltage, $V(d)$, and the current $I(d)$ can be written as follows:

$$V(d) = V_A\, e^{j\beta d}\left(1 + \frac{V_B}{V_A} e^{-2j\beta d}\right)$$
$$= V_A\, e^{j\beta d}\left[1 + \Gamma(d)\right] \quad (7.41)$$

whilst

$$I(d) = \frac{V_A}{Z_o} e^{j\beta d}\left[1 - \Gamma(d)\right] \quad (7.42)$$

and the impedance, $Z_{in}(d)$, as a function of the distance d, is given by the following expression:

$$Z_{in}(d) = \frac{V(d)}{I(d)}$$
$$= Z_o \frac{1 + \Gamma(d)}{1 - \Gamma(d)} \quad (7.43)$$

For $d = 0$ it is known that $Z_{in}(0) = Z_L$, that is;

$$Z_L = Z_o \frac{1 + \Gamma(0)}{1 - \Gamma(0)}$$

Solving the last eqn in order to obtain $\Gamma(0)$, we get:

$$\Gamma_o \triangleq \Gamma(0) = \frac{Z_L - Z_o}{Z_L + Z_o} \quad (7.44)$$

The reflection coefficient given by eqn 7.44 effectively 'measures' the difference between the load Z_L and the transmission line characteristic impedance Z_o. If $\Gamma_o = 0$ then $Z_L = Z_o$. When $\Gamma_o \neq 0$ these two impedances are different. The greater the value for $|\Gamma_o|$ the greater the difference between Z_L and Z_o. Setting $d = 0$ in eqn 7.40 and using eqn 7.44 we have

$$\Gamma(d) = \Gamma_o\, e^{-2j\beta d} \quad (7.45)$$

At this point we introduce the transmission coefficient, $T(d)$, which is defined as the ratio between the voltage wave $V(d)$ and the voltage of the incident wave (see also eqn 7.40):

$$T(d) = \frac{V_A\, e^{j\beta d} + V_B\, e^{-j\beta d}}{V_A\, e^{j\beta d}}$$
$$= 1 + \frac{V_B\, e^{-j\beta d}}{V_A\, e^{j\beta d}} \quad (7.46)$$

that is

$$T(d) = 1 + \Gamma(d) \tag{7.47}$$

The transmission coefficient is often presented, in dB, as the 'insertion loss', $IL(d)$:

$$IL(d) = 20 \log_{10} |T(d)| \tag{7.48}$$

Using eqns 7.43, 7.44 and 7.45, the input impedance $Z_{in}(d)$ can be written (see problem 7.3) as a function of Z_L, Z_o and βd:

$$Z_{in}(d) = Z_o \frac{Z_L + j Z_o \tan(\beta d)}{Z_o + j Z_L \tan(\beta d)} \tag{7.49}$$

βd represents an angle which is commonly referred to as the 'electrical length', where the angle $\beta d = 2\pi$ corresponds to a single wavelength. Equation 7.49 shows that, at different locations on the transmission line, the input impedance varies between being capacitive and inductive depending on the value of the load Z_L.

Example 7.3.2 Consider a transmission line with $Z_o = 50$ Ω. The load impedance is $Z_L = 10$ Ω. Determine the line input impedance for:

1. An electrical length of 45°;
2. An electrical length of 90°;
3. An electrical length of 135°.

Solution: Using eqn 7.49 for the required electrical lengths, βd;

1. $\beta d = \pi/4$, $Z_{in} = 19.2 + j\, 46.2$ Ω.
2. $\beta d = \pi/2$, $Z_{in} = 250$ Ω.
3. $\beta d = 3\pi/4$, $Z_{in} = 19.2 - j\, 46.2$ Ω.

7.3.1 Voltage Standing Wave Ratio – VSWR

When a transmission line is terminated by a load which is different from the characteristic impedance there is a wave reflected from the load. With such a reflection, we effectively have two waves travelling in opposite directions along the transmission line; the incident wave and the reflected wave. The addition of these two waves produces a standing wave pattern along the transmission line which is characterised by the 'Voltage Standing Wave Ratio' (VSWR). The VSWR is defined as the ratio between the absolute value of the maximum voltage in the transmission line and the absolute value of the minimum voltage in the transmission line:

$$\text{VSWR} = \frac{|V(d)|_{max}}{|V(d)|_{min}} \tag{7.50}$$

7. RF circuit analysis techniques

From eqn 7.41 we can write

$$|V(d)|_{max} = |V_A|\,(1+|\Gamma_o|)$$
$$|V(d)|_{min} = |V_A|\,(1-|\Gamma_o|)$$

and the VSWR can be written as:

$$\text{VSWR} = \frac{1+|\Gamma_o|}{1-|\Gamma_o|} \tag{7.51}$$

We now consider three very important cases:

- The matched transmission line;
- The open-circuit transmission line;
- The short-circuited transmission line.

The matched transmission line

The matched transmission line has been discussed above. For this situation $Z_L = Z_o$, $\Gamma_o = 0$ and $Z_{in}(d) = Z_o$. It follows that the VSWR is constant at its minimum value: 1. This is expected since there is no reflected wave. Therefore, the wave pattern resulting from the incident wave is constant, as shown in figure 7.6 a). For this situation the voltage signal can be written as

$$V(d) = V_A e^{j\beta d} \tag{7.52}$$
$$v(t,d) = \text{Real}\left[V_A\, e^{j\beta d}\, e^{j\omega t}\right]$$
$$= V_A \cos(\omega t + \beta d) \tag{7.53}$$

The open-circuit transmission line

For this case we have

$$Z_L \to \infty$$
$$\Gamma_o = \lim_{Z_L \to \infty} \frac{1-\frac{Z_o}{Z_L}}{1+\frac{Z_o}{Z_L}} = 1$$
$$\text{VSWR} \to \infty$$
$$Z_{in}(d) = \lim_{Z_L \to \infty} Z_o \frac{1+j\frac{Z_o \tan(\beta d)}{Z_L}}{\frac{Z_o}{Z_L}+j\tan(\beta d)}$$
$$= \frac{Z_o}{j\tan(\beta d)} \tag{7.54}$$

From eqn 7.54 it can be seen that the transmission line has a capacitive nature for $0 < d < \lambda/4$. For this situation the voltage signal can be written as:

$$V(d) = V_A\left(e^{j\beta d}+e^{-j\beta d}\right)$$
$$= 2V_A \cos(\beta d) \tag{7.55}$$
$$v(t,d) = \text{Real}\left[V_A\left(e^{j\beta d}+e^{j\beta d}\right)e^{j\omega t}\right]$$
$$= 2V_A \cos(\beta d)\cos(\omega t) \tag{7.56}$$

Figure 7.6 b) shows this waveform.

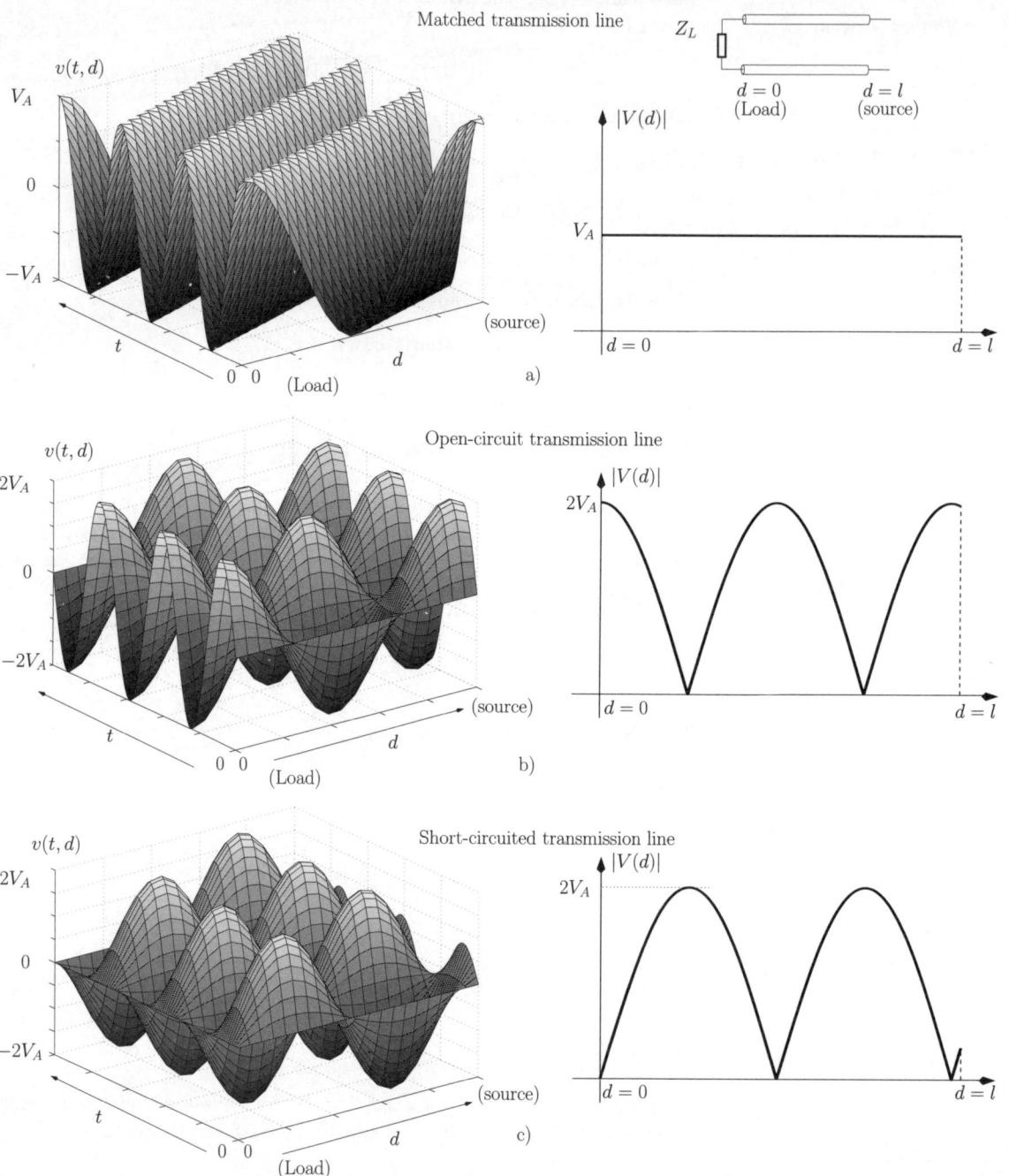

Figure 7.6: *Voltage patterns on a transmission line showing standing waves versus time and distance (left) and magnitude of the standing wave voltage (right). a) Matched transmission line (no standing wave). b) Open-circuit transmission line. c) Short-circuited transmission line.*

7. RF circuit analysis techniques

The short-circuited transmission line

For a short-circuited transmission line we have:

$$Z_L = 0$$
$$\Gamma_o = -1$$
$$\text{VSWR} \to \infty$$
$$Z_{in}(d) = j Z_o \tan(\beta d) \tag{7.57}$$

From eqn 7.57 it can be seen that the transmission line has an inductive nature for $0 < d < \lambda/4$. For this situation the voltage signal can be written as:

$$\begin{aligned} V(d) &= V_A \left(e^{j\beta d} - e^{-j\beta d} \right) \\ &= 2 j V_A \sin(\beta d) \end{aligned} \tag{7.58}$$

$$\begin{aligned} v(t,d) &= \text{Real}\left[V_A \left(e^{j\beta d} - e^{j\beta d} \right) e^{j\omega t} \right] \\ &= 2 V_A \sin(\beta d) \cos(\omega t + \pi/2) \end{aligned} \tag{7.59}$$

Figure 7.6 c) shows this waveform.

Example 7.3.3 V_s is a DC voltage source with a resistive output impedance, Z_s, applied to an open-circuit transmission line with characteristic impedance Z_o. Assuming that the source is switched-on at $t = 0$, show that the voltage at the output of the transmission line tends to V_s as $t \to \infty$.

Solution: We refer to figure 7.7 where we illustrate the following 'transient analysis'; When the source is switched-on the voltage source V_s only 'sees' the voltage divider formed by Z_s and Z_o since the voltage waveform has not yet travelled along the transmission line. Hence, the voltage at the input of the transmission line ($x = 0$) is given by

$$\begin{aligned} V(x=0) &= V_s \frac{Z_o}{Z_o + Z_s} \\ &= V_s \frac{1}{1+r} \end{aligned} \tag{7.60}$$

where $r = Z_s/Z_o$. The voltage described by eqn 7.60 propagates along the transmission line. T_P is the propagation time which is the time taken by the voltage to travel from $x = 0$ to $x = l$ (or from $x = l$ to $x = 0$). Since the line is terminated by an open-circuit ($\Gamma_o = 1$), this voltage is totally reflected back towards the source as illustrated in figure 7.8. Now, at a time instant immediately after T_P, i.e. $t = T_P^+$, the voltage at $x = l$ is

$$V(x=l) = 2V_s \frac{1}{1+r} \tag{7.61}$$

When the voltage propagating back to the signal source reaches $x = 0$ it sees a reflection coefficient Γ_s given by

$$\begin{aligned} \Gamma_s &= \frac{Z_s - Z_o}{Z_s + Z_o} \\ &= \frac{r-1}{r+1} \end{aligned} \tag{7.62}$$

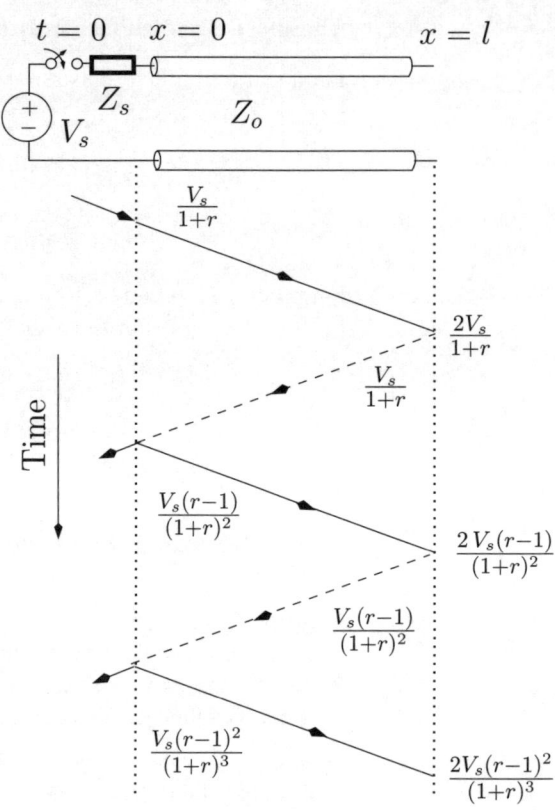

Figure 7.7: *Open-circuit transmission line driven by a voltage source. Transient analysis.* — *Transmission;* - - - *Reflection.*

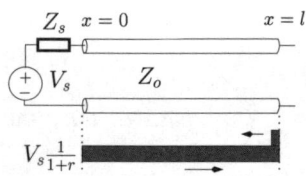

Figure 7.8: *Voltage versus the distance at* $t = T_P^+$.

Hence, a fraction of this voltage is reflected back towards the open-circuit with a value given by

$$V_s \frac{1}{1+r} \frac{r-1}{r+1} = V_s \frac{r-1}{(r+1)^2} \qquad (7.63)$$

and so on, as shown in figure 7.60. The total voltage at $x = l$, as t increases, is given by the addition of the partial voltages, that is:

$$\begin{aligned} V(x=l) &= 2V_s \left(\frac{1}{r+1} + \frac{r-1}{(r+1)^2} + \frac{(r-1)^2}{(r+1)^3} + \ldots \right) \\ &= \frac{2V_s}{r+1} \sum_{k=0}^{\infty} \left(\frac{r-1}{r+1} \right)^k \end{aligned} \qquad (7.64)$$

The last eqn is the sum of an infinite geometric series (see appendix A) whose value is given by:

$$\begin{aligned} V(x=l) &= \frac{2V_s}{r+1} \frac{r+1}{2} \\ &= V_s \end{aligned}$$

7. RF circuit analysis techniques

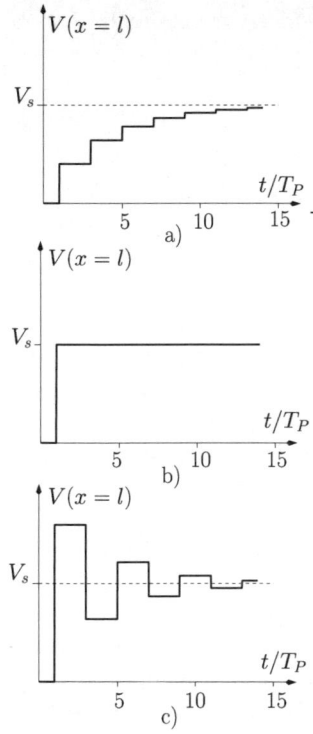

Figure 7.9: *Load voltage at $x = l$ versus the time normalised to T_P. a) $r = 4$. b) $r = 1$. c) $r = 0.25$.*

Figure 7.9 shows the voltage at $x = l$ versus the time normalised to T_P for three values of r; 4, 1 and 0.25. We observe that in all situations the voltage tends to V_s as predicted by our discussion above. Note that for $r = 1$ (corresponding to $Z_s = Z_o$) the voltage $V(x = l)$ equals V_s for $t > T_P$. This is because the reflected voltage wave is totally absorbed by Z_s since the line is matched to the signal source impedance.

Example 7.3.4 We want to determine the location of a failure in a coaxial cable with a length of 430 metres. In order to identify this location we send a 5 μs pulse through the cable and we monitor the reflection. Figure 7.10 a) shows the waveform monitored at the input of the cable. Determine the location of the fault knowing that the cable has an inductance per metre of 250 nH and a capacitance per metre of 100 pF.

Solution: The waveform of figure 7.10 a) can be seen as the sum of two pulses as shown in figure 7.10 b) where we clearly identify the pulse which was sent and the one reflected back. Since the reflected pulse has the opposite polarity of that transmitted we conclude that the fault is a short-circuit ($\Gamma_o = -1$).

The propagation velocity is $v_p = (LC)^{-1/2}$ and therefore we calculate the location of the fault at point x given by:

$$\begin{aligned} x &= v_p T_P \\ &= 300 \text{ m} \end{aligned}$$

This technique of fault diagnosis is commonly used and is known as Time Domain Reflectometry (TDR).

7.3.2 The $\lambda/4$ transformer

Another very important transmission line case, widely used in practical applications, is the quarter-wavelength transmission line also known as the quarter-wave transformer. This transmission line has an electrical length of $\beta d =$

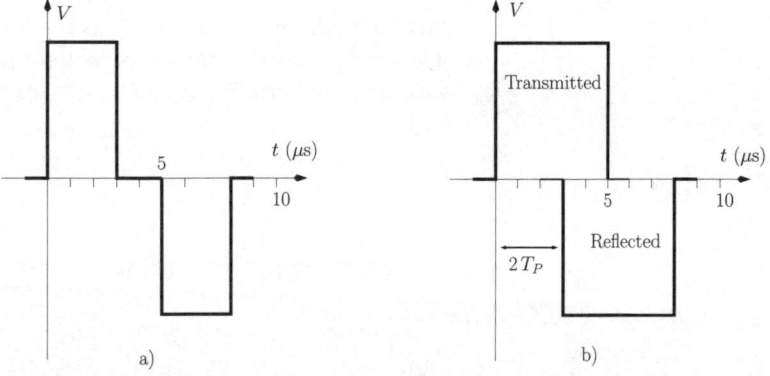

Figure 7.10: *Waveforms monitored at the input of a faulty cable.*

$(2\pi/\lambda) \times (\lambda/4) = \pi/2$. The input impedance for this transmission line can be calculated by rewriting eqn 7.49 as:

$$Z_{in}(d) = Z_o \frac{\frac{Z_L}{\tan(\beta d)} + jZ_o}{\frac{Z_o}{\tan(\beta d)} + jZ_L} \tag{7.65}$$

and if we let $\beta d \to \pi/2$ we get:

$$Z_{in}(d = \lambda/4) = \frac{Z_o^2}{Z_L} \tag{7.66}$$

Equation 7.66 reveals the importance of the quarter wave transformer which is the ability of transforming a real impedance (Z_L) into another real impedance given by Z_o^2/Z_L. This result is very important since it allows the matching of a load (Z_L) to a transmission line with a characteristic impedance Z_{oA} different from Z_L. The matching is achieved using a quarter-wave transformer with a characteristic impedance $Z_o = \sqrt{Z_{oA} Z_L}$ as shown in figure 7.11. In order to understand how this matching process is achieved we refer now to figure 7.12 where we perform a 'transient analysis': Let us suppose a normalised travelling voltage wave along the transmission line, with $Z_o = Z_{oA}$, towards the load Z_L. For the sake of simplicity we consider the phase of the voltage wave to be zero at $d = 0$. When this wave reaches the quarter-wave transformer with $Z_o = Z_{oB}$ for the *first time* it 'sees' only the impedance Z_{oB} since it has not travelled along the quarter-wave transformer and it has not reached the load Z_L. Hence, a partial reflection, Γ_{AB}, and a partial transmission, T_{AB}, take place at the interface between these two transmission lines:

$$\Gamma_{AB} = \frac{Z_{oB} - Z_{oA}}{Z_{oB} + Z_{oA}} \tag{7.67}$$

$$T_{AB} = 1 + \Gamma_{AB} = \frac{2Z_{oB}}{Z_{oB} + Z_{oA}} \tag{7.68}$$

This partial transmitted wave travels a distance $d = \lambda/4$ to the load where a fraction is reflected back towards the line Z_{oA}:

$$\Gamma_{BL} = \frac{Z_L - Z_{oB}}{Z_L + Z_{oB}} \tag{7.69}$$

This last reflected wave arrives to the transmission line Z_{oA} with an amplitude $-T_{AB} \Gamma_{BL}$. It should be noted that the round trip along the quarter-wave transformer corresponds to 180 degrees (or π) phase shift. A fraction

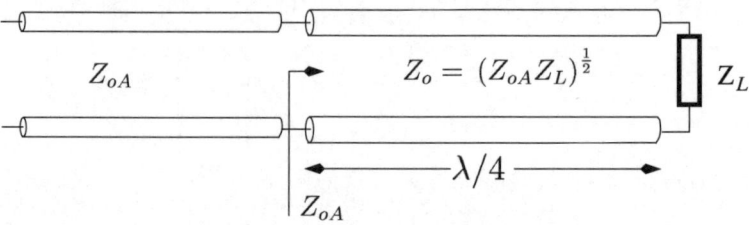

Figure 7.11: *Load matching using a quarter-wave transformer.*

7. RF circuit analysis techniques

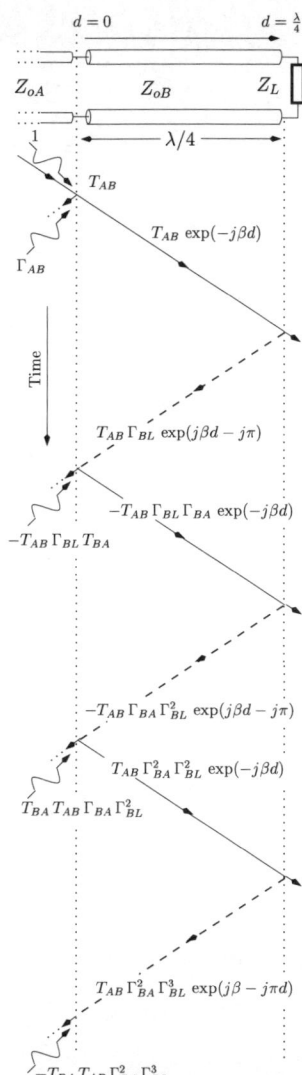

Figure 7.12: *The quarter-wave transformer load matching: transient analysis.*

$-T_{AB}\, \Gamma_{BL}\, T_{BA}$ of this wave travels back to the signal source while a fraction $-T_{AB}\, \Gamma_{BL}\, \Gamma_{BA}$ is reflected back to the load. For this analysis the following eqns apply;

$$\Gamma_{BA} = -\Gamma_{AB} \tag{7.70}$$

$$T_{BA} = \frac{2Z_{oA}}{Z_{oB} + Z_{oA}} \tag{7.71}$$

$$\Gamma_{BL} = \frac{Z_L - Z_{oB}}{Z_L + Z_{oB}} \tag{7.72}$$

The total reflected wave at the boundary between the transmission line Z_{oA} and the quarter wave transformer can be calculated by summing all the partial reflections (see also figure 7.12):

$$\begin{aligned}\Gamma_{tot} &= \Gamma_{AB} - T_{AB}\, T_{BA}\Gamma_{BL}(1 - \Gamma_{BL}\Gamma_{BA} + \Gamma_{BL}^2\Gamma_{BA}^2 - \ldots) \\ &= \Gamma_{AB} - T_{AB}\, T_{BA}\Gamma_{BL}\sum_{k=0}^{\infty}(-\Gamma_{BL}\Gamma_{BA})^k \end{aligned} \tag{7.73}$$

Equation 7.73 represents the sum of an infinite geometric series (see appendix A) whose value is

$$\begin{aligned}\Gamma_{tot} &= \Gamma_{AB} - \frac{\Gamma_{BL}T_{AB}T_{BA}}{1 + \Gamma_{BL}\Gamma_{BA}} \\ &= \frac{\Gamma_{AB}(1 + \Gamma_{BL}\Gamma_{BA}) - \Gamma_{BL}T_{AB}T_{BA}}{1 + \Gamma_{BL}\Gamma_{BA}} \end{aligned} \tag{7.74}$$

It can be shown (see problem 7.7) that the last eqn can be expressed as:

$$\Gamma_{tot} = \frac{Z_{oB}^2 - Z_{oA}\, Z_L}{Z_{oB}^2 + Z_{oA}\, Z_L} \tag{7.75}$$

which vanishes if

$$Z_{oB} = \sqrt{Z_{oA}\, Z_L} \tag{7.76}$$

In other words, if $Z_{oB} = \sqrt{Z_{oA}\, Z_L}$, the reflections at the boundary between the transmission line Z_{oA} and the quarter wave transformer add to zero and the transmission line with $Z_o = Z_{oA}$ is matched.

Example 7.3.5 Consider the load matching problem shown in figure 7.12. Determine an expression for the incident wave and the resulting reflected wave within the quarter-wave transformer for $Z_{oB} = \sqrt{Z_{oA}\, Z_L}$.

Solution: The total incident voltage, $V_i^+(d)$, in the load Z_L can be obtained by summing the partial incident voltages:

$$\begin{aligned}V_i^+(d) &= e^{-j\beta d}\, T_{AB}\, [1 - \Gamma_{BA}\Gamma_{BL} + (\Gamma_{BA}\Gamma_{BL})^2 - \ldots] \\ &= e^{-j\beta d}\, T_{AB}\sum_{k=0}^{\infty}(-\Gamma_{BA}\Gamma_{BL})^k \quad , \, 0 \le d \le \lambda/4 \end{aligned} \tag{7.77}$$

It is known that

$$\sum_{k=0}^{\infty} r^k = \frac{1}{1-r} \quad , \; |r| < 1 \quad (7.78)$$

Since $|-\Gamma_{BA}\Gamma_{BL}| < 1$ we can write

$$V_i^+(d) = e^{-j\beta d} \frac{T_{AB}}{1+\Gamma_{BA}\Gamma_{BL}} \quad , \; 0 \leq d \leq \lambda/4 \quad (7.79)$$

where

$$\frac{T_{AB}}{1+\Gamma_{BA}\Gamma_{BL}} = \frac{2Z_{oB}(Z_L+Z_{oB})}{(Z_L+Z_{oB})(Z_{oA}+Z_{oB})+(Z_L-Z_{oB})(Z_{oA}-Z_{oB})}$$

$$= \frac{Z_{oB}(Z_L+Z_{oB})}{Z_L Z_{oA}+Z_{oB}^2} \quad (7.80)$$

Since $Z_{oB} = \sqrt{Z_{oA} Z_L}$,

$$\frac{T_{AB}}{1+\Gamma_{BA}\Gamma_{BL}} = \frac{1}{2}\left(1+\sqrt{\frac{Z_L}{Z_{oA}}}\right) \quad (7.81)$$

and the incident wave can be written as:

$$V_i^+(d) = e^{-j\beta d} \frac{1}{2}\left(1+\sqrt{\frac{Z_L}{Z_{oA}}}\right) \quad , \; 0 \leq d \leq \lambda/4 \quad (7.82)$$

Similarly, the total voltage reflected from the load is obtained by summing all the partial reflections:

$$V_i^-(d) = e^{j\beta d - j\pi} T_{AB}\Gamma_{BL}[1-\Gamma_{BA}\Gamma_{BL}+(\Gamma_{BA}\Gamma_{BL})^2-\ldots]$$

$$= e^{j\beta d - j\pi} T_{AB}\Gamma_{BL} \sum_{k=0}^{\infty}(-\Gamma_{BA}\Gamma_{BL})^k \quad , \; 0 \leq d \leq \lambda/4 \quad (7.83)$$

Using eqn 7.78 we obtain:

$$V_i^-(d) = e^{j\beta d - j\pi} \frac{T_{AB}\Gamma_{BL}}{1+\Gamma_{BA}\Gamma_{BL}}$$

$$= e^{j\beta d - j\pi} \frac{Z_L-Z_{oB}}{2Z_{oB}}$$

$$= e^{j\beta d} \frac{1}{2}\left(1-\sqrt{\frac{Z_L}{Z_{oA}}}\right) \quad , \; 0 \leq d \leq \lambda/4 \quad (7.84)$$

7.3.3 Lossy transmission lines

In practical transmission lines there is power dissipation when a wave signal travels along a transmission line. These dissipative phenomena are usually

7. RF circuit analysis techniques

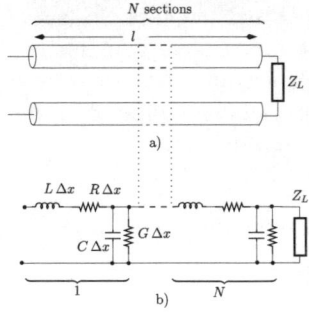

Figure 7.13: *Electrical model for a lossy transmission line.*

due to the finite conductivity of transmission lines and losses in the dielectric around them. A more realistic model for a transmission line than that of figure 7.3 is given in figure 7.13 where the resistance per section $R\Delta x$ and the conductance per section $G\Delta x$ account for the finite conductivity and the dielectric losses, respectively. If we assume that the transmission line is matched, that is, that all the power arriving to the load, Z_L is absorbed, it is possible to show that (see example 7.3.6 and exercise 7.8)

$$V(x) = e^{-\gamma x} V(0) \qquad (7.85)$$

$$\gamma = \alpha + j\beta = \sqrt{(R + j\omega L)(G + j\omega C)} \qquad (7.86)$$

$$I(x) = e^{-\gamma x} \frac{V(0)}{Z_o} \qquad (7.87)$$

$$Z_o = \sqrt{\frac{R + j\omega L}{G + j\omega C}} \qquad (7.88)$$

where γ is called the complex propagation constant, α is the attenuation constant (in nepers[3] per metre) and β is the propagation constant, as before. In general, the characteristic impedance and the complex propagation constant are frequency dependent. It can be shown (see problem 7.9) that at low frequencies where $R >> \omega L$ and $G >> \omega C$, we can write

$$\alpha_{LF} \simeq \sqrt{RG} \qquad (7.89)$$

$$\beta_{LF} \simeq \omega \frac{1}{2} \left(C \sqrt{\frac{R}{G}} + L \sqrt{\frac{G}{R}} \right) \qquad (7.90)$$

$$Z_{o_{LF}} \simeq \sqrt{\frac{R}{G}} \qquad (7.91)$$

and at high frequencies where $R << \omega L$ and $G << \omega C$ we can write:

$$\alpha_{HF} \simeq \frac{1}{2} \left(R\sqrt{\frac{C}{L}} + G\sqrt{\frac{L}{C}} \right) \qquad (7.92)$$

$$\beta_{HF} \simeq \omega \sqrt{LC} \qquad (7.93)$$

$$Z_{o_{HF}} \simeq \sqrt{\frac{L}{C}} \qquad (7.94)$$

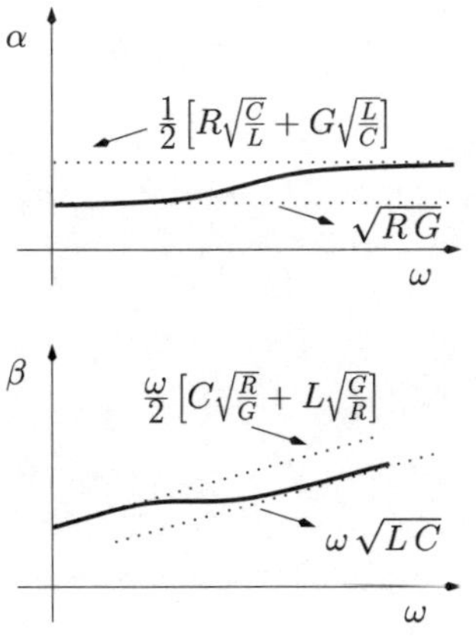

Figure 7.14 shows typical variations of the complex propagation constants (α and β) with the angular frequency, ω. Since the propagation velocity (ω/β) and the attenuation constant (α) are frequency dependent we can expect, in addition to amplitude attenuation, linear signal distortion (see figure 7.15) in a lossy transmission line. This is because the different frequency components of a propagating signal will travel at different speeds and will experience different delays when arriving at the load. Also, further distortion can arise from different frequency components experiencing different attenuation levels (see also section 3.3.4).

Figure 7.14: *a) Attenuation constant versus the angular frequency. b) Propagation constant versus the angular frequency.*

---

[3]Neper is a unit expressing the ratio of two numbers as a natural logarithm where the attenuation in nepers is 1/2 ln(output/input). Attenuation of one neper approximately equals 13.5% $\simeq$ -8.7 dB.

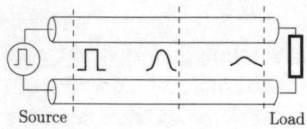

Figure 7.15: *Signal distortion in a lossy transmission line.*

Fortunately, most practical transmission lines exhibit low loss. In fact, if these losses were very high the transmission line would be of very limited use. Hence, in practice the propagation constant and the characteristic impedance can be characterised by their high frequency approximations expressed in eqns 7.92 to 7.94, for which the signal propagation is distortion free.

Example 7.3.6 Show that for a matched lossy transmission line we have

$$V(x) = e^{-\gamma x} V(0) \qquad (7.95)$$

Solution: Assuming that the transmission line is terminated by a load Z_L such that the impedance looking into each section is also given by Z_L we can write:

$$Z_L = Z_1' + \frac{Z_2' Z_L}{Z_2' + Z_L} \qquad (7.96)$$

with

$$Z_1' = (R + j\omega L)\Delta x$$
$$Z_2' = \frac{1}{(G + j\omega C)\Delta x}$$

Solving eqn 7.96 in order to obtain Z_L we get

$$\begin{aligned} Z_L &= \frac{Z_1' + \sqrt{Z_1'^2 + 4Z_2'Z_1'}}{2} \\ &= \frac{(R+j\omega L)\Delta x}{2} + \frac{\sqrt{(R+j\omega L)^2 \Delta x^2 + 4\frac{R+j\omega L}{G+j\omega C}}}{2} \end{aligned} \qquad (7.97)$$

Under the assumptions mentioned above, the voltage transfer function between any two adjacent sections is equal and given by the impedance voltage divider:

$$\frac{V(k\Delta x)}{V([k-1]\Delta x)} = \frac{\frac{Z_2' Z_L'}{Z_2' + Z_L'}}{\frac{Z_2' Z_L'}{Z_2' + Z_L'} + Z_1'} \qquad (7.98)$$

with $k = 1, 2, \ldots, N$. From eqn 7.96 it is known that:

$$\frac{Z_2' Z_L}{Z_2' + Z_L} = Z_L' - Z_1' \qquad (7.99)$$

Using the result of eqn 7.99 on eqn 7.98 the voltage transfer function between any two adjacent sections can be written as:

$$\begin{aligned} \frac{V(k\Delta x)}{V([k-1]\Delta x)} &= \frac{Z_L - Z_1'}{Z_L} \\ &= \frac{\sqrt{(R+j\omega L)^2 \Delta x^2 + 4\frac{R+j\omega L}{G+j\omega C}} - (R+j\omega L)\Delta x}{\sqrt{(R+j\omega L)^2 \Delta x^2 + 4\frac{R+j\omega L}{G+j\omega C}} + (R+j\omega L)\Delta x} \end{aligned}$$

7. RF circuit analysis techniques

The last eqn can be written as:

$$\frac{V(k\Delta x)}{V([k-1]\Delta x)} = \frac{\left(\sqrt{1+\frac{4}{(R+j\omega L)(G+j\omega C)\Delta x^2}}-1\right)}{\left(\sqrt{1+\frac{4}{(R+j\omega L)(G+j\omega C)\Delta x^2}}+1\right)} \quad (7.100)$$

We use now the variable $W = (R+j\omega L)(G+j\omega C)/4$ to simplify the analysis. Hence, we can write eqn 7.100 as follows:

$$\frac{V(k\Delta x)}{V([k-1]\Delta x)} = \frac{\sqrt{1+\frac{1}{W\Delta x^2}}-1}{\sqrt{1+\frac{1}{W\Delta x^2}}+1}$$
$$= (\sqrt{W\Delta x^2+1} - \sqrt{W\Delta x^2})^2 \quad (7.101)$$

The voltage transfer function considered in a particular section of the line, $x = k\Delta x$, is given as:

$$\frac{V(x=k\Delta x)}{V(x=0)} = \left(\frac{V(k\Delta x)}{V([k-1]\Delta x)}\right)^k$$
$$= (\sqrt{W\Delta x^2+1} - \sqrt{W\Delta x^2})^{2k} \quad (7.102)$$

The last eqn can be expanded as a Maclaurin series as follows:

$$\frac{V(x=k\Delta x)}{V(x=0)} = 1 - 2k\sqrt{W}\Delta x + 4k^2 \frac{(\sqrt{W}\Delta x)^2}{2!}$$
$$+ (2k - 8k^3)\frac{(\sqrt{W}\Delta x)^3}{3!} + (16k^4 - 16k)\frac{(\sqrt{W}\Delta x)^4}{4!}$$
$$+ (80k^3 - 32k^4 - 18k^2)\frac{(\sqrt{W}\Delta x)^5}{5!} + \ldots \quad (7.103)$$

If we increase the number of sections, that is, $N \to \infty$, and if we decrease the length of each section, $\Delta x \to 0$, in such a way that the product $l = \Delta x\, N$ is kept constant, then eqn 7.19 applies and the last eqn can be written as shown below:

$$\lim_{\substack{k\to\infty \\ \Delta x\to 0}} \frac{V(x=k\Delta x)}{V(x=0)} = 1 - 2\sqrt{W}x + 4\frac{(\sqrt{W}x)^2}{2!} + (-8)\frac{(\sqrt{W}x)^3}{3!}$$
$$+ (16)\frac{(\sqrt{W}x)^4}{4!} + (-32)\frac{(\sqrt{W}x)^5}{5!} + \ldots$$
$$= \sum_{n=0}^{\infty} \frac{(-2\sqrt{W}x)^n}{n!}$$
$$= e^{-2\sqrt{W}\,x}$$
$$= e^{-\gamma x} \quad (7.104)$$

with γ given by eqn 7.86. Hence we have

$$V(x) = e^{-\gamma x} V(0) \quad (7.105)$$

Example 7.3.7 Prove that for a lossy transmission line with $RC = LG$ the signal propagation is distortionless.

Solution: The complex propagation constant for a lossy transmission line can be expressed as:

$$\begin{aligned}\gamma &= \sqrt{(j\omega L)(j\omega C)\left(1 + \frac{R}{j\omega L}\right)\left(1 + \frac{G}{j\omega C}\right)} \\ &= j\omega\sqrt{LC}\sqrt{1 - j\left(\frac{R}{\omega L} + \frac{G}{\omega C}\right) - \frac{RG}{\omega^2 LC}}\end{aligned}$$

Using the condition $RC = LG$ we get

$$\begin{aligned}\gamma &= j\omega\sqrt{LC}\sqrt{1 - 2j\frac{R}{\omega L} - \frac{R^2}{\omega^2 L^2}} \\ &= j\omega\sqrt{LC}\sqrt{\left(1 - j\frac{R}{\omega L}\right)^2} \\ &= j\omega\sqrt{LC}\left(1 - j\frac{R}{\omega L}\right)\end{aligned} \quad (7.106)$$

that is

$$\alpha = R\sqrt{\frac{C}{L}} \quad (7.107)$$

$$\beta = \omega\sqrt{LC} \quad (7.108)$$

and because α is constant and does not depend on the frequency and β varies linearly with the frequency, all signal frequency components are equally attenuated and they all travel at the same propagation velocity, $(LC)^{-1/2}$, effectively resulting in distortionless transmission.

When a lossy transmission line is terminated with a load impedance Z_L, the solution for the voltage and current wave signals, at any position of the transmission line, d, is given by:

$$V(d) = \underbrace{V_A\, e^{\gamma d}}_{\text{incident wave}} + \underbrace{V_B\, e^{-\gamma d}}_{\text{reflected wave}} \quad (7.109)$$

$$I(d) = I_A\, e^{\gamma d} - I_B\, e^{-\gamma d} \quad (7.110)$$

$$= \frac{V_A}{Z_o} e^{\gamma d} - \frac{V_B}{Z_o} e^{-\gamma d} \quad (7.111)$$

The reflection coefficient at any point d on the lossy transmission line follows the definition presented in eqn 7.40, that is:

$$\Gamma(d) = \frac{V_A e^{-\gamma d}}{V_B e^{\gamma d}} = \Gamma_o e^{-2\gamma d} \quad (7.112)$$

$$\Gamma_o = \frac{Z_L - Z_o}{Z_L + Z_o} \quad (7.113)$$

7. RF circuit analysis techniques

and the input impedance is now given by (see problem 7.10):

$$Z_{in}(d) = Z_o \frac{Z_L + Z_o \tanh(\gamma\, d)}{Z_o + Z_L \tanh(\gamma\, d)} \tag{7.114}$$

7.3.4 Microstrip transmission lines

There is a large variety of transmission lines including coaxial cables, striplines, and several types of waveguides. However, the microstrip line is one of the most popular types because it is easily fabricated using printed-circuit techniques and because it is easily integrated with other active and passive microwave devices such as integrated circuits operating at high frequency (RF), microwave connectors, etc. Here we introduce the reader to the basic concepts of microstrip transmission lines. Detailed studies of these lines and other transmission line structures have been presented by Edwards and Steer [6] and also by Fooks and Zakarvicius [7].

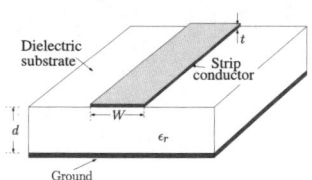

Figure 7.16: *Geometry of microstrip lines.*

The geometry of microstrip transmission lines is illustrated in figure 7.16. A strip conductor of width W and thickness t is printed on a grounded dielectric of thickness d and relative permittivity ϵ_r. When the thickness of the strip conductor is small ($t/d < 0.005$) the characteristic impedance can be calculated, given the physical dimensions of the microstrip line, as follows:

$$Z_o \simeq \begin{cases} \frac{60}{\sqrt{\epsilon_e}} \ln\left(8\frac{d}{W} + 0.25\frac{W}{d}\right) & \text{if } \frac{W}{d} < 1 \\[2ex] \frac{120\pi/\sqrt{\epsilon_e}}{W/d + 1.393 + 0.667 \ln(W/d + 1.444)} & \text{if } \frac{W}{d} \geq 1 \end{cases} \tag{7.115}$$

where ϵ_e is the effective dielectric permittivity expressed as:

$$\epsilon_e \simeq \frac{\epsilon_r + 1}{2} + \frac{\epsilon_r - 1}{2} \times \frac{1}{\sqrt{1 + 12\, d/W}} \tag{7.116}$$

The effective dielectric permittivity accounts for the fact that whilst part of the wave propagation takes place within the dielectric substrate ($\epsilon_r\, \epsilon_o$) some occurs through the air (ϵ_o).

For a given value of d it is necessary to calculate W to achieve correct electrical parameters. Hence, for a given characteristic impedance, Z_o, and dielectric constant ϵ_r, the W/d ratio can be calculated as follows:

$$\frac{W}{d} \simeq \frac{8\, e^A}{e^{2A} - 2} \qquad \text{if } \frac{W}{d} < 2 \tag{7.117}$$

$$\frac{W}{d} \simeq \frac{2}{\pi}\left[B - 1 - \ln(2B - 1)\right.$$
$$\left. + \frac{\epsilon_r - 1}{2\epsilon_r}\left(\ln(B - 1) + 0.39 - \frac{0.61}{\epsilon_r}\right)\right] \qquad \text{if } \frac{W}{d} > 2 \tag{7.118}$$

where

$$A = \frac{Z_o}{60}\sqrt{\frac{\epsilon_r + 1}{2}} + \frac{\epsilon_r - 1}{\epsilon_r + 1}\left(0.23 + \frac{0.11}{\epsilon_r}\right) \tag{7.119}$$

and

$$B = \frac{377\,\pi}{2\,Z_o\,\sqrt{\epsilon_r}} \qquad (7.120)$$

The propagation constant β can be calculated according to;

$$\beta = \frac{\omega}{c}\sqrt{\epsilon_e} \qquad (7.121)$$

with c representing the speed of light in vacuum.

The attenuation due to the dielectric loss can be determined as follows:

$$\alpha_d \simeq \frac{\omega\,\epsilon_r(\epsilon_e - 1)\tan(\delta)}{2\,c\,\sqrt{\epsilon_e}(\epsilon_r - 1)} \text{ nepers/m} \qquad (7.122)$$

with $\tan(\delta)$ representing the 'loss tangent' of the dielectric given by:

$$\tan(\delta) = \frac{\sigma_d}{\omega\,\epsilon_r\,\epsilon_o} \qquad (7.123)$$

where σ_d is called the total effective conductivity of the dielectric.

The attenuation due to losses in the stripline conductor can be determined as follows:

$$\alpha_c \simeq \frac{\sqrt{\omega\mu_o/(2\sigma_c)}}{Z_o\,W} \text{ nepers/m} \qquad (7.124)$$

where $\mu_o = 4\pi \cdot 10^{-7}$ Henry/m is the permeability of free-space and σ_c is the conductivity of the stripline conductor.

Example 7.3.8 Determine W for a microstrip transmission line to give $Z_o = 50\,\Omega$. The substrate thickness is 0.127 cm and the relative permittivity is 2.20. For such a line determine also the effective dielectric permittivity.

Solution: Taking the initial guess that $W/d > 2$ we use eqns 7.118 and 7.120 to obtain;

$$B = 7.99 \quad , \quad W/d = 3.08 \qquad (7.125)$$

Note that the value obtained for W/d is greater than two. Otherwise, it would be necessary to use eqn 7.117, valid for $W/d < 2$. $W = 3.08\,d = 0.39$ cm. From eqn 7.116 we obtain $\epsilon_e = 1.87$.

7.4 Scattering parameters

Scattering parameters (S-parameters) were developed in the early 1960s for the purposes of high-frequency transistor assessment and measurements. These parameters are defined according to the voltage and current wave signal definitions presented previously in the context of lossless (ideal) transmission lines.

7. RF circuit analysis techniques

Let us consider the voltage and current wave signals propagating in an ideal transmission line as shown in figure 7.17, for which we shall use the following notation:

$$V^+(x) = V_A \, e^{-j\beta x} \tag{7.126}$$
$$V^-(x) = V_A \, e^{+j\beta x} \tag{7.127}$$
$$I^+(x) = \frac{V_A}{Z_o} e^{-j\beta x} \tag{7.128}$$
$$I^-(x) = \frac{V_A}{Z_o} e^{+j\beta x} \tag{7.129}$$

where $V^+(x)$ and $V^-(x)$ represent the incident and reflected voltage waves, respectively. $I^+(x)$ and $I^-(x)$ represent the incident and reflected current waves, respectively. Z_o is the characteristic impedance of the lossless transmission line. It is now possible to write

$$V(x) = V^+(x) + V^-(x) \tag{7.130}$$
$$I(x) = I^+(x) - I^-(x) = \frac{V^+(x)}{Z_o} - \frac{V^-(x)}{Z_o} \tag{7.131}$$
$$\Gamma(x) = \frac{V^-(x)}{V^+(x)} = \frac{I^-(x)}{I^+(x)} \tag{7.132}$$

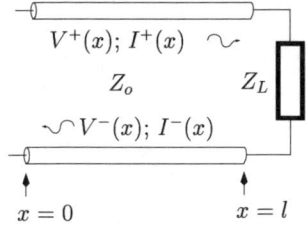

Figure 7.17: *Incident and reflected waves in a transmission line.*

Using the following normalisations

$$v(x) = \frac{V(x)}{\sqrt{Z_o}} \tag{7.133}$$
$$i(x) = \sqrt{Z_o} I(x) \tag{7.134}$$

and defining $a(x)$ and $b(x)$ as indicated below:

$$a(x) = \frac{V^+(x)}{\sqrt{Z_o}} = \sqrt{Z_o} I^+(x) \tag{7.135}$$
$$b(x) = \frac{V^-(x)}{\sqrt{Z_o}} = \sqrt{Z_o} I^-(x) \tag{7.136}$$

we can write the following set of eqns

$$v(x) = a(x) + b(x) \tag{7.137}$$
$$i(x) = a(x) - b(x) \tag{7.138}$$
$$b(x) = \Gamma(x) \, a(x) \tag{7.139}$$

or

$$a(x) = \frac{1}{2}[v(x) + i(x)] = \frac{1}{2\sqrt{Z_o}}[V(x) + Z_o I(x)] \tag{7.140}$$

and

$$b(x) = \frac{1}{2}[v(x) - i(x)] = \frac{1}{2\sqrt{Z_o}}[V(x) - Z_o I(x)] \tag{7.141}$$

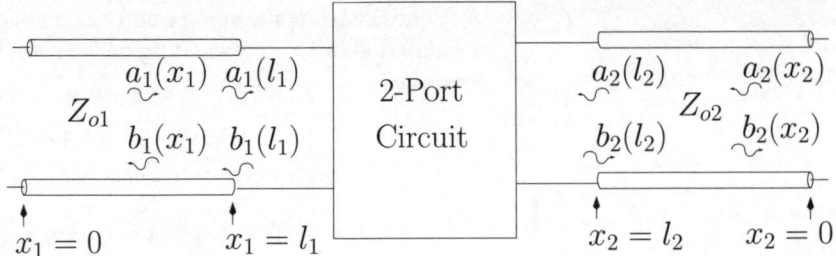

Figure 7.18: *S-parameter definition for a two-port circuit.*

For a two-port circuit (see figure 7.18) we have now $a_1(l_1)$ and $b_1(l_1)$ representing the incident and reflected waves, respectively, at port one located at $x_1 = l_1$. Similarly, $a_2(l_2)$ and $b_2(l_2)$ represent the incident and reflected waves, respectively, at port two located at $x_2 = l_2$. We can relate the incident and the reflected waves in port one and port two by generalising eqn 7.139 for the characterisation of a two-port circuit, like in figure 7.18, as follows:

$$b_1(l_1) = S_{11}\, a_1(l_1) + S_{12}\, a_2(l_2) \tag{7.142}$$
$$b_2(l_2) = S_{21}\, a_1(l_1) + S_{22}\, a_2(l_2) \tag{7.143}$$

These last two eqns can be written in matrix form:

$$\begin{bmatrix} b_1(l_1) \\ b_2(l_2) \end{bmatrix} = \begin{bmatrix} S_{11} & S_{12} \\ S_{21} & S_{22} \end{bmatrix} \begin{bmatrix} a_1(l_1) \\ a_2(l_2) \end{bmatrix} \tag{7.144}$$

where $a_1(l_1)$, $b_1(l_1)$, $a_2(l_2)$, $b_2(l_2)$ represent the normalised values for the incident and reflected waves at $x_1 = l_1$ and $x_2 = l_2$ as illustrated in figure 7.18. The S-parameters represent the reflection and transmission coefficients for the two-port circuit. From eqns 7.142 and 7.143 we can define each parameter as follows:

$$S_{11} = \left.\frac{b_1(l_1)}{a_1(l_1)}\right|_{a_2(l_2)=0} \quad \text{Input reflection coefficient} \tag{7.145}$$

$$S_{12} = \left.\frac{b_1(l_1)}{a_2(l_2)}\right|_{a_1(l_1)=0} \quad \text{Reverse gain coefficient} \tag{7.146}$$

$$S_{21} = \left.\frac{b_2(l_2)}{a_1(l_1)}\right|_{a_2(l_2)=0} \quad \text{Forward gain coefficient} \tag{7.147}$$

$$S_{22} = \left.\frac{b_2(l_2)}{a_2(l_2)}\right|_{a_1(l_1)=0} \quad \text{Output reflection coefficient} \tag{7.148}$$

Figure 7.19 a) shows an experimental set-up for the measurement or calculation of S_{11} and S_{21}. From this figure it can be observed that to ensure that

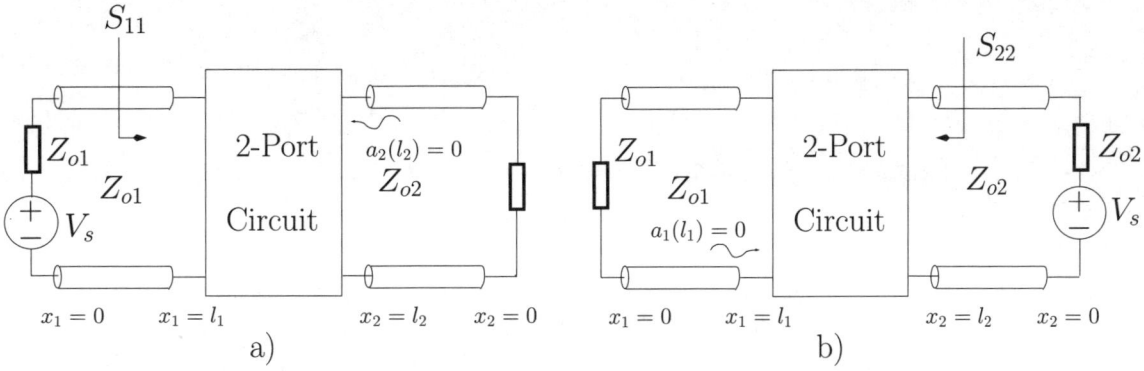

Figure 7.19: *Two-port circuit. a) Calculation of S_{11} and of S_{21}. b) Calculation of S_{12} and of S_{22}.*

$a_2(l_2) = 0$ the output transmission line is terminated with a load equal to its characteristic impedance, that is, S_{11} and S_{21} are determined with the output transmission line matched to Z_{o2}. S_{12} and S_{22} are determined with the input transmission line matched to Z_{o1} (see figure 7.19 b)).

Example 7.4.1 Determine the S-parameters of a series impedance, Z, in a Z_o system.

Solution: S_{11} is calculated as follows (see also figure 7.20):

$$S_{11} = \frac{Z_{IN} - Z_o}{Z_{IN} + Z_o} \tag{7.149}$$

Since $Z_{IN} = Z + Z_o$ we have

$$S_{11} = \frac{Z}{Z + 2Z_o} \tag{7.150}$$

S_{21} is calculated as:

$$S_{21} = \left.\frac{b_2(l_2)}{a_1(l1)}\right|_{a_1(l2)=0}$$
$$= \frac{V_2(l_2) - Z_o\, I_2(l_2)}{V_1(l_1) + Z_o\, I_1(l_1)} \tag{7.151}$$

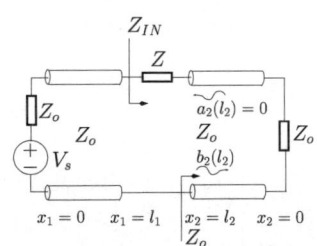

Figure 7.20: *Set-up for the calculation of S_{11} and of S_{21} of a series impedance Z.*

It is known that:

$$\begin{aligned} V_1(l_1) &= Z_{IN}\, I_1(l_1) \\ &= (Z + Z_o)\, I_1(l_1) \end{aligned} \tag{7.152}$$

Also, $V_2(l_2)$ can be related to $V_1(l_1)$ by the voltage divider expression:

$$V_2(l_2) = \frac{Z_o}{Z + Z_o} V_1(l_1) \tag{7.153}$$

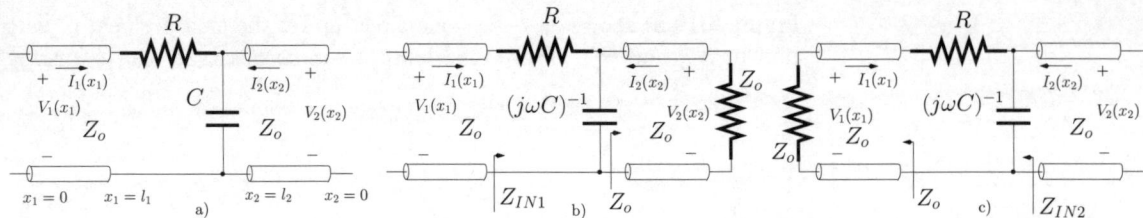

Figure 7.21: *a) RC low-pass filter. b) Set-up for the calculation of S_{11} and S_{21}. c) Set-up for the calculation of S_{12} and S_{22}.*

$a_2(l_2) = 0$ implying that $V_2(l_2) = -Z_o I_2(l_2)$ (see also eqn 7.140). Hence we have

$$S_{21} = \frac{2 V_2(l_2)}{V_1(l_1)\left(1 + \frac{Z_o}{Z_o + Z}\right)}$$

$$= \frac{2Z_o}{Z + 2Z_o} \quad (7.154)$$

From symmetry considerations it is straightforward to conclude that $S_{22} = S_{11}$ and that $S_{12} = S_{21}$.

Example 7.4.2 Determine the S-parameters of the low-pass filter shown in figure 7.21 a) in a Z_o system.

Solution: S_{11} and S_{21} are calculated from the circuit of 7.21 b) where we consider that the output transmission line is matched. From its definition S_{11} can be calculated as

$$S_{11} = \frac{Z_{IN1} - Z_o}{Z_{IN1} + Z_o} \quad (7.155)$$

with Z_{IN1} given by

$$Z_{IN1} = R + \left(Z_o \| \frac{1}{j\omega C}\right)$$

$$= R + \frac{Z_o}{1 + j\omega C Z_o} \quad (7.156)$$

Therefore we can write

$$S_{11} = \frac{R + j\omega C Z_o (R - Z_o)}{R + 2Z_o + 2j\omega C Z_o (R + Z_o)} \quad (7.157)$$

S_{21} is given by

$$S_{21} = \left.\frac{b_2(l_2)}{a_1(l1)}\right|_{a_1(l2)=0}$$

$$= \frac{V_2(l_2) - Z_o I_2(l_2)}{V_1(l_1) + Z_o I_1(l_1)} \quad (7.158)$$

7. RF circuit analysis techniques

Since $a_2(l_2) = 0 \Rightarrow V_2(l_2) = -Z_o I_2(l_2)$ (see also eqn 7.140), the last eqn can be written as

$$S_{21} = \frac{2 V_2(l_2)}{V_1(l_1) + Z_o I_1(l_1)} \qquad (7.159)$$

$V_2(l_2)$ can be related to $V_1(l_1)$ using the impedance voltage divider formula:

$$V_2(l_2) = V_1(l_1) \frac{\frac{Z_o}{1+j\omega C Z_o}}{R + \frac{Z_o}{1+j\omega C Z_o}}$$

$$= V_1(l_1) \frac{Z_o}{Z_o + R + j\omega C Z_o R} \qquad (7.160)$$

Using the result of the last eqn and taking also that $V_1(l_1) = Z_{IN1} I_1(l_1)$ we can calculate S_{21} as follows:

$$S_{21} = \frac{2 Z_o}{2 Z_o + R + j\omega C Z_o (R + Z_o)} \qquad (7.161)$$

S_{22} and S_{12} are calculated considering the circuit of 7.21 c) where we have the input transmission line terminated by a load equal to its characteristic impedance. Z_{IN2} is given by

$$Z_{IN2} = (R + Z_o) || \frac{1}{j\omega C}$$

$$= \frac{R + Z_o}{1 + j\omega C (Z_o + R)} \qquad (7.162)$$

Therefore we can write

$$S_{22} = \frac{R - j\omega C Z_o (Z_o + R)}{R + 2 Z_o + j\omega C Z_o (Z_o + R)} \qquad (7.163)$$

S_{12} is given by

$$S_{12} = \left. \frac{b_1(l_1)}{a_2(l2)} \right|_{a_1(l1)=0}$$

$$= \frac{V_1(l_1) - Z_o I_1(l_1)}{V_2(l_2) + Z_o I_2(l_2)} \qquad (7.164)$$

Since $a_1(l_1) = 0 \Rightarrow V_1(l_1) = -Z_o I_1(l_1)$ (see also eqn 7.140), the last eqn can be written as

$$S_{12} = \frac{2 V_1(l_1)}{V_2(l_2) + Z_o I_2(l_2)} \qquad (7.165)$$

$V_1(l_1)$ can be related to $V_2(l_2)$ using the impedance voltage divider formula:

$$V_1(l_1) = V_2(l_2) \frac{Z_o}{R + Z_o}$$

Using the result of the last eqn and since $V_2(l_2) = Z_{IN2} I_2(l_2)$, we can calculate S_{12} as follows:

$$S_{12} = \frac{2 Z_o}{2 Z_o + R + j\omega C Z_o (R + Z_o)} \qquad (7.166)$$

Reference planes

The practical measurement of S-parameters requires the usage of connecting cables, which are effectively transmission lines, between measurement instruments and the circuit to be measured. Thus, these measurements probe the S-parameters at the inputs of these connecting cables; s'_{ij}. Since the transmission lines impose a phase shift which depends on their physical lengths and the measurement frequency, it is possible to relate the measured S-parameters with the S-parameters of the circuit under test as follows:

$$\begin{bmatrix} S_{11} & S_{12} \\ S_{21} & S_{22} \end{bmatrix} = \begin{bmatrix} S'_{11} e^{2j\theta_1} & S'_{12} e^{j(\theta_1+\theta_2)} \\ S'_{21} e^{j(\theta_1+\theta_2)} & S'_{22} e^{2j\theta_2} \end{bmatrix} \qquad (7.167)$$

where θ_1 and θ_2 represent the electrical lengths of the transmission lines at the circuit input and output, respectively (see figure 7.22).

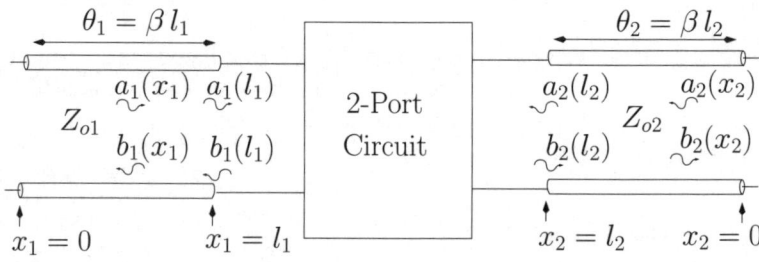

Figure 7.22: *Measurement of the S-parameters for a two-port circuit. θ_1 and θ_2 are the electrical lengths imposed by the input and output connecting cables, respectively.*

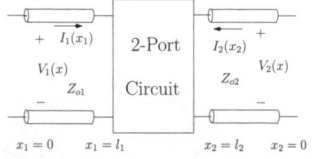

Figure 7.23: *S-parameters and travelling waves (voltage and current).*

7.4.1 S-parameters and power waves

Using eqns 7.126 to 7.139 for the circuit shown in figure 7.23 we can write the following general eqns:

$$a_i(x_i) = \frac{1}{2\sqrt{Z_{oi}}} [V_i(x_i) + Z_{oi} I_i(x_i)] \qquad (7.168)$$

$$b_i(x_i) = \frac{1}{2\sqrt{Z_{oi}}} [V_i(x_i) - Z_{oi} I_i(x_i)] \qquad (7.169)$$

with $i = 1$ for port one and $i = 2$ for port two. The average power in port one at $x_1 = 0$, $P_1(0)$, is equal to;

$$P_1(0) = \frac{1}{2} \text{Real} [V_1(0) I_1^*(0)] \qquad (7.170)$$

It can be shown (see example 7.4.3) that $P_1(0)$ can be expressed in terms of $a_1(0)$ and $b_1(0)$ as follows:

$$P_1(0) = \frac{1}{2}|a_1(0)|^2 - \frac{1}{2}|a_2(0)|^2 \qquad (7.171)$$

where $1/2\,|a_1(0)|^2$ and $1/2\,|b_1(0)|^2$ are usually termed the incident power, $P_1^+(0)$, and the reflected power, $P_1^-(0)$, at the input port, respectively. $P_1^+(0)$ and $P_1^-(0)$ can also be expressed as:

$$
\begin{aligned}
P_1^+(0) &= \frac{1}{2}\text{Real}\left[V_1^+(0)I_1^{+*}(0)\right] \\
&= \frac{1}{2}\frac{|V_1^+(0)|^2}{Z_{o1}}
\end{aligned}
\tag{7.172}
$$

$$
\begin{aligned}
P_1^-(0) &= \frac{1}{2}\text{Real}\left[V_1^-(0)I_1^{-*}(0)\right] \\
&= \frac{1}{2}\frac{|V_1^-(0)|^2}{Z_{o1}}
\end{aligned}
\tag{7.173}
$$

For the circuit in figure 7.24, and for $x_1 = 0$ we can write:

$$
\begin{aligned}
V_1(0) &= V_s - Z_s\, I_1(0) \\
&= V_s - Z_{o1}\, I_1(0)
\end{aligned}
\tag{7.174}
$$

$$
\begin{aligned}
a_1(0) &= \frac{1}{2\sqrt{Z_{o1}}}[V_1(0) + Z_{o1}\, I_1(0)] \\
&= \frac{1}{2\sqrt{Z_{o1}}}V_s
\end{aligned}
\tag{7.175}
$$

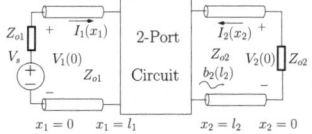

Figure 7.24: *Calculation of S_{11} and of S_{21}. The travelling wave concept.*

that is,

$$
|a_1(0)|^2 = \frac{1}{4}\frac{|V_s|^2}{Z_{o1}}
\tag{7.176}
$$

and for this situation we have the incident power wave given as:

$$
P_1^+(0) = \frac{1}{8}\frac{|V_s|^2}{Z_{o1}}
\tag{7.177}
$$

This power is the maximum power that a source can deliver to a load. It is known as the *available power* (P_{AV}). For a load impedance to absorb P_{AV} the source impedance, Z_s, must be equal to Z_{o1}.

For a lossless transmission line we have

$$
a_1(l_1) = a_1(0)\, e^{-j\beta_1 l_1}
\tag{7.178}
$$
$$
b_1(l_1) = b_1(0)\, e^{-j\beta_1 l_1}
\tag{7.179}
$$

with β_1 representing the propagation constant of the input transmission line. Therefore, we have that $P_1^+(l_1) = P_1^+(0)$, $P_1^-(l_1) = P_1^-(0)$ and $P_1(l_1) = P_1(0)$. Hence, we can write $P_1(l_1)$ as

$$
\begin{aligned}
P_1(l_1) &= P_1^+(l_1) - P_1^-(l_1) \\
&= P_{AV} - P_1^-(l_1) \\
&= P_{AV} - \frac{1}{2}|b_1(l_1)|^2
\end{aligned}
\tag{7.180}
$$

and we can write

$$|S_{11}|^2 = \left.\frac{|b_1(l_1)|^2}{|a_1(l_1)|^2}\right|_{a_2(l_2)=0}$$

$$= \left.\frac{\frac{1}{2}|b_1(l_1)|^2}{\frac{1}{2}|a_1(l_1)|^2}\right|_{a_2(l_2)=0}$$

$$= \frac{P_{AV} - P_1(l_1)}{P_{AV}} \quad (7.181)$$

and eqn 7.180 can be expressed as

$$P_1(l_1) = P_{AV}(1 - |S_{11}|^2) \quad (7.182)$$

We conclude that $|S_{11}|^2$ represents the ratio of the reflected power to the available power in port one.

Example 7.4.3 Show that $P_1(0)$ given by eqn 7.170, in figure 7.24, can be expressed in terms of $a_1(0)$ and $b_1(0)$ as follows:

$$P_1(0) = \frac{1}{2}|a_1(0)|^2 - \frac{1}{2}|a_2(0)|^2 \quad (7.183)$$

Solution: Using eqns 7.168 and 7.169 we can write eqn 7.183 as

$$P_1(0) = \frac{1}{8Z_{o1}}[V_1(0) + Z_{o1} I_1(0)][V_1^*(0) + Z_{o1} I_1^*(0)]$$

$$- \frac{1}{8Z_{o1}}[V_1(0) - Z_{o1} I_1(0)][V_1^*(0) - Z_{o1} I_1^*(0)]$$

$$= \frac{1}{8Z_{o1}} 4\, \text{Real}\,[Z_{o1} V_1(0) I_1^*(0)]$$

$$= \frac{1}{2}\text{Real}\,[V_1(0)I_1^*(0)] \quad (7.184)$$

The average power in port one at $x_2 = 0$, that is the power delivered to the load, $P_2(0)$, is equal to;

$$P_2(0) = \frac{1}{2}\text{Real}\,[-V_2(0)I_2^*(0)] \quad (7.185)$$

It can be shown (see example 7.4.4) that $P_2(0)$ can be expressed in terms of $a_2(0)$ and $b_2(0)$ as follows:

$$P_2(0) = \frac{1}{2}|b_2(0)|^2 - \frac{1}{2}|a_1(0)|^2 \quad (7.186)$$

For the circuit in figure 7.24 we can write:

$$V_2(0) = -Z_{o2} I_2(0) \quad (7.187)$$

so that

$$a_2(0) = \frac{1}{2\sqrt{Z_{o2}}}[V_2(0) + Z_{o2} I_2(0)] = 0 \qquad (7.188)$$

The result that $a_2(0) = 0$ is expected since the output transmission line is matched with $Z_L = Z_{o2}$. Using the result of the last eqn we can write $b_2(0)$ as

$$\begin{aligned} b_2(0) &= \frac{1}{2\sqrt{Z_{o2}}}[V_2(0) - Z_{o2} I_2(0)] \\ &= \frac{V_2(0)}{\sqrt{Z_{o2}}} \\ &= -\sqrt{Z_{o2}} I_2(0) \end{aligned} \qquad (7.189)$$

From eqns 7.186 and 7.188 we observe that the power delivered to the load $Z_L = Z_{o2}$, $P_2(0)$, is

$$P_2(0) = \frac{1}{2}|b_2(0)|^2$$

The transmission coefficient S_{21} is given by:

$$S_{21} = \left. \frac{b_2(l_2)}{a_1(l_1)} \right|_{a_2(l_2)=0} \qquad (7.190)$$

Since the transmission lines are lossless we have

$$\begin{aligned} a_1(l_1) &= a_1(0)e^{-j\beta_1 l_1} \\ &= \frac{1}{2\sqrt{Z_{o2}}} V_s e^{-j\beta_1 l_1} \end{aligned} \qquad (7.191)$$

$$b_2(l_2) = b_2(0)e^{-j\beta_2 l_2} \qquad (7.192)$$

where β_1 and β_2 are the propagation constants of the input and output transmission lines, respectively. $|S_{21}|^2$ can be written as

$$\begin{aligned} |S_{21}|^2 &= \left. \frac{\frac{1}{2}|b_2(l_2)|^2}{\frac{1}{2}|a_1(l_1)|^2} \right|_{a_2(l_2)=0} \\ &= \frac{P_2(0)}{P_{AVs}} \end{aligned} \qquad (7.193)$$

and we conclude that $|S_{21}|^2$ represents the ratio of the power delivered to the load $Z_L = Z_{o2}$ to the available power, P_{AV}. It follows that $|S_{21}|^2$ represents a power gain, G_T, named the 'Transducer Power Gain'. Note that if the source impedance and the load impedance are not equal to the characteristic impedances Z_{o1} and Z_{o2}, respectively, the power gain is different from that given by eqn 7.193.

Similar analysis of the circuit of figure 7.25 shows us that $|S_{22}|^2$ represents the ratio of the reflected power to the available power in port two while $|S_{12}|^2$ is the reverse transducer power gain. This analysis is similar to that used to obtain $|S_{11}|^2$ and $|S_{21}|^2$. However, now we apply the source V_s to port two.

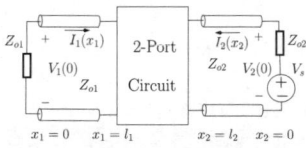

Figure 7.25: *Calculation of S_{12} and of S_{22}.*

Example 7.4.4 Show that $P_2(0)$ given by eqn 7.185, in figure 7.24, can be expressed in terms of $a_2(0)$ and $b_2(0)$ as follows:

$$P_2(0) = \frac{1}{2}|b_2(0)|^2 - \frac{1}{2}|a_1(0)|^2 \qquad (7.194)$$

<u>Solution</u>: Using eqns 7.168 and 7.169 we can write eqn 7.194 as

$$\begin{aligned} P_1(0) &= \frac{1}{8Z_{o1}}[V_2(0) - Z_{o2}I_2(0)][V_2^*(0) - Z_{o2}I_2^*(0)] \\ &\quad - \frac{1}{8Z_{o1}}[V_2(0) + Z_{o2}I_2(0)][V_2^*(0) + Z_{o2}I_2^*(0)] \\ &= -\frac{1}{8Z_{o2}}4\,\text{Real}\,[Z_{o2}V_2(0)I_2^*(0)] \\ &= \frac{1}{2}\text{Real}\,[-V_2(0)I_2^*(0)] \qquad (7.195)\end{aligned}$$

7.4.2 Power waves and generalised S-parameters

The representation of the voltage and current in terms of incident and reflected waves is quite natural when we deal with transmission lines. The generalisation of this concept to circuits described by lumped elements is attractive specially when such circuits are considered together with distributed circuits. This is made possible by generalising the concept of power waves.

Power waves

Let us consider the impedance voltage divider of figure 7.26. For this circuit it is not possible to normalise the waveforms to the characteristic impedance since this impedance is meaningless for this circuit. It is possible, however, to consider new waveforms which can be normalised to the source impedance, Z_s. Such waveforms are called 'power waves':

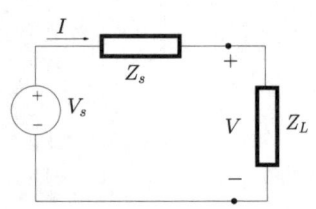

Figure 7.26: *Impedance voltage divider.*

$$a_p = \frac{1}{2\sqrt{R_s}}(V + Z_s I) \qquad (7.196)$$

$$b_p = \frac{1}{2\sqrt{R_s}}(V - Z_s^* I) \qquad (7.197)$$

$$R_s = \text{Real}\,[Z_s] \qquad (7.198)$$

The average power delivered to the load can be expressed as:

$$\begin{aligned} P_L &= \frac{1}{2}\text{Real}\,[V I^*] = \frac{1}{2}\text{Real}\,[Z_L I I^*] = \frac{1}{2}|I|^2\text{Real}\,[Z_L] \\ &= \frac{1}{2}\left|\frac{V_s}{Z_L + Z_s}\right|^2 \text{Real}\,[Z_L] \\ &= \frac{|V_s|^2}{8R_s} \quad \text{if} \quad Z_L = Z_s^* \qquad (7.199)\end{aligned}$$

7. RF circuit analysis techniques

that is, if $Z_L = Z_s^*$ the power absorbed by the load is maximum and equal to the available power, P_{AV}.

Example 7.4.5 Show that the available power can be expressed as:

$$P_{AV} = \frac{1}{2}|a_p|^2 \tag{7.200}$$

and that the power delivered to the load can be expressed as:

$$P_L = \frac{1}{2}|a_p|^2 - \frac{1}{2}|b_p|^2 \tag{7.201}$$

Solution: From eqn 7.196 we can write:

$$\frac{1}{2}|a_p|^2 = \frac{1}{8R_s}|V + Z_s I|^2 \tag{7.202}$$

and from figure 7.26 we can write $V_s = V + I Z_s$. Thus, eqn 7.202 can be written as

$$\frac{1}{2}|a_p|^2 = \frac{|V_s|^2}{8R_s} = P_{AV} \tag{7.203}$$

From eqns 7.196 and 7.197 we can write:

$$\frac{1}{2}|a_p|^2 = \frac{1}{8R_s}(V + Z_s I)(V^* + Z_s^* I^*) \tag{7.204}$$

$$\frac{1}{2}|b_p|^2 = \frac{1}{8R_s}(V - Z_s^* I)(V^* - Z_s I^*) \tag{7.205}$$

and

$$\begin{aligned}
\frac{1}{2}|a_p|^2 - \frac{1}{2}|b_p|^2 &= \frac{1}{8R_s}(2\text{Real}[V I^* Z_s^*] + 2\text{Real}[V I^* Z_s]) \\
&= \frac{1}{8R_s}(4\text{Real}[V I^*]\,\text{Real}[Z_s]) \\
&= \frac{1}{2}\text{Real}[V I^*] = P_L
\end{aligned} \tag{7.206}$$

It should be noted that $1/2|b_p|^2$ represents the reflected power. If $Z_L = Z_s^*$ then $b_p = 0$ as expected.

The reflection coefficient for power waves can be defined as:

$$\Gamma_p = \frac{b_p}{a_p} = \frac{V - Z_s^* I}{V + Z_s I} = \frac{Z_L - Z_s^*}{Z_L + Z_s} \tag{7.207}$$

It is possible to write:

$$\begin{aligned}
P_L &= \frac{1}{2}|a_p|^2\left(1 - \frac{|b_p|^2}{|a_p|^2}\right) \\
&= P_{AV}(1 - |\Gamma_p|^2)
\end{aligned} \tag{7.208}$$

This is consistent with our earlier discussion showing that maximum power is delivered to the load when perfect matching is achieved ($\Gamma_p = 0$).

Generalised S Parameters

From eqns 7.196 and 7.197 we can write:

$$V = \frac{1}{\sqrt{R_s}}[Z_s^* a_p + Z_s b_p] \tag{7.209}$$

$$I = \frac{1}{\sqrt{R_s}}[a_p - b_p] \tag{7.210}$$

Now it is possible to define incident and reflected voltage and current waves as follows

$$V = V_p^+ + V_p^- \tag{7.211}$$

$$I = I_p^+ - I_p^- \tag{7.212}$$

with

$$V_p^+ = \frac{Z_s^* a_p}{\sqrt{R_s}} \tag{7.213}$$

$$V_p^- = \frac{Z_s b_p}{\sqrt{R_s}} \tag{7.214}$$

$$I_p^+ = \frac{a_p}{\sqrt{R_s}} \tag{7.215}$$

$$I_p^- = \frac{b_p}{\sqrt{R_s}} \tag{7.216}$$

so that

$$V_p^+ = Z_s^* I_p^+ \tag{7.217}$$

and

$$V_p^- = Z_s I_p^- \tag{7.218}$$

The reflection coefficients for voltage and current waves are defined as follows

$$\Gamma_V = \frac{V_p^-}{V_p^+} = \frac{Z_s}{Z_s^*}\Gamma_p \tag{7.219}$$

$$\Gamma_I = \frac{I_p^-}{I_p^+} = \Gamma_p \tag{7.220}$$

with Γ_p given by eqn 7.207. When the impedance Z_s is a positive real quantity, the expressions for a_p and b_p are identical to those derived for a and b in the context of transmission lines. For this situation we have, $Z_s = Z_s^* = Z_o$. Therefore:

$$\Gamma_o = \Gamma_V = \Gamma_I = \Gamma_p = \frac{Z_L - Z_o}{Z_L + Z_o} \tag{7.221}$$

With these definitions and the normalisations presented above, we are in a position to determine the generalised S-parameters for a two-port circuit denoted below as S_{pij} (see figure 7.27).

$$b_{p1} = S_{p11} a_{p1} + S_{p12} a_{p2} \tag{7.222}$$

$$b_{p2} = S_{p21} a_{p1} + S_{p22} a_{p2} \tag{7.223}$$

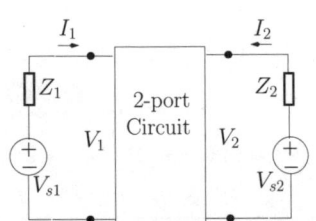

Figure 7.27: *Calculation of S_p of a two-port circuit.*

7. RF circuit analysis techniques

with

$$a_{pi} = \frac{1}{2\sqrt{R_i}}(V_i + Z_i I_i), \quad i = 1, 2 \qquad (7.224)$$

$$b_{pi} = \frac{1}{2\sqrt{R_i}}(V_i - Z_i^* I_i), \quad i = 1, 2 \qquad (7.225)$$

with Z_i representing the *reference* impedances shown in figure 7.27. $R_i = \text{Real}[Z_i]$.

S_{p11} and S_{p21} are calculated after setting $V_{s2} = 0$ in figure 7.27. From eqn 7.207, S_{p11} is calculated as follows:

$$S_{p11} = \frac{Z_{IN1} - Z_1^*}{Z_{IN1} + Z_1} \qquad (7.226)$$

where Z_{IN1} represents the circuit input impedance when port two is terminated by Z_2.

For the calculation of S_{p21} we can determine the power delivered to the load P_L which is given by;

$$P_L = \frac{1}{2}|b_{p2}|^2 = \frac{1}{2}|S_{p21}|^2|a_{p1}|^2, \quad \text{and if } a_{p2} = 0 \qquad (7.227)$$

Hence,

$$|S_{p21}|^2 = \frac{\frac{1}{2}|b_{p2}|^2}{\frac{1}{2}|a_{p1}|^2}$$

$$= \frac{P_L}{P_{AV}} \qquad (7.228)$$

$|S_{p21}|^2$ is also called a transducer power gain, G_T. Note the similarity of the last eqn with eqn 7.193.

The calculations of S_{p22} and of S_{p12} are similar to the calculation of S_{p11} and of S_{p21}, respectively, but now setting $V_{s1} = 0$. It should be noted that if $Z_1 = Z_2 = Z_o$ (real), the results obtained for the S_p-parameters are identical to those obtained for the S-parameters.

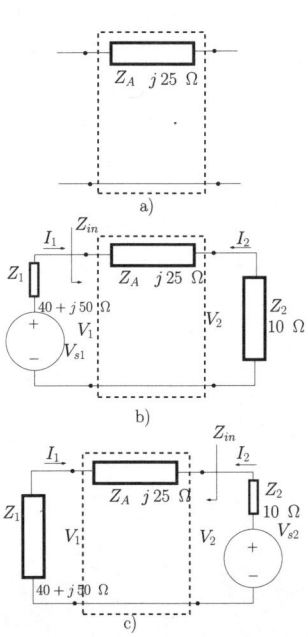

Figure 7.28: *Calculation of S_{p11} and of S_{p21} of a two-port circuit.*

Example 7.4.6 Consider the two-port network of figure 7.28 a)

1. Determine S_{p11} and S_{p21} of the two-port circuit using figure 7.28 b).

2. Determine S_{p12} and S_{p22} of the two-port circuit using figure 7.28 c).

<u>Solution</u>:

1. Applying Ohm's law to the circuit of figure 7.28 b) we find that:

$$I_1 = \frac{V_{s1}}{Z_1 + Z_2 + Z_A}$$
$$= V_{s1}\, 0.01\, e^{-j\, 0.98} \text{ A}$$
$$I_2 = -I_1$$

262 7. RF circuit analysis techniques

$$\begin{aligned}
&= V_{s1}\,0.01\,e^{j\,2.16}\text{ A}\\
V_1 &= (Z_A + Z_2)\,I_1\\
&= V_{s1}\,0.30\,e^{j\,0.21}\text{ V}\\
V_2 &= Z_2\,I_1\\
&= V_{s1}\,0.11\,e^{-j\,0.98}\text{ V}\\
Z_{in} &= (Z_A + Z_2)\\
&= 26.93\,e^{j\,1.19}\;\Omega
\end{aligned}$$

Using eqns 7.224 and 7.225 we obtain

$$\begin{aligned}
a_{p1} &= V_{s1}\,0.08\,e^{j\,0} = V_{s1}\,0.08\\
b_{p2} &= V_{s1}\,0.04\,e^{-j\,0.98}
\end{aligned}$$

and we have

$$\begin{aligned}
S_{p11} &= \frac{Z_{in} - Z_1^*}{Z_{in} + Z_1}\\
&= 0.90\,e^{-j\,0.97}\\
S_{p21} &= \left.\frac{b_{p2}}{a_{p1}}\right|_{a_{p2}=0}\\
&= 0.44\,e^{-j\,0.98}
\end{aligned}$$

2. Applying Ohm's law to the circuit of figure 7.28 c) we find that:

$$\begin{aligned}
I_2 &= \frac{V_{s2}}{Z_1 + Z_2 + Z_A}\\
&= V_{s2}\,0.01\,e^{-j\,0.98}\text{ A}\\
I_1 &= -I_2\\
&= V_{s2}\,0.01\,e^{j\,2.16}\text{ A}\\
V_2 &= (Z_A + Z_2)\,I_2\\
&= V_{s2}\,0.94\,e^{j\,0.98}\text{ V}\\
V_1 &= Z_1\,I_2\\
&= V_{s2}\,0.71\,e^{-j\,0.09}\text{ V}\\
Z_{in} &= (Z_A + Z_1)\\
&= 85\,e^{j\,1.08}\;\Omega
\end{aligned}$$

Using eqns 7.224 and 7.225 we obtain

$$\begin{aligned}
a_{p2} &= V_{s2}\,0.16\,e^{j\,0} = V_{s2}\,0.16\\
b_{p1} &= V_{s2}\,0.07\,e^{-j\,0.98}
\end{aligned}$$

and we have

$$S_{p22} = \frac{Z_{in} - Z_2^*}{Z_{in} + Z_2}$$

7. RF circuit analysis techniques

$$S_{p12} = \left. \frac{b_{p1}}{a_{p2}} \right|_{a_{p1}=0} \begin{aligned} &= 0.90\, e^{j\, 21} \\ &\\ &= 0.44\, e^{-j\, 0.98} \end{aligned}$$

7.4.3 Conversions between different two-port parameters

S-parameters are normally obtained from RF measurements and are quoted by manufacturers of high frequency devices and circuits. On the other hand, for circuit analysis, design and optimisation it is, sometimes, more convenient to use other two-port circuit parameters such as those studied in Chapter 5. It is possible to convert between S-parameters and other circuit parameters using elementary matrix algebra. For example, the conversion between Z-parameters and S-parameters can be obtained using the following mathematical manipulation:

$$[\boldsymbol{V}] = [\boldsymbol{Z}][\boldsymbol{I}] \tag{7.229}$$

where

$$[\boldsymbol{Z}] = \begin{bmatrix} Z_{11} & Z_{12} \\ Z_{21} & Z_{22} \end{bmatrix} \tag{7.230}$$

$$[\boldsymbol{V}] = \begin{bmatrix} V_1 \\ V_2 \end{bmatrix} \tag{7.231}$$

$$[\boldsymbol{I}] = \begin{bmatrix} I_1 \\ I_2 \end{bmatrix} \tag{7.232}$$

Equation 7.229 can be generalised as follows

$$[\boldsymbol{V}^+] + [\boldsymbol{V}^-] = [\boldsymbol{Z}]\,([\boldsymbol{I}^+] - [\boldsymbol{I}^-]) \tag{7.233}$$

to include incident and reflected quantities. Recall that

$$[\boldsymbol{V}^+] = [\boldsymbol{Z_o}][\boldsymbol{I}^+] \tag{7.234}$$

and that

$$[\boldsymbol{V}^-] = [\boldsymbol{Z_o}][\boldsymbol{I}] \tag{7.235}$$

with

$$[\boldsymbol{Z_o}] = Z_o\,[\boldsymbol{1}] \tag{7.236}$$

where $[\boldsymbol{1}]$ represents the identity matrix, that is

$$[\boldsymbol{1}] = \begin{bmatrix} 1 & 0 \\ 0 & 1 \end{bmatrix} \tag{7.237}$$

Equation 7.233 becomes

$$([\boldsymbol{Z_o}] + [\boldsymbol{Z}])\,[\boldsymbol{I}^-] = ([\boldsymbol{Z}] - [\boldsymbol{Z_o}])\,[\boldsymbol{I}^+] \tag{7.238}$$

$$[S] = \frac{[V^-]}{[V^+]} = \frac{[I^-]}{[I^+]}$$
$$= ([Z] + [Z_o])^{-1} ([Z] - [Z_o]) \quad (7.239)$$

It should be noted that
$$\frac{[A]}{[B]} = [B]^{-1} [A] \quad (7.240)$$

and that
$$[S] = \begin{bmatrix} S_{11} & S_{12} \\ S_{21} & S_{22} \end{bmatrix} \quad (7.241)$$

Similarly, conversions can be made between the other electrical parameters such as impedance, chain (or ABCD), etc. In appendix C we present a table with the conversions between the main electrical parameters including the S-parameters.

Example 7.4.7 Derive the S-parameters from the admittance parameters.

Solution:
$$[I] = [Y][V] \quad (7.242)$$

where
$$[Y] = \begin{bmatrix} Y_{11} & Y_{12} \\ Y_{21} & Y_{22} \end{bmatrix} \quad (7.243)$$

Equation 7.242 can be expanded as:
$$[I^+] - [I^-] = [Y]([V^+] + [V^-]) \quad (7.244)$$

Using eqns 7.234 and 7.235 we can write:
$$([1] + [Y][Z_o])[I^-] = ([1] - [Y][Z_o])[I^+] \quad (7.245)$$

Finally, the S-parameters are expressed as
$$[S] = \frac{[I^-]}{[I^+]} = ([1] - [Y][Z_o])^{-1} ([1] + [Y][Z_o]) \quad (7.246)$$

7.5 The Smith chart

The analysis of impedance and transmission line matching problems using analytical eqns can be cumbersome. The Smith chart is a powerful tool which provides a graphical analysis of such problems.

7.5.1 The impedance and the reflection coefficient planes

In essence the Smith chart is a graphical representation of impedances in a plane called 'reflection coefficient plane' – the Γ plane. This representation is

7. RF circuit analysis techniques

valid for all values of Z (usually for Real $[Z] \geq 0$). The normalised impedance plane is defined according to:

$$z = \frac{Z}{Z_o} = \frac{R + jX}{Z_o} = r + jx \qquad (7.247)$$

where Z_o is a real (non-complex) number representing either the characteristic impedance or a reference impedance as discussed previously for S- and generalised S-parameters, respectively. The reflection coefficient plane can now be defined as follows:

$$\begin{aligned}
\Gamma &= \frac{Z - Z_o}{Z + Z_o} \\
&= \frac{\frac{Z}{Z_o} - 1}{\frac{Z}{Z_o} + 1} \\
&= \frac{z - 1}{z + 1} = U + jV \qquad (7.248)
\end{aligned}$$

Using the same normalisation as used for impedance (eqn 7.247) we can write:

$$\Gamma = \frac{r - 1 + jx}{r + 1 + jx} = U + jV \qquad (7.249)$$

that is

$$U = \frac{r^2 - 1 + x^2}{(r + 1)^2 + x^2} \qquad (7.250)$$

$$V = \frac{2x}{(r + 1)^2 + x^2} \qquad (7.251)$$

The last two eqns allow for the transformation from the normalised impedance z-plane to the reflection coefficient Γ-plane and allows the variables r and x to be mapped as circles in the Γ plane as explained below.

Constant resistance circles

If we solve eqn 7.250 in order to obtain x we have

$$x = \pm\sqrt{\frac{r^2 - 1 - U(r + 1)^2}{U - 1}} \qquad (7.252)$$

If we now substitute x in eqn 7.251, we can show that:

$$\left(U - \frac{r}{r + 1}\right)^2 + V^2 = \frac{1}{(r + 1)^2} \qquad (7.253)$$

Note that in this procedure x is eliminated as a variable from the eqns (7.250 and 7.251) that define the transformation to the Γ-plane. This effectively allows us to obtain a representation of impedances with a constant real part (constant resistance) in the Γ plane. Such a representation, given by eqn 7.253, describes a family of circles with centres on the U axis at the points $(r/(r + 1), 0)$ and with radii of $(1 + r)^{-1}$.

Figure 7.29 a) presents impedances of constant resistance in the z-plane while figure 7.29 b) shows these constant resistance impedances mapped into the Γ-plane as given by eqn 7.253.

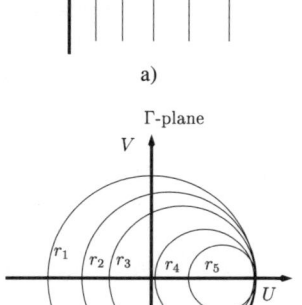

Figure 7.29: *a) Constant resistance impedances. b) Constant resistance circles.*

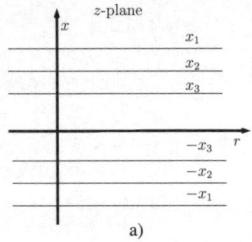

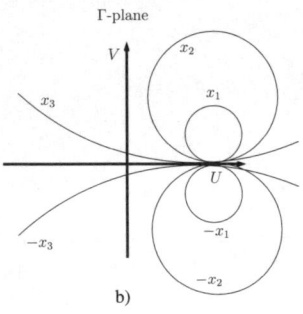

Figure 7.30: *a) Constant reactance impedances. b) Constant reactance circles.*

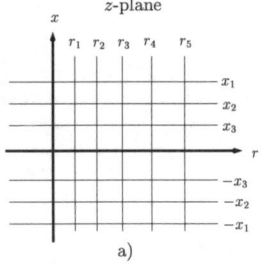

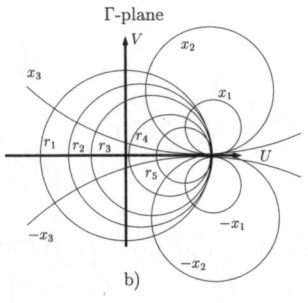

Figure 7.31: *a) Constant reactance and constant resistance impedances. b) The Smith chart.*

Constant reactance circles

If we eliminate r from eqns 7.250 and 7.251 we obtain a representation of complex impedances with a constant reactance part in the Γ plane defined by the following eqn:

$$(U-1)^2 \left(V - \frac{1}{x}\right)^2 = \frac{1}{x^2} \qquad (7.254)$$

which also represents a family of circles. The centres of these circles are now at the points $(1, x^{-1})$ and they have radii of x^{-1}. Figure 7.30 a) presents examples of constant reactance impedances in the z plane and figure 7.30 b) shows the same impedances mapped into the Γ plane.

The combined representation of constant resistance and constant reactance circles is called the Smith chart and is illustrated in figure 7.31 b). The upper part of the chart represents positive reactive (inductive) impedances while the lower part represents negative reactive (capacitive) impedances. The U axis which separates these two regions represents pure resistances.

The Smith chart can also be used to represent admittances by considering another plane, Γ_y, such that:

$$\Gamma_y = \frac{y-1}{y+1} \qquad (7.255)$$

where y represents the admittances normalised to $Y_o = Z_o^{-1}$:

$$y = \frac{Y}{Y_o} = Y Z_o \qquad (7.256)$$

It is left to the reader, as an exercise, to show that:

$$\Gamma_y = -\Gamma = \Gamma e^{j\pi} \qquad (7.257)$$

that is, the admittance map is obtained by rotating the impedance map by 180 degrees.

7.5.2 Representation of impedances

The representation of impedances in the Smith chart is straightforward given the graphical nature of the Γ plane where the constant resistance and constant reactance circles are clearly indicated. Figure 7.32 shows the representation of the following impedances[4] normalised to 50 Ω:

$$z_1 = 1 + j \qquad z_2 = 0.4 + j\,0.5$$
$$z_3 = 3 - j\,3 \qquad z_4 = 0.2 - j\,0.6$$
$$z_5 = 0 \qquad z_6 = 1$$

It is also possible to determine and to represent an impedance given the corre-

---
[4]Note that the Smith chart of figure 7.32 allows us to represent impedances with a real part greater than or equal to zero.

7. RF circuit analysis techniques

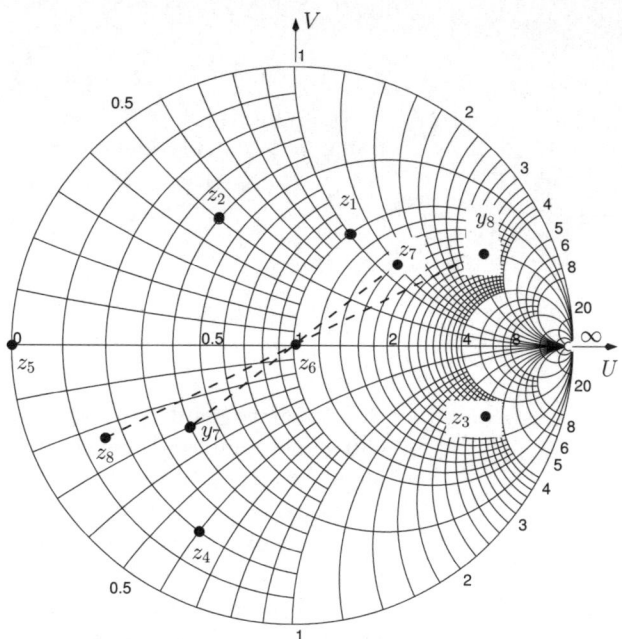

Figure 7.32: *Impedance representation in the Smith chart.*

sponding admittance $y = z^{-1}$. The procedure to determine such an impedance is as follows: first represent the admittance in the impedance chart as if we were representing an impedance. Then, rotate this representation by 180 degrees (around the centre of the chart, see also eqn 7.257) to find the corresponding impedance representation[5]. Figure 7.32 also illustrates the application of this procedure to find the impedances, normalised to 50 Ω, corresponding to the following normalised admittances:

$$y_7 = 0.4 - j\,0.3 \qquad y_8 = 2 + 3\,j$$

From figure 7.32 we can read

$$z_7 = 1.6 + j\,1.2 \qquad z_8 = 0.15 - j\,0.23$$

It is important to note that the normalisation of the admittance is obtained by dividing the admittance by $Y_o = Z_o^{-1}$ while the normalisation of the impedances is obtained by dividing the impedance by Z_o.

In addition to the direct representation of impedances, the Smith chart, by its very nature, allows a straightforward representation of the reflection coefficient and, therefore, allows graphical solutions of the eqns discussed previously in the context of transmission lines (see also figure 7.33). These are:

$$\Gamma_o = \frac{Z_L - Z_o}{Z_L + Z_o} \tag{7.258}$$

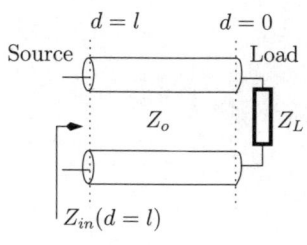

Figure 7.33: *Transmission line.*

---

[5] Recall that the numbers indicated in figure 7.32 indicate the value of constant resistance (conductance) and constant reactance (susceptance) circles.

$$\Gamma_{in}(d) = \Gamma_o e^{-j2\beta d} \quad (7.259)$$

$$Z_{in}(d) = Z_o \frac{e^{j\beta d} + \Gamma_o e^{-j\beta d}}{e^{j\beta d} - \Gamma_o e^{-j\beta d}} \quad (7.260)$$

Using the normalisations $z_L = Z_L/Z_o$ and $z_{in}(d) = Z_{in}(d)/Z_o$ these eqns can be written as:

$$\Gamma_o = \frac{z_L - 1}{z_L + 1} \quad (7.261)$$

$$\Gamma_{in}(d) = \Gamma_o e^{-j2\beta d} \quad (7.262)$$

$$z_{in}(d) = \frac{1 + \Gamma_{in}(d)}{1 - \Gamma_{in}(d)} \quad (7.263)$$

Recall that the Smith chart is plotted in the Γ plane. It should be noted that travelling on a complete circle around the Smith chart corresponds to an electrical length of $2\beta d = 2\pi$ (or $\beta d = \pi$, see eqn 7.259) which, in turn, corresponds to a physical length of the transmission line equal to $\lambda/2$. Recall that $\beta = 2\pi/\lambda$. The direction of movement around the chart is important; 'travelling' towards the signal source corresponds to a clockwise rotation while 'travelling' towards the load corresponds to a counter-clockwise rotation (see also eqn 7.262 and figure 7.33). To illustrate these points consider figure 7.34. Here we show how to determine the input impedance, $Z_{in}(d = l)$, and the reflection coefficient, Γ_o, for a transmission line with a length $l = \lambda/8$ (corresponding to an electrical length $2\beta l = \pi/2$) and terminated with a load impedance $Z_L = 50 + j\,50$ Ω. First, we represent the normalised load impedance $z_L = 1 + j$ in the Smith

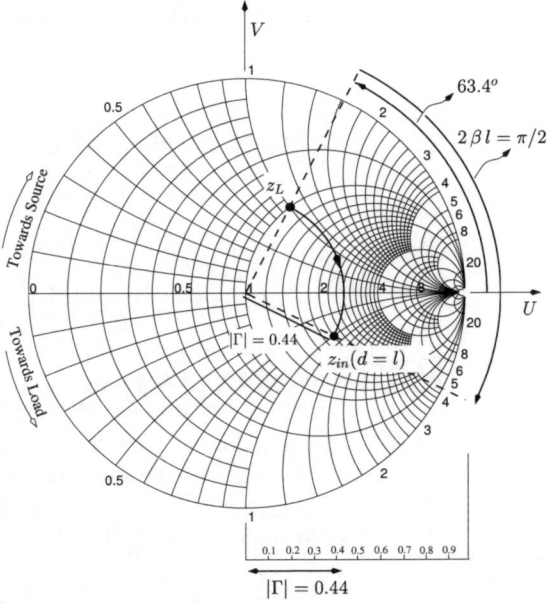

Figure 7.34: *Calculation of transmission line input impedance and reflection coefficient.*

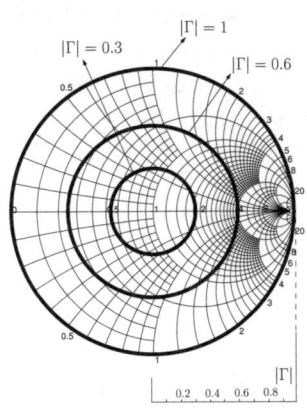

Figure 7.35: *a) Constant Γ circles.*

chart, taken to be at distance $d = 0$. We find the normalised input impedance, $z_{in}(d = l)$ by 'travelling' $2\beta l = \pi/2$, towards the signal source, on a constant $|\Gamma|$ circle. Constant $|\Gamma|$ circles (see figure 7.35) are centred at the centre of the chart ($\Gamma = 0$, corresponding to $z = 1 + j\,0$). Each constant $|\Gamma|$ circle has a radius equal to the magnitude of the reflection coefficient under consideration. For our example the reflection coefficient is Γ_o as defined in eqn 7.261 and is found to be $0.44 \angle 63.4°$. The angle of Γ_o, also represented in figure 7.34, is measured from the U axis, of course. After the rotation described above we obtain $Z_{in}(\lambda/8) = (2 - j) \times 50\,\Omega = 100 - j\,50\,\Omega$.

Example 7.5.1 Determine the length of a short-circuited transmission line, l_{sc}, and the length of an open-circuit transmission line, l_{oc}, such that the input impedance for these transmission lines is $Z_{in} = j\,100\,\Omega$.

Solution:

1. <u>Short-circuited line</u>: First we represent the normalised impedance $z_{in} = j\,100/50 = j\,2$ in the Smith chart as shown in Figure 7.36. Then, travelling from $z_L = 0$, towards the signal source, to $z_{in} = j\,2$ in a constant $|\Gamma|$ circle, we determine the angle $2\beta l_{sc} = 127°$. Knowing that $\beta = 2\pi/\lambda$ we get $l_{sc} = 0.176\lambda$.

2. <u>Open-circuit line</u>: The calculation of l_{oc} is similar to the calculation of l_{sc}. The main difference is that now $z_L = \infty$ (see also Figure 7.36). $l_{oc} = 0.25\lambda + l_{sc} = 0.426\lambda$.

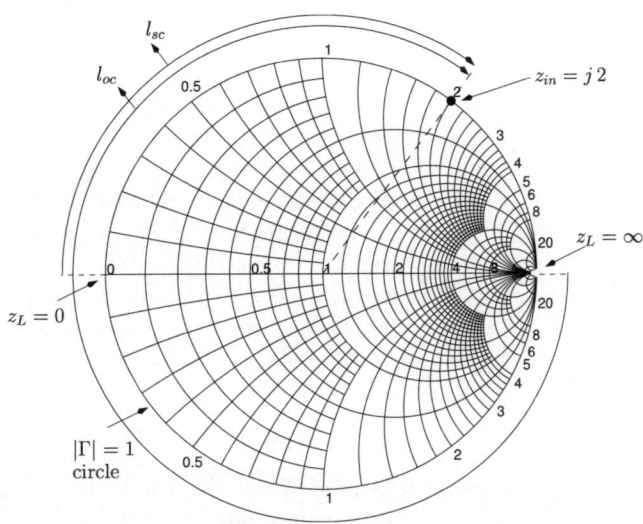

Figure 7.36: *Calculation of the electrical lengths of short-circuited and open-circuit transmission lines. Example 7.5.1.*

The Smith chart is also very useful in representing the impedance or admittance versus frequency of various circuits. Figure 7.37 illustrates the representation of the impedance versus frequency of a resistor-inductor (RL) and resistor-capacitor (RC) series combinations. These representations are done for a frequency range of f_a to f_b. It is clear that for each of the circuits the impedance follows a constant resistance circle, as expected. Note that as the frequency is increased (from f_a to f_b) the reactance of the RL circuit increases while it decreases for the RC circuit.

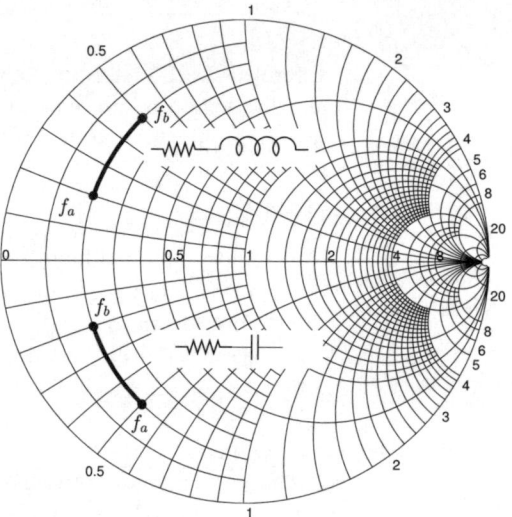

Figure 7.37: *Representation of the impedance versus frequency of a resistor-inductor and resistor-capacitor series combinations.* $f_b > f_a$.

It is also possible to represent the impedance of a *parallel* combination of passive elements. However, such a representation is not so straightforward as the series combinations. Figure 7.38 illustrates the procedure for the case of a resistor in parallel with a capacitor. First we determine the equivalent admittance values for the circuit. For this example, the value of the admittance at frequency f_a is $y(f_a) = 0.3 + j\,0.1$ and the value of the admittance at frequency f_b is $y(f_b) = 0.3 + j\,0.6$. All the admittance values are represented in the Smith chart in a dashed line. Rotating these values by 180 degrees we find the correspondent impedance representation from $z(f_a)$ to $z(f_b)$.

This procedure can be generalised for representing the impedance of any parallel combination of passive elements.

Example 7.5.2 Determine the impedance of a capacitor in parallel with the series connection of a resistor with an inductor. Consider the frequency range 200 MHz $\leq f \leq 900$ MHz. $L = 2$ nH, $C = 12$ pF and R=21.5 Ω.

<u>Solution</u>: we refer now to Fig. 7.39.

7. RF circuit analysis techniques

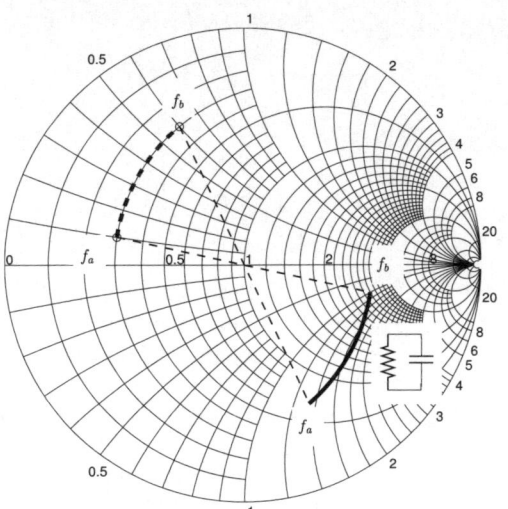

Figure 7.38: *Representation of the impedance versus frequency of a resistor in parallel with a capacitor. $f_b > f_a$.*

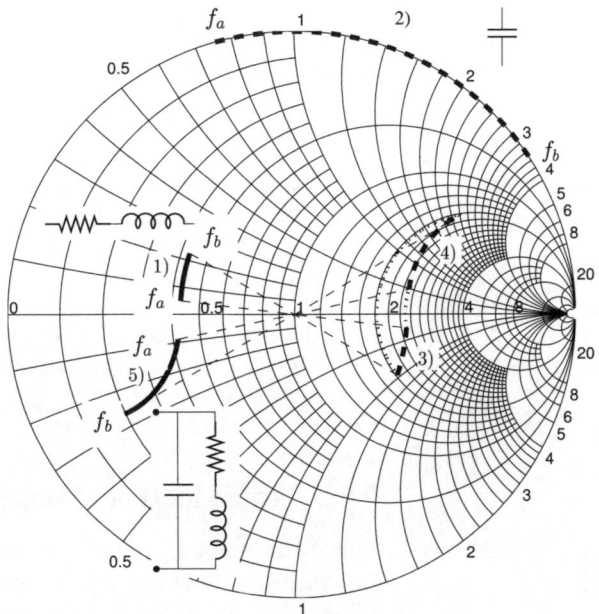

Figure 7.39: *Representation of the impedance versus frequency of a capacitor in parallel with the series connection of a resistor and an inductor. $f_b > f_a$. Example 7.5.2.*

1. First we normalise the admittances and impedances to 50 Ω. This gives

$$y_{cap}(f) \;=\; (j\,2\,\pi\,f\,C)\,50$$

$$
\begin{aligned}
&= j2\pi f 6 \times 10^{-10} \\
z_{ind}(f) &= \frac{j2\pi f L}{50} \\
&= j2\pi f 4 \times 10^{-11} \\
r &= \frac{R}{50} \\
&= 0.43
\end{aligned}
$$

2. Then we represent the normalised impedance resulting from the series combination of the resistor with the inductor for the frequency range mentioned above

$$0.43 + j2\pi f 4 \times 10^{-11}$$

This representation corresponds to curve 1) where $f_a = 200$ MHz and $f_b = 900$ MHz.

3. Then we represent the frequency response of the normalised admittance associated with the capacitor – curve 2)

$$j2\pi f 6 \times 10^{-10}$$

4. Then we determine the equivalent admittance response for the RL combination. This is done by rotating the curve 1) by 180 degrees to obtain curve 3).

5. Now we have the admittance representation of the two parallel branches (RL and C). To obtain the overall admittance we simply add, point by point, the admittances of curves 2) and 3). This gives the overall admittance of curve 4).

6. Finally, by rotating curve 4) by 180 degrees we obtain curve 5) which is the impedance of the overall network.

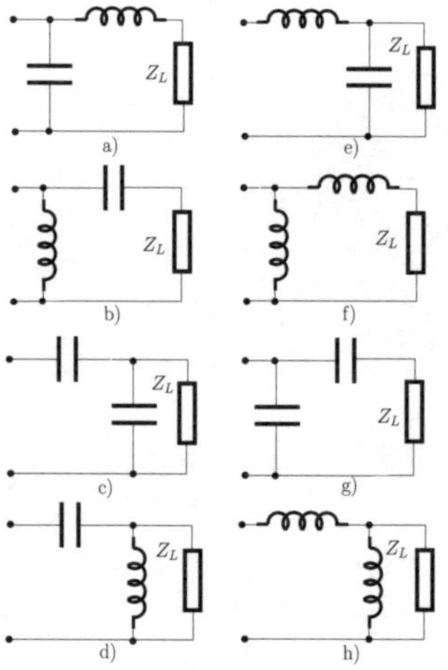

7.5.3 Introduction to impedance matching

Impedance matching is a very important issue in microwave engineering where, in a wide variety of applications such as amplification, the main objective is to achieve maximum power transfer to a load as described in section 7.4.1. The impedance matching can be achieved using many different circuit topologies. However, L-section circuits, illustrated in Fig. 7.40, result in very simple and practical solutions for this problem. It should be noted that there are no dissipative elements in any of the L-section circuits.

Let us consider the problem of matching a load Z_L to a signal source with an output impedance $R_s = 50$ Ω using L-section circuits, as illustrated in figure 7.41. The matching is to be effected at 500 MHz. The load Z_L is the series combination of an inductor $L_L = 3.18$ nH with a resistor $R_L = 10$ Ω.

Figure 7.40: *L-section circuits for impedance matching with the load Z_L.*

7. RF circuit analysis techniques

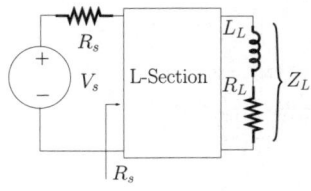

Figure 7.41: *Impedance matching with L-section circuits.*

Figure 7.42 shows two solutions for this matching problem. At 500 MHz the load is $Z_L = 10 + j\,10\ \Omega$. Normalising to $R_s = 50\ \Omega$ we get $z_L = 0.2 + j\,0.2$. It should be noted that, by using an L-section circuit, we aim to obtain a normalised unit impedance (or admittance) which is represented in the centre of the Smith chart.

Figure 7.42 shows a circle (dotted line) which represents all the admittances (or impedances) corresponding to the constant unit conductance (resistance) circle. This circle plays a major role in obtaining the solution for the L-section as we will show. Let us consider the solution a). First we represent the normalised load impedance, z_L in the Smith chart. By inserting an inductor, with a normalised impedance of $j0.2$, in series with z_L, the resulting impedance increases along a constant resistance ($r = 0.2$) circle until it arrives to point a) on the dotted circle. Hence, at point a), the normalised impedance is $z_a = 0.2 + j\,0.4$. The corresponding normalised admittance can be obtained by rotating point a) by 180 degrees resulting in point a1) with a normalised admittance $y_a = 1 - j\,2$. Finally, in order to get a unit normalised admittance (or impedance) we need to reduce the negative susceptance of y_a to zero. This is obtained by adding a positive susceptance of $+j\,2.0$ which corresponds to a move from point a1) to the centre of the Smith chart along the unit conductance circle. The addition of a positive susceptance of $+j\,2.0$ corresponds to the addition of a capacitor, with normalised admittance of $j\,2$, in parallel with the series combination of the inserted inductor and the load. It is now possible to determine the values of the inductor and capacitor as follows:

$$j\frac{2\pi\,500\times 10^6\,L}{50} = j\,0.2$$

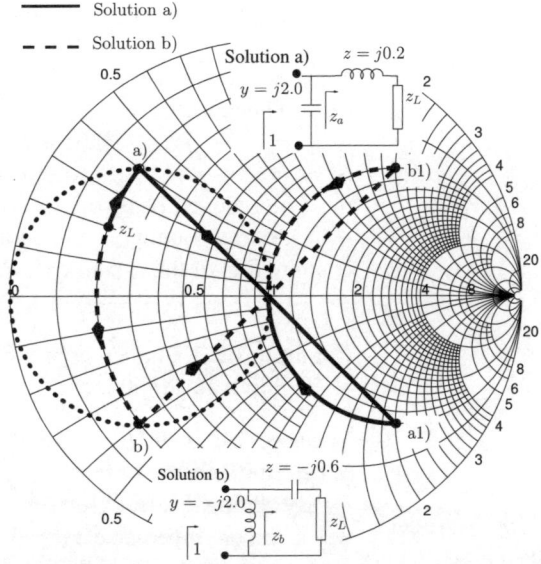

Figure 7.42: *Impedance matching using L-section circuits.*

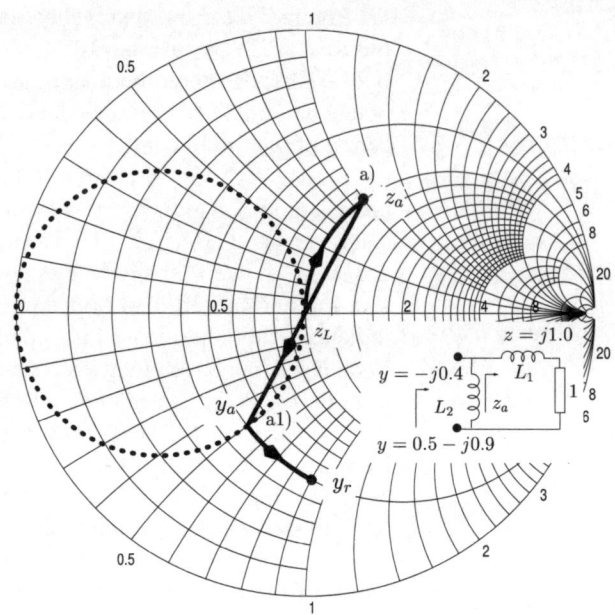

Figure 7.43: *Impedance matching using an L-section circuit.*

$$\begin{aligned} L &= 3.18 \text{ nH} \\ j\,50 \times 2\pi 500 \times 10^6\, C &= j\,2.0 \\ C &= 12.7 \text{ pF} \end{aligned}$$

A different solution b) can also be reached, following a similar procedure. This is represented by the dashed lines. Matching is achieved here firstly by adding a negative reactance corresponding to inserting a capacitor in series with the load. The value of this reactance is $-j\,0.6$. Hence, the impedance at point b) is $z_b = 0.2 - j\,0.4$. The corresponding admittance, represented in point $b1$), is $y_b = 1 + j\,2.0$. In order to obtain $y = 1$ the positive susceptance of y_b must be reduced to zero. This is obtained by adding a negative susceptance $-j\,2.0$ corresponding to the addition of an inductor in parallel with the series combination of the capacitor and load.

Example 7.5.3 Find an appropriate L-section circuit which transforms a load of 50 Ω into an admittance $Y_r = 10 - j\,1.8$ mS.

<u>Solution</u>: The normalised admittance $y_r = Y_r \times 50 = 0.5 - j\,0.9$ is represented in the Smith chart of figure 7.43.
  By adding a normalised impedance of $j\,1$ to the normalised unit impedance we obtain an impedance $z_a = 1 + j$. This is equivalent to adding an inductor L_1 in series with the load. $z_a = 1 + j$ corresponds to an admittance of $y_a = 0.5 - j\,0.5$. To get the desired admittance we need to add now a susceptance of $-j\,0.4$ which corresponds to adding another inductor L_2, with $y = -j\,0.4$, but

this time in parallel with the impedance resulting from the series connection of the inductor L_1 and the load.

Impedance matching with transmission lines

It is possible to provide impedance matching using transmission lines instead of using lumped elements. For this purpose, short-circuited and open-circuit

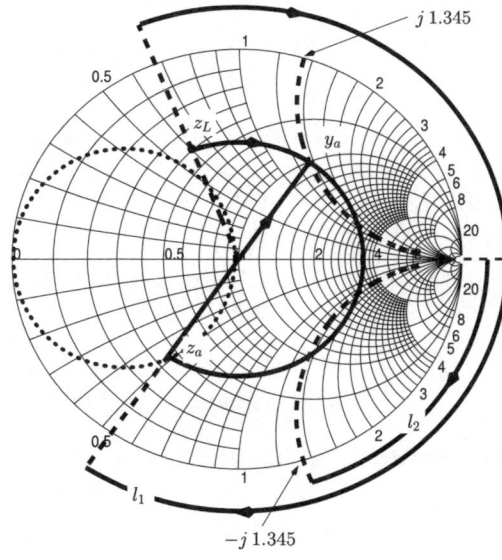

Figure 7.44: *Impedance matching with transmission lines using Smith chart calculations.*

transmission lines play a significant role. Recall that a short-circuited transmission line presents an inductive input impedance (for lengths less than $\lambda/4$) while the open-circuit transmission line presents a capacitive input impedance (for lengths less than $\lambda/4$). The matching procedure using transmission lines is very similar to the matching procedure using L-sections. The capacitors are now replaced by open-circuited transmission lines and the inductors are replaced by short-circuited transmission lines.

Figure 7.44 illustrates a transmission line calculation, using the Smith chart, to transform an impedance $Z_L = 20 + j\,30\ \Omega$ into a 50 Ω impedance. First, the load impedance is normalised to 50 Ω, that is, $z_L = 0.4 + j\,0.6$. By adding a transmission line with a length $l_1 = 0.325\lambda$ we obtain an impedance z_a such that the real part of its admittance, y_a, is equal to one. Now it is necessary to eliminate the susceptance of y_a which is $j\,1.345$. This is obtained by adding (in parallel) a susceptance of $-j\,1.345$ provided by a short-circuited transmission line with a length $l_2 = 0.1\,\lambda$. Figure 7.45 shows the matching circuit designed here. It is left to the reader to verify that matching can also be achieved if the short-circuited line was replaced by an open-circuit line with length $l_2 = 0.1\,\lambda + \lambda/4$.

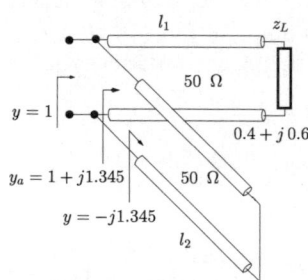

Figure 7.45: *Impedance matching circuit.*

The method of matching described above is known as single stub-tuning and is commonly used in RF and microwave circuits. More complex methods of matching are also used and some of these are described in the references of this chapter.

7.6 Bibliography

1. G.F. Miner, *Lines and Electromagnetic Fields for Engineers*, 1996 (Oxford University Press).

2. E. da Silva, *High Frequency and Microwave Engineering*, 2001 (Butterworth-Heinemann).

3. G. Gonzalez, *Microwave Transistor Amplifiers, Analysis and Design*, 1997 (Prentice Hall), 2nd edition.

4. D.M. Pozar, *Microwave Engineering*, 1998, (Wiley), 2nd edition.

5. J.D. Kraus, *Electromagnetics with applications*, 1999 (McGraw-Hill) 5th edition.

6. T.E. Edwards, M.B. Steer, *Foundations of microstrip design*, 2000 (John Wiley) 3rd edition.

7. E.H. Fooks, R.A. Zakarvicius, *Microwave Engineering using microstrip circuits*, 1990 (Prentice Hall).

7.7 Problems

7.1 Show that the transfer function $V(x)/V(x=0)$ of an ideal transmission line, terminated by a load Z_L which allows for maximum power transfer, can be expressed by eqn 7.22.

7.2 Show that the current $I(x)$ in an ideal transmission line, terminated by a load Z_L which allows for maximum power transfer, can be expressed by eqn 7.26.

7.3 Show that the input impedance of an ideal transmission line terminated by a load Z_L can be expressed by eqn 7.49.

7.4 Consider a transmission line with an inductance per metre of 550 nH and a capacitance per metre of 100 pF. This line has a length $l = 13$ m and is terminated by a load $Z_L = 25$ Ω. Determine the line input impedance for the following frequencies:

1. $\omega = 2\pi\, 3 \times 10^2$ rad/s;

2. $\omega = 2\pi\, 5 \times 10^8$ rad/s.

7.5 Plot eqn 7.49 for $0 < \beta d < 2\pi$ and the following situations:

1. $Z_L/Z_o = 0.1$

2. $Z_L/Z_o = 1$

3. $Z_L/Z_o = 10$

7. RF circuit analysis techniques

7.6 A circuit with a 30 Ω output impedance is to be matched to a 50 Ω load, at 500 MHz, using a quarter-wave transformer. Determine the characteristic impedance of the quarter-wave transformer.

7.7 Show that the eqn 7.74 can be expressed by eqn 7.75.

7.8 Show that if a lossy transmission line is matched then $I(x)$ and Z_o can be expressed by eqns 7.87 and 7.88, respectively.

7.9 Consider a lossy transmission line.

1. Show that for low frequencies the attenuation, α, the propagation constant, β, and the characteristic impedance, Z_o, can be approximated by eqns 7.89, 7.90 and 7.91, respectively.

2. Show that for high frequencies the attenuation, α, the propagation constant, β, and the characteristic impedance, Z_o, can be approximated by eqns 7.92, 7.93 and 7.94, respectively.

7.10 Show that the input impedance of a lossy transmission line terminated by a load Z_L can be expressed by eqn 7.114.

7.11 A microstrip material with $\epsilon_r = 8.1$ and $d = 1.3$ mm is used to build a transmission line. Determine the width of the microstrip for a 50 Ω characteristic impedance.

7.12 Determine the S-parameters of a CR circuit (high-pass filter).

7.13 Determine the S-parameters of the circuits of figure 7.46.

7.14 Determine the S-parameters of the high-frequency model for the field-effect transistor.

7.15 For the previous problem assume $k_n W/L = 40$ mA/V^2, $C_{gs} = 3$ pF, $C_{gd} = 1.5$ pF, $V_A = 60$ V and $I_D = 10$ mA. $Z_o = 50$ Ω. Plot the S-parameters for a frequency range 1 MHz–10 GHz.

7.16 Derive the chain parameters from the S-parameters.

7.17 Represent the following impedances on the Smith chart and determine the corresponding admittances

1. $10 - j\,30$ Ω
2. $75 + j\,20$ Ω
3. $60 - j\,40$ Ω
4. $5 - j\,70$ Ω
5. $j\,50$ Ω

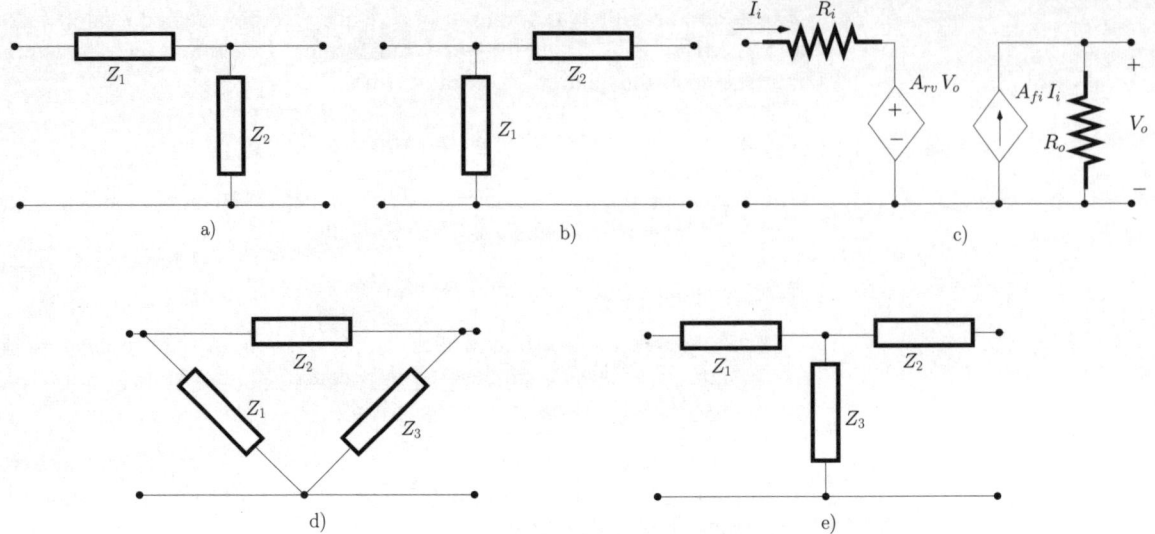

Figure 7.46: *Circuits of problem 7.13.*

6. $-j\,180\ \Omega$

7.18 Represent the following admittances on the Smith chart and determine the corresponding impedances

1. $j\,4 \times 10^{-3}$ S
2. $8 \times 10^{-2} - j\,6 \times 10^{-2}$ S
3. $0.2 + j\,4 \times 10^{-2}$ S

7.19 Convert a 50 Ω impedance into an impedance of $25 - j\,15$ Ω by using L-sections.

7.20 Sketch the S-parameters of the RC circuit considered in problem 7.12 on a Smith chart for $0.1/(RC) < \omega < 10/(RC)$. Use values of $R = 60$ Ω, $C = 1$ pF.

7.21 Find an L-section circuit which converts a $60 + j\,20$ Ω impedance into an impedance of $40 + j\,30$ Ω. *Hint: Normalise the impedances to 40 Ω.*

7.22 Design a transmission line circuit to convert a complex impedance of $30 + j\,45$ Ω into a real 45 Ω impedance.

8 Noise in electronic circuits

8.1 Introduction

Electronic noise is defined as a signal that either corrupts, masks or interferes with the desired signal which is being processed by an electronic circuit. There is a very broad range of noise sources that can be present in such circuits. Here we divide these noise sources into two major classes. The first refers to noise sources which are intrinsic to electronic devices and arise from fundamental physical effects. Such noise sources, sometimes known as intrinsic, are thermal (or Johnson) noise, electronic shot-noise, and $1/f$ noise. The second class encompasses all coupled noise sources that arise from interactions between the electronic circuit and the surrounding environment. Examples of extrinsic noise sources are atmospheric-based noise, glitches induced by fast switching digital circuits, coupling from nearby electrical circuits, etc. In this chapter we shall address only the first class of noise sources.

We start by revising very important statistical and probability concepts which are fundamental to modelling electronic noise. This is done in the next section and in section 8.3, where the main mathematical properties of random variables and of stochastic (or random) processes are discussed. In section 8.4 we present models for the various sources of intrinsic noise in active and passive devices. Also, we present a method to address the performance of electronic amplifiers in terms of equivalent input noise sources and the noise figure. Finally, in section 8.5 we present a matrix-based method suitable for a computer-aided analysis of noise performance of linear circuits.

8.2 Random variables

In the previous chapters we have studied the current–voltage relationships of electronic circuits and, for that purpose, we have treated current and voltage as deterministic (non-random) quantities; that is, they were characterised by precise values following deterministic models. These deterministic models, although very useful for the analysis and design of a large variety of circuits, do not account for the randomness associated with currents and voltages. For example, when the measurement of a DC current through a resistor reads 3.4 amperes this value corresponds to the *average value* of this DC current as there is always some randomness associated with the flow of electronic charges.

In order to understand some of the statistical properties of noise we consider the following experiment where the current passing through N identical resistances is measured in a 'very precise manner'. The resistances have no voltage applied to their terminals. According to Ohm's law we would expect no current flow. In fact, although the average (net) current is zero there is a random motion of free electronic charges in the resistors as illustrated in fig-

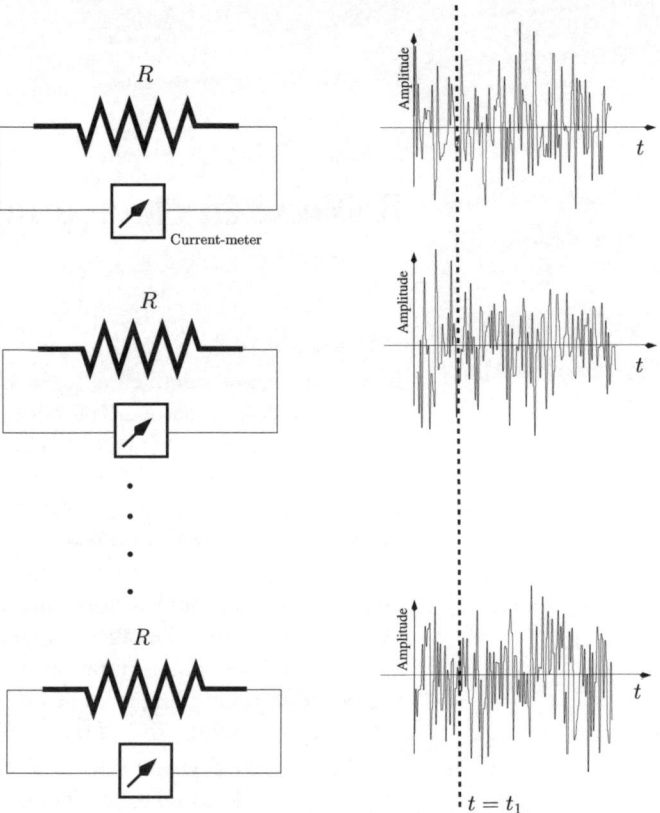

Figure 8.1: *Current measured in N resistances.*

ure 8.1. Such a random motion results in noise termed the 'current noise'. If we take an instant of time $t = t_1$ and we record the N measurements taken, we can construct a histogram which tell us the likelihood or the probability of the measured current being in a certain amplitude interval. The histogram is constructed by dividing the amplitude of the measured current into intervals and by plotting the number of measured amplitudes in each interval relative to the total number of measurements, N, as in figure 8.2 a). It should be noted that the amplitude of the current taken at time t_1 is a random variable (r.v.). In this chapter the random variables are represented by capital letters and the possible outcomes of a random variable are represented by lower cases. The random variable associated with the current at time t_1 is represented here by I_1 and the outcomes of I_1 are represented by i_1.

From this histogram it is already possible to extract the following information; the average current amplitude is zero as expected. Also, it is more likely to measure amplitudes near zero than further away from this average (or mean) value.

If we increase the number of measurements and if we decrease the amplitude intervals we get a more refined histogram, such as that depicted in figure

8. Noise in electronic circuits

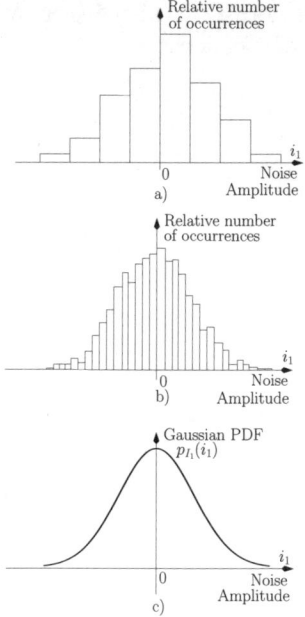

Figure 8.2: *a) Histogram of the noise associated with the current. b) Histogram of the noise associated with the current. c) Gaussian Probability Density Function.*

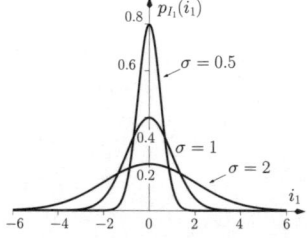

Figure 8.3: *Gaussian PDF.*

8.2 b), which characterises more accurately the probability of the measured current, in time t_1, being in a certain amplitude interval.

In figure 8.2 c) we present the theoretical model for the Probability Density Function (PDF) of the current measured at t_1. This PDF would result from a non-discrete continuous histogram where the number of measurements tends towards infinity while the amplitude interval tends towards zero. This PDF is called the 'Gaussian' PDF, or the 'Gaussian distribution', and can be described mathematically as follows:

$$p_{I_1}(i_1) = \frac{1}{\sqrt{2\pi}\sigma} \exp\left(-\frac{(i_1-\mu)^2}{2\sigma^2}\right) \qquad (8.1)$$

where μ is the mean, or the average value, of the distribution. σ is the *standard deviation* and it accounts for the 'amount of randomness' of the random variable. If we take the square value of the standard deviation (σ^2) we obtain what is defined as the *variance*. Basically, the greater the value of σ the more likely it is to have occurrences of the random variable further away from the mean value μ. Figure 8.3 illustrates the Gaussian PDF with $\mu = 0$ and with $\sigma = 0.5$, $\sigma = 1$ and $\sigma = 2$. Clearly, as the the value of σ increases the PDF gets broader. The area under $p_{I_1}(i_1)$ is unity:

$$\int_{-\infty}^{\infty} p_{I_1}(i_1)\, di_1 = 1 \qquad (8.2)$$

In other words, when σ increases, resulting in a broader PDF, the maximum amplitude of the PDF decreases in order to maintain the area constant and equal to one. Conversely, when σ decreases the PDF gets narrower and the density around the mean value, $\mu = 0$, increases. When σ tends to zero the PDF tends to a Dirac delta function with unity area centred at the mean value:

$$\lim_{\sigma \to 0} \frac{1}{\sqrt{2\pi}\sigma} \exp\left(-\frac{(i_1-\mu)^2}{2\sigma^2}\right) = \delta(i_1-\mu) \qquad (8.3)$$

It follows that the PDF of a completely deterministic event is the Dirac delta function located at the mean value with unity area. If the random fluctuations (noise) associated with the current measured at t_1 were non-existent then its PDF would be a Dirac delta function centred at zero.

Probability calculation

The calculation of the probability of I_1 being in a given interval $[i_a, i_b]$, with $i_a < i_b$, is defined as follows:

$$P[i_a < I_1 < i_b] \triangleq \int_{i_a}^{i_b} p_{I_1}(i_1)\, di_1 \qquad (8.4)$$

The interpretation of this eqn is illustrated in figure 8.4. It corresponds to determining the area under $p_{I_1}(i_1)$ between the values i_a and i_b. The calculation of probabilities when the distribution of the random variable is Gaussian cannot be effected in a direct manner since the Gaussian PDF does not have an analytic primitive, that is, it cannot be integrated analytically. However, numerical methods have been devised to generate tables of the normalised integrals below (see also appendix A) which are also known as error functions[1]:

$$Q(x) \triangleq \frac{1}{\sqrt{2\pi}} \int_x^\infty e^{-\lambda^2/2} \, d\lambda$$

$$\text{erf}(x) \triangleq \frac{2}{\sqrt{\pi}} \int_0^x e^{-\lambda^2} \, d\lambda$$

$$\text{erfc}(x) \triangleq \frac{2}{\sqrt{\pi}} \int_x^\infty e^{-\lambda^2} \, d\lambda$$

$$= 1 - \text{erf}(x)$$

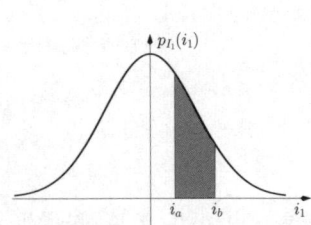

Figure 8.4: *The calculation of probabilities as the calculation of an area.*

The calculation of the probability of the Gaussian r.v. X exceeding a specific value a is expressed as $P[X > a]$. The calculation of $P[X > a]$ when X has a mean value of μ and a standard deviation of σ, can be performed as follows:

$$P[X > a] = \int_a^\infty \frac{1}{\sqrt{2\pi}\sigma} e^{-\frac{(x-\mu)^2}{2\sigma^2}} \, d\lambda \tag{8.5}$$

Considering $a > 0$ we use the following change of variable (see example 8.2.2 for $a < 0$)

$$\lambda = \frac{x - \mu}{\sigma} \tag{8.6}$$

we can write

$$d\lambda = dx \frac{1}{\sigma} \tag{8.7}$$

$$x = a \quad ; \quad \lambda = \frac{a - \mu}{\sigma} \tag{8.8}$$

$$x \to \infty \quad ; \quad \lambda \to \infty \tag{8.9}$$

and eqn 8.5 can be written as

$$P[X > a] = \frac{1}{\sqrt{2\pi}} \int_{(a-\mu)/\sigma}^\infty e^{-\lambda^2/2} \, d\lambda$$

$$= Q\left(\frac{a - \mu}{\sigma}\right) \tag{8.10}$$

Example 8.2.1 Show that if I_1 is a Gaussian r.v. with mean value μ and variance σ^2 then eqn 8.4 can be written as

$$P[i_a < I_1 < i_b] = Q\left(\frac{i_a - \mu}{\sigma}\right) - Q\left(\frac{i_b - \mu}{\sigma}\right) \tag{8.11}$$

---

[1]The function erfc($\cdot$) is also known as the complementary error function.

Solution: Since I_1 is a Gaussian r.v. then we have:

$$P_{I_1}(i_1) = \frac{1}{\sqrt{2\pi}\sigma} \exp\left(-\frac{(i_1-\mu)^2}{2\sigma^2}\right) \qquad (8.12)$$

and $P[i_a < I_1 < i_b]$ can be calculated according to eqn 8.4, that is;

$$P[i_a < I_1 < i_b] = \int_{i_a}^{i_b} \frac{1}{\sqrt{2\pi}\sigma} \exp\left(-\frac{(i_1-\mu)^2}{2\sigma^2}\right) di_1 \qquad (8.13)$$

The last eqn can be written as follows:

$$\begin{aligned} P[i_a < I_1 < i_b] &= \int_{i_a}^{\infty} \frac{1}{\sqrt{2\pi}\sigma} \exp\left(-\frac{(i_1-\mu)^2}{2\sigma^2}\right) di_1 \\ &- \int_{i_b}^{\infty} \frac{1}{\sqrt{2\pi}\sigma} \exp\left(-\frac{(i_1-\mu)^2}{2\sigma^2}\right) di_1 \end{aligned} \qquad (8.14)$$

Using the mathematical manipulation described in eqns 8.6–8.10 we can write this eqn as

$$\begin{aligned} P[i_a < I_1 < i_b] &= \frac{1}{\sqrt{2\pi}} \int_{(i_a-\mu)/\sigma}^{\infty} e^{-\lambda_1^2/2} \, d\lambda_1 \\ &- \frac{1}{\sqrt{2\pi}} \int_{(i_b-\mu)/\sigma}^{\infty} e^{-\lambda_2^2/2} \, d\lambda_2 \\ &= Q\left(\frac{i_a-\mu}{\sigma}\right) - Q\left(\frac{i_b-\mu}{\sigma}\right) \end{aligned} \qquad (8.15)$$

Example 8.2.2 Consider a Gaussian r.v. X with $\mu = 3$ and $\sigma = 2$. Determine $P[X < -7]$.

Solution:

$$P[X < -7] = \int_{-\infty}^{-7} \frac{1}{\sqrt{2\pi}\sigma} \exp\left(-\frac{(x-\mu)^2}{2\sigma^2}\right) dx \qquad (8.16)$$

Using the change of variable

$$x = -y \qquad (8.17)$$
$$dx = -dy \qquad (8.18)$$
$$x = -7 \quad ; \quad y = 7 \qquad (8.19)$$
$$x \to -\infty \quad ; \quad y \to +\infty \qquad (8.20)$$

eqn 8.16 can now be written as follows:

$$\begin{aligned} P[X < -7] &= \int_{+\infty}^{+7} \frac{1}{\sqrt{2\pi}\sigma} \exp\left(-\frac{(-y-\mu)^2}{2\sigma^2}\right) (-dy) \\ &= \int_{7}^{\infty} \frac{1}{\sqrt{2\pi}\sigma} \exp\left(-\frac{(y+\mu)^2}{2\sigma^2}\right) dy \end{aligned} \qquad (8.21)$$

Finally by using another change of variable

$$\lambda = \frac{y+\mu}{\sigma} \tag{8.22}$$

$$d\lambda = dy\frac{1}{\sigma} \tag{8.23}$$

$$y = 7 \quad ; \quad \lambda = \frac{7+\mu}{\sigma} \tag{8.24}$$

$$y \to \infty \quad ; \quad \lambda \to \infty \tag{8.25}$$

eqn 8.21 can be calculated as follows

$$\begin{aligned} P[X < -7] &= Q\left(\frac{7+\mu}{\sigma}\right) \\ &= 2.9 \times 10^{-7} \end{aligned}$$

As an exercise, repeat the above for $P[X > 13]$. You will find that this will give an identical answer due to the symmetry of the Gaussian distribution around the mean ($\mu = 3$ in this case).

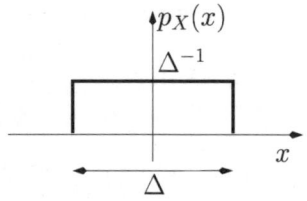

Figure 8.5: *Uniform PDF. Example.*

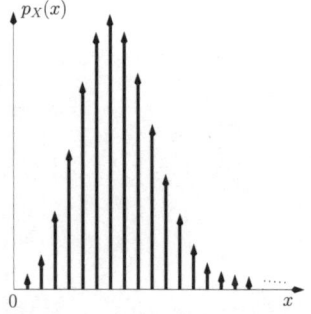

Figure 8.6: *Poisson distribution. Example.*

Other distributions

There is a large variety of PDFs which have been devised as appropriate models for a broad range of random phenomena. Among these distributions the following are commonly found in electronic and communication systems:

- **Uniform.** This continuous distribution is characterised by a PDF which can be expressed by the following equation:

$$p_X(x) = \frac{1}{\Delta}\text{rect}\left(\frac{x-\mu}{\Delta}\right) \tag{8.26}$$

where $\text{rect}(\cdot)$ represents the rectangular function (see appendix A) and μ is the mean value. Figure 8.5 illustrates a zero mean uniform PDF with range Δ. Note that, unlike the Gaussian distribution with infinite tails, the uniform distribution has a finite range of possible occurrences for the random variable. Note that the area of $p_X(x)$ is one, as expected.

- **Poisson.** This distribution differs substantially from the two previous distributions (Gaussian and uniform) since it is categorised as a (non-continuous) random variable. It is used to model a very broad range of random phenomena including shot noise in electronic devices, a subject that will be discussed further in section 8.4.2. Discrete distributions are often characterised by probability frequency functions. However, using the Dirac delta function, it is possible to describe these discrete distributions using probability density functions. For the Poisson distribution we can write:

$$p_X(x) = e^{-\mu} \sum_{k=0}^{\infty} \frac{\mu^k}{k!} \delta(x-k) \tag{8.27}$$

where μ represents the mean of the distribution.

8.2.1 Moments of a random variable

The moments of a random variable are, in essence, weighted averaging operations over the PDF. These moments provide valuable information about this r.v. Mathematically the n-th order moment of a random variable X is defined as

$$\mathrm{E}\left[X^n\right] \triangleq \int_{-\infty}^{\infty} x^n \, p_X(x) \, dx \tag{8.28}$$

where the operator $\mathrm{E}\left[\cdot\right]$ is called the 'expectation' or 'averaging' operator.

The average value

Of particular interest are the first and the second moment. The first moment represents the average value, or the mean, of the random variable:

$$\mu = \mathrm{E}\left[X\right] = \int_{-\infty}^{\infty} x \, p_X(x) \, dx \tag{8.29}$$

Figure 8.7 illustrates the mathematical operation of eqn 8.29. In figure 8.7 a) we illustrate the zero mean PDF, $p_X(x)$, and the linear function $f(x) = x$.

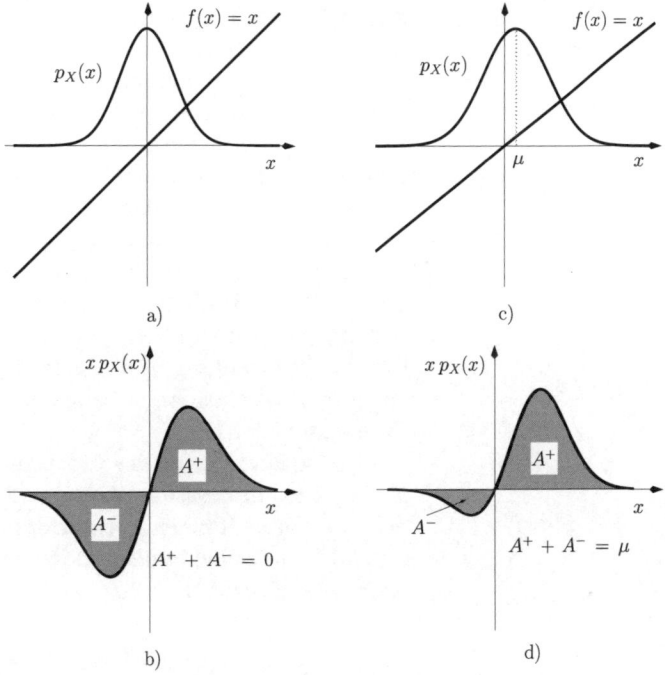

Figure 8.7: *Calculation of the mean of a PDF. a) Representation of a zero mean $p_X(x)$ and of $f(x) = x$. b) Representation of the area of the product of x with the zero mean $p_X(x)$. c) Representation of a non-zero mean (μ) $p_X(x)$ and of $f(x) = x$. d) Representation of the area of the product of x with $p_X(x)$.*

Figure 8.7 b) shows the product of x with $p_X(x)$. It can be seen that, for this situation, the positive area, A^+, is equal to the negative area, A^-. Thus, the overall area, that is the sum of $A^+ + A^-$ is zero as expected. In Figure 8.7 c) we see the PDF with a mean value $\mu \neq 0$ and the linear function $f(x) = x$. Figure 8.7 d) represents the product of x with $p_X(x)$. Now the positive area, A^+ is greater that the negative area, A^-. In fact the sum of $A^+ + A^-$ is equal to μ, the mean value of the PDF.

The variance

The second moment is called the mean-square value of the random variable:

$$\mathrm{E}\left[X^2\right] = \int_{-\infty}^{\infty} x^2 \, p_X(x) \, dx \tag{8.30}$$

The mean-square value can be be expressed as $\mathrm{E}\left[x^2\right] = \mu^2 + \sigma^2$ where σ^2 is the variance of the random variable. σ^2 is also called the centred second moment and can be calculated as follows:

$$\sigma^2 = \mathrm{E}\left[(X-\mu)^2\right] = \int_{-\infty}^{\infty} (x-\mu)^2 \, p_X(x) \, dx \tag{8.31}$$

When the mean of the PDF is zero then $\mathrm{E}\left[x^2\right] = \sigma^2$. Figure 8.8 illustrates the mathematical operation of eqn 8.31 for two zero mean distributions with different variances.

Figure 8.8 a) represents the distribution $p_X(x)$ and the parabolic function $f(x) = x^2$. The role of the function x^2 is as follows: after multiplying x^2 with $p_X(x)$, as shown in figure 8.8 b), the parabolic function attenuates the importance of those values of $p_X(x)$ near the mean value while it enhances the values of $p_X(x)$ further away from the mean value. Thus, the calculation of the area resulting from this product provides a measure of how broad (or scattered) a PDF is and the result is called the variance. In figure 8.8 c) we show another distribution $p_X(x)$ with a larger variance represented together with the parabolic function $f(x) = x^2$. It can be seen, in a qualitative manner, that the area resulting from the product of $p_X(x)$ with x^2 for this situation is larger than the situation presented in figure 8.8 a) a result that is also apparent from figure 8.8 d).

The square root of the variance is the standard deviation, σ, which has already been discussed in the context of the Gaussian distribution. The interpretation for σ, although presented in the context of the Gaussian distribution, is valid for any kind of distribution. In fact Chebyshev's inequality states that, regardless of the PDF $p_X(x)$, we have

$$P(|X - \mu| \geq a\,\sigma) \leq \frac{1}{a^2} \tag{8.32}$$

where μ and σ are the mean and the standard deviation of $p_X(x)$. Equation 8.32 states that the probability of observing any outcome of a random variable X outside $\pm a$ times the standard deviation of its average value is never greater than $1/a^2$, regardless of its distribution.

8. Noise in electronic circuits

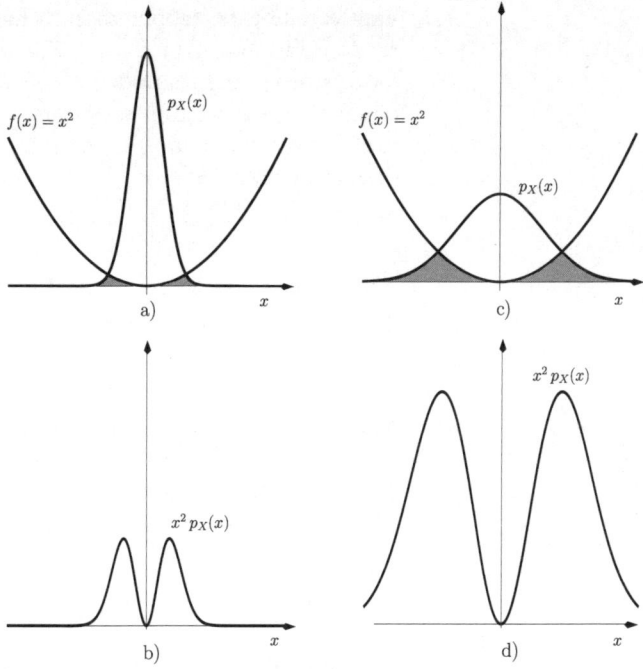

Figure 8.8: *Calculation of the variance of different zero mean PDFs. a) Representation of $p_X(x)$ and of $f(x) = x^2$. b) The product of x^2 with $p_X(x)$. c) Representation of $p_X(x)$, with larger variance, and of $f(x) = x^2$. d) The product of x^2 with $p_X(x)$ with the larger variance.*

Example 8.2.3 Determine the variance of a zero mean uniform distribution.

<u>Solution</u>: The variance of a zero mean uniform distribution coincides with the second moment and can be calculated as follows:

$$\begin{aligned}
\sigma^2 &= \int_{-\infty}^{\infty} x^2 \frac{1}{\Delta} \text{rect}\left(\frac{x}{\Delta}\right) dx \\
&= \int_{-\Delta/2}^{\Delta/2} x^2 \frac{1}{\Delta} dx = \frac{x^3}{3\Delta}\bigg|_{-\Delta/2}^{\Delta/2} \\
&= \frac{\Delta^2}{12}
\end{aligned} \quad (8.33)$$

Multivariate expectations

The term multivariate expectations refers to the calculation of expectations of more than one random variable and the calculation of functions of multiple random variables. However, in this chapter we only consider simple but important multivariate expectations cases. The first situation refers to the product of N random variables X_1, X_2, ..., X_N. If these random variables are statistically independent, such as those arising from different noise sources in electronic circuits, the expectation of this product is equal to the product of the individual expectation values, that is, the mean value of the product is equal to the product of each mean value;

$$\text{E}[X_1 X_2 \ldots X_N] = \text{E}[X_1]\text{E}[X_2] \ldots \text{E}[X_N] \qquad (8.34)$$

The last eqn can be generalised for higher order moments provided that the random variables are independent:

$$\text{E}[X_1^r X_2^r \ldots X_N^r] = \text{E}[X_1^r]\text{E}[X_2^r] \ldots \text{E}[X_N^r] \qquad (8.35)$$

where r represents the moment order and is a positive integer.

The second case refers to the sum of N random variables X_1, X_2, ..., X_N. The expectation of this sum is equal to the sum of all individual expectation values, *regardless* of whether the random variables are independent or not independent:

$$\text{E}[X_1 + X_2 + \ldots X_N] = \text{E}[X_1] + \text{E}[X_2] + \ldots + \text{E}[X_N] \qquad (8.36)$$

As before, the last eqn can be generalised for higher order moments as

$$\text{E}[X_1^r + X_2^r + \ldots X_N^r] = \text{E}[X_1^r] + \text{E}[X_2^r] + \ldots + \text{E}[X_N^r] \qquad (8.37)$$

8.2.2 The characteristic function

The characteristic function of a random variable X is defined by the following eqn

$$C_X(\lambda) \triangleq \text{E}\left[e^{-j\lambda X}\right] = \int_{-\infty}^{\infty} p_X(x)\, e^{-j\lambda X}\, dx \qquad (8.38)$$

with $j = \sqrt{-1}$ and λ representing an independent variable. Equation 8.38 is easily recognised as a Fourier integral. In fact, if we let $\lambda = 2\pi\alpha$ we can write:

$$C_X(2\pi\alpha) = \mathfrak{F}\left[p_X(x)\right] \qquad (8.39)$$

and consequently

$$p_X(x) = \mathfrak{F}^{-1}\left[C_X(2\pi\alpha)\right] = \int_{-\infty}^{\infty} C_X(2\pi\alpha)\, e^{j\,2\pi x \alpha}\, d\alpha \qquad (8.40)$$

The last two eqns reveal that the characteristic function and the PDF of a random variable form a Fourier transform pair.

The moments of a random variable can be calculated from its characteristic function according to:

$$\text{E}[X^n] = \frac{1}{(-j\,2\pi)^n} \left.\frac{d}{d\alpha^n} C_X(2\pi\alpha)\right|_{\alpha=0} \qquad (8.41)$$

Example 8.2.4 Calculate the second moment of a zero mean uniform distribution using eqn 8.41 and show that it is equal to $\Delta^2/12$, where Δ represents the range of the r.v.

Solution: The PDF of a zero mean uniform distribution can be written as:

$$p_x(x) = \frac{1}{\Delta}\text{rect}\left(\frac{x}{\Delta}\right) \tag{8.42}$$

and the characteristic function is its Fourier transform, that is, (see appendix A):

$$C_X(\alpha) = \text{sinc}(\alpha\Delta) \tag{8.43}$$

From eqn 8.41 we have:

$$\begin{aligned}
\mathrm{E}[X^2] &= \frac{1}{(-j\,2\pi)^2}\left.\frac{d}{d\alpha^2}\text{sinc}(\alpha\Delta)\right|_{\alpha=0} \\
&= \lim_{\alpha\to 0}\frac{1}{(-j\,2\pi)^2}\left[-\pi\Delta\frac{\sin(\pi\alpha\Delta)}{\alpha} - 2\frac{\cos(\pi\alpha\Delta)}{\alpha^2} + 2\frac{\sin(\pi\alpha\Delta)}{\pi\Delta\alpha^3}\right] \\
&= \frac{1}{(-j\,2\pi)^2}\left[-\pi^2\Delta^2 + \frac{2}{3}\pi^2\Delta^2\right] \\
&= \frac{\Delta^2}{12}
\end{aligned}$$

Note that above result is identical to that obtained in example 8.2.3.

The characteristic function is very useful when dealing with sums of independent variables. Considering the sum of two independent random variables $Y = X_1 + X_2$, the characteristic function of Y is calculated as:

$$\mathrm{E}\left[e^{j\lambda Y}\right] = \mathrm{E}\left[e^{j\lambda(X_1+X_2)}\right] = \mathrm{E}\left[e^{j\lambda X_1}\right]\mathrm{E}\left[e^{j\lambda X_2}\right] \tag{8.44}$$

that is

$$C_Y(2\pi\alpha) = C_{X_1}(2\pi\alpha)\,C_{X_2}(2\pi\alpha) \tag{8.45}$$

and consequently, from the convolution theorem,

$$p_Y(z) = p_{X_1}(z) * p_{X_2}(z) \tag{8.46}$$

That is, the PDF resulting from the sum of two independent random variables is given by the convolution of the PDFs of each random variable.

Example 8.2.5 Show that the sum of two independent Gaussian random variables is also a Gaussian random variable.

Solution: Let us consider two Gaussian random variables X_1 and X_2 each one characterised by a PDF

$$p_{X_i}(x) = \frac{1}{\sqrt{2\pi}\,\sigma_i} \exp\left(-\frac{(x_i - \mu_i)^2}{2\sigma_i^2}\right) \qquad i = 1, 2 \tag{8.47}$$

where μ_i and σ_i^2 are the mean value and the variance of X_i, $i = 1, 2$, respectively. Taking the Fourier transform (see appendix A) of each PDF we get:

$$C_{X_1}(2\pi\alpha) = e^{-j\,2\pi\alpha\mu_1 - \alpha^2\,(2\pi\sigma_1)^2} \tag{8.48}$$

$$C_{X_2}(2\pi\alpha) = e^{-j\,2\pi\alpha\mu_2 - \alpha^2(2\pi\sigma_2)^2} \tag{8.49}$$

The characteristic function of $Z = X_1 + X_2$ is given by

$$\begin{aligned}
C_Z(2\pi\alpha) &= e^{-j\,2\pi\alpha\mu_1 - \alpha^2\,(2\pi\sigma_1)^2} \times e^{-j\,2\pi\alpha\mu_2 - \alpha^2\,(2\pi\sigma_2)^2} \\
&= e^{-j\,2\pi\alpha\,(\mu_1 + \mu_2) - \alpha^2(2\pi)^2\,(\sigma_1^2 + \sigma_2^2)} \\
&= e^{-j2\pi\alpha\mu_z - \alpha^2(2\pi\sigma_z)^2}
\end{aligned} \tag{8.50}$$

where $\mu_z = \mu_1 + \mu_2$ and $\sigma_z^2 = \sigma_1^2 + \sigma_2^2$. Taking the inverse Fourier transform we get;

$$p_Z(z) = \frac{1}{\sqrt{2\pi}\,\sigma_z} \exp\left(-\frac{(z - \mu_z)^2}{2\,\sigma_z^2}\right) \tag{8.51}$$

The last eqn represents a Gaussian PDF.

Equations 8.45 and 8.46 can be generalised for the sum of N independent random variables. Hence if $Y = X_1 + X_2 + \ldots X_N$ then,

$$C_Y(2\pi\alpha) = \prod_{k=1}^{N} C_{X_k}(2\pi\alpha) \tag{8.52}$$

and,

$$p_Y(z) = p_{X_1}(z) * p_{X_2}(z) * \ldots * p_{X_N}(z) \tag{8.53}$$

8.2.3 The central limit theorem

The central limit theorem states that the sum of N independent and general (meaning regardless of their distribution) random variables, X_i, tends towards a Gaussian distribution with a mean, μ given by:

$$\mu = \sum_{i=1}^{N} \mu_i \tag{8.54}$$

and a variance given by

$$\sigma^2 = \sum_{i=1}^{N} \sigma_i^2 \qquad (8.55)$$

where μ_i and σ_i^2 represent the mean value and the variance, respectively, of each random variable X_i. This theorem is very important since it allows the characterisation of the PDF of a r.v., which results from a large sum of r.v.s, as a Gaussian distribution. It should be noted that there is no need to know the PDF of any of these individual random variables. All that is required is that the mean and the variance of each of these individual random variables be known.

To illustrate this important theorem we consider the sum of three identical and uniform distributions. Figure 8.9 a) shows a zero mean uniform distribution with range $2A$ and a zero mean Gaussian PDF with the same variance that is, $\sigma^2 = (2A)^2/12$. From this figure it can be observed that these two distributions are quite different. If we sum two uniform random variables $Y = X_1 + X_2$ its PDF is triangular as shown in figure 8.9 b). It can be seen

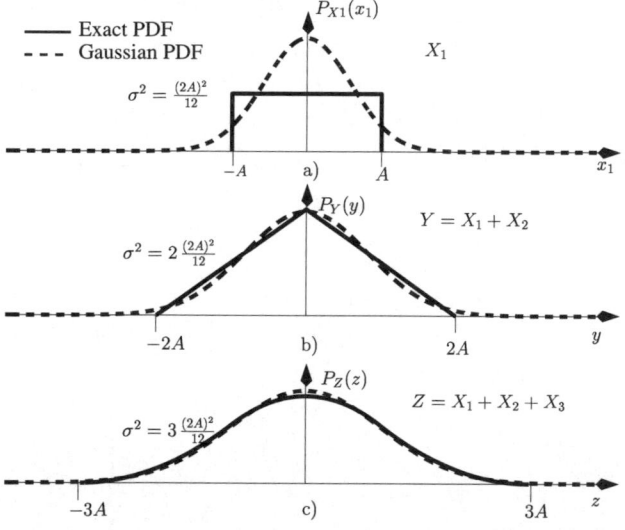

Figure 8.9: *Comparison between the exact PDF of the sum of N uniform random variables (X_i, $i = 1, 2, 3$), which are independent identically distributed, with the Gaussian approximation given by the central limit theorem. a) $N = 1$. b) $N = 2$. c) $N = 3$.*

that the range of the PDF of Y is now $4A$. Figure 8.9 b) also shows the Gaussian approximation to the PDF of Y, as stated by the central limit theorem. It can be observed that the difference between these two PDFs is not as large as in the previous situation described by figure 8.9 a). In figure 8.9 c) we show the PDF for the random variable Z resulting from the sum of three uniform random variables. The range of this distribution is now $6A$ and we observe that the piecewise parabolic shape of this distribution starts to resemble the

Gaussian PDF. In fact, from figure 8.9 c) it can be seen that the difference between these two PDFs is small. Clearly, the PDF of the r.v. resulting from the addition of more than one uniform random variable tends to a Gaussian distribution as stated by the central limit theorem. However, we emphasise that the distribution resulting from the sum of three uniform distributions has a finite range while the Gaussian PDF exhibits infinite tails. Therefore, some caution must be exercised using the Gaussian PDF to approximate the PDF of Z when dealing with very small probabilities! The following example illustrates this idea.

Example 8.2.6 Consider the random variable Z resulting from the sum of three independent, identically distributed, uniform random variables X_i, $i = 1, 2, 3$:

$$p_{X_i}(x) = \frac{1}{2A} \text{rect}\left(\frac{x}{2A}\right) \tag{8.56}$$

with $2A = 1$.

1. Determine the value z_a such that the $P(Z > z_a) = 0.3$.

2. Consider the Gaussian approximation to the PDF of Z and determine an estimate for the value z_a such that the $P(Z > z_a) = 0.3$. Compare the value of z_a with that obtained above.

3. Repeat the last two questions but now for z_b such that $P(Z > z_b) = 2.1 \times 10^{-5}$.

Solution:

1. The piecewise parabolic PDF is given by the convolution of the three uniform PDFs and can be expressed by;

$$p_Z(z) = \begin{cases} \dfrac{(z+3A)^2}{16A^3} & -3A \leq z < -A \\[2mm] -\dfrac{(z^2 - 3A^2)}{8A^3} & -A \leq z \leq A \\[2mm] \dfrac{(z-3A)^2}{16A^3} & A < z \leq 3A \\[2mm] 0 & \text{elsewhere} \end{cases} \tag{8.57}$$

The value z_a satisfies the following equation:

$$\int_{z_a}^{\infty} p_Z(z)\, dz = 0.3$$

that is

$$\int_{z_a}^{3A} \frac{(z-3A)^2}{16A^3}\, dz = 0.3$$

$$\Leftrightarrow \frac{9}{16} - \frac{1}{6}z_a^3 + \frac{3}{4}z_a^2 - \frac{9}{8}z_a = 0.3$$
$$\Leftrightarrow z_a = 0.2836$$

2. For each uniform PDF the variance is $\sigma_x^2 = 1/12$. Hence the Gaussian approximation for Z is

$$p_Z(z) = \frac{1}{\sqrt{2\pi}\,\sqrt{3}\sigma_x} e^{\frac{z^2}{2\times 3\sigma_x^2}} \tag{8.58}$$

$P(Z > z_a) = 0.3$ can be written as

$$Q\left(\frac{z_a}{\sqrt{3}\sigma_x}\right) = 0.3$$

From the tables of the Gaussian error function (see appendix A), z_a is estimated as 0.2620. The error for z_a given by the true PDF and by the Gaussian approximation is relatively small; about 7.6%.

3. The true value of z_b satisfies the following equation:

$$\int_{z_b}^{\infty} P_Z(z)\,dz = 2.1 \times 10^{-5}$$

that is

$$\frac{9}{16} - \frac{1}{6}z_b^3 + \frac{3}{4}z_b^2 - \frac{9}{8}z_b = 2.1 \times 10^{-5}$$
$$\Leftrightarrow z_b = 1.4499$$

The Gaussian approximation for z_b is such that:

$$Q\left(\frac{z_b}{\sqrt{3}\sigma_x}\right) = 2.1 \times 10^{-5}$$
$$\Leftrightarrow z_b = 2.05$$

It should be noted that, for this case, there is a significant error in the prediction of z_b by the Gaussian approximation. Also, since the true PDF of Z has a limited range $[-1.5, 1.5]$ the outcome $z_b = 2.05$ is meaningless in the context of the random variable Z.

8.2.4 Bivariate Gaussian distributions

We consider now the joint PDF of two Gaussian random variables, X and Y, which are statistically *dependent*. The interdependence of X and Y is quantified by the *covariance* of X and Y which is defined as follows:

$$\rho_{XY} \triangleq \frac{1}{\sigma_x \sigma_y} \mathrm{E}\left[(X - \mu_x)(Y - \mu_y)\right] \tag{8.59}$$

where μ_x and σ_x^2 are the mean and the variance of X, respectively. μ_y and σ_y^2 are the mean and the variance of Y, respectively. If the two random variables are statistically independent then $\rho_{XY} = 0$. If the two random variables are totally correlated then $\rho_{XY} = \pm 1$. Figure 8.10 illustrates the outcomes of a random variable Y versus the outcomes of a random variable X for $\rho_{XY} = 0$, $\rho_{XY} = 0.7$, $\rho_{XY} = -0.8$, $\rho_{XY} = 1$. When the two random variables are uncorrelated, $\rho_{XY} = 0$, it can be seen that the outcomes of Y do not relate

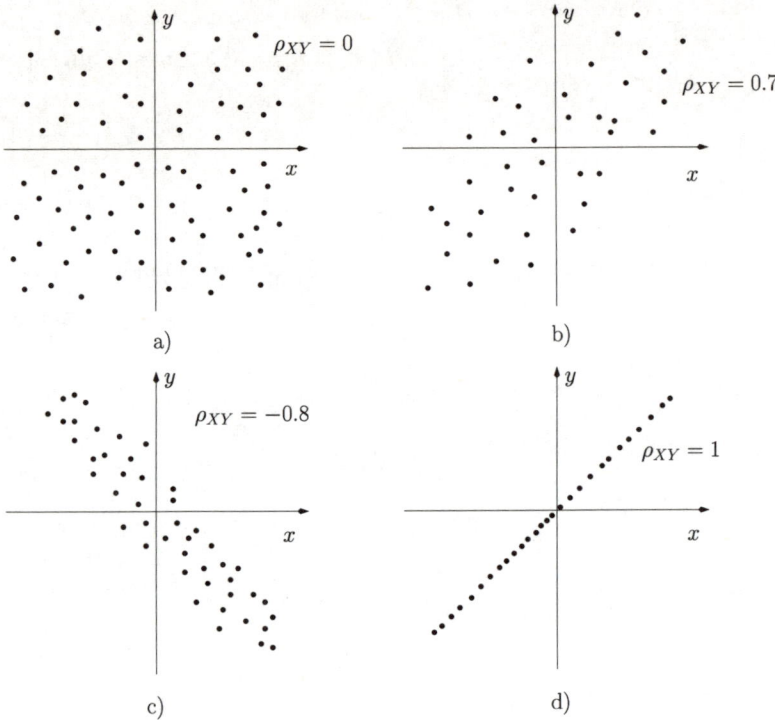

Figure 8.10: *Correlation between X and Y. a) $\rho_{XY} = 0$. b) $\rho_{XY} = 0.7$. c) $\rho_{XY} = -0.8$. d) $\rho_{XY} = 1$.*

to the outcomes of X. However, as $|\rho_{XY}|$ increases it can be seen that the outcomes of Y and of X become increasingly dependent on each other.

The bivariate Gaussian PDF can be written as follows:

$$p_{XY}(x,y) = \frac{1}{2\pi\sigma_x\sigma_y\sqrt{1-\rho_{XY}^2}}$$
$$\times \exp\left\{-\frac{(x-\mu_x)^2}{2\sigma_x^2(1-\rho_{XY}^2)} - \frac{(y-\mu_y)^2}{2\sigma_y^2(1-\rho_{XY}^2)} - \frac{\rho(x-\mu_x)(y-\mu_y)}{\sigma_x\sigma_y(1-\rho_{XY}^2)}\right\}$$
(8.60)

$p_{XY}(x,y)$ is also called the joint probability distribution (or joint PDF) of X and Y.

8.3 Stochastic processes

Let us consider again the measurements of the noise current in each resistor illustrated in figure 8.1. Each measured noise current waveform is also called a sample function or a sample waveform. Previously we have considered a fixed time instant t_1 for which we considered a random variable, I_1. If we consider any other instant of time, $t = t_2$ for example, we have another random variable, I_2. Therefore, a random process can be viewed as a family of random variables considered at different time instants.

8.3.1 Ensemble averages

If the PDF of the noise current, for *any* instant of time t, is represented by $p_I(i(t))$ then the mean value for $i(t)$ can be written as:

$$\mathrm{E}\left[I(t)\right] \triangleq \int_{-\infty}^{\infty} i(t)\, p_I(i(t))\, di(t) \tag{8.61}$$

It should be noted that the time variable t in eqn 8.61 is treated like a constant. Hence, eqn 8.61 is often written without the explicit time dependency, that is:

$$\mathrm{E}\left[I(t)\right] = \int_{-\infty}^{\infty} i\, p_I(i)\, di \tag{8.62}$$

Equation 8.61 or 8.62 represents an ensemble average, that is, an average over the ensemble of current waveforms (or sample functions) for any given instant of time. Its meaning is similar to the first moment (or average value) discussed in the context of a single random variable. The main difference is that for a random process the value of $\mathrm{E}\left[I(t)\right]$ might be time dependent.

Correlation functions

Another very important ensemble average is the autocorrelation function. This function, like the covariance defined in eqn 8.59, is a measure of the relatedness or dependence between random variables I_1 and I_2, considered at time instants t_1 and t_2 respectively. As it will be shown latter on, the measure of such a dependence can provide valuable information about the bandwidth and about the power of the noise which is modelled as a random process. The autocorrelation function is defined as follows:

$$R_i(t_1,t_2) \triangleq \mathrm{E}\left[I_1(t_1)\, I_2(t_2)\right] = \int_{-\infty}^{\infty}\int_{-\infty}^{\infty} i_1\, i_2\, p_{I_1\, I_2}(i_1,i_2)\, di_1\, di_2 \tag{8.63}$$

where the dependency of I_1 and of I_2 on the time variable has been dropped for reasons of simplicity. $p_{I_1\, I_2}(i_1,i_2)$ is the joint probability distribution of the random variables I_1 and I_2 at $t = t_1$ and at $t = t_2$, respectively.

If I_1 is statistically independent of I_2 then $R_i(t_1,t_2) = \mathrm{E}\left[I(t_1)\right]\mathrm{E}\left[I(t_2)\right]$. On the other hand, $R_i(t,t) = \mathrm{E}\left[I(t)^2\right]$, that is $R_i(t,t)$ is the mean square value of $I(t)$ as a function of time.

If I_1 and I_2 are both zero mean random variables then

$$R_i(t_1,t_2) = \sigma_{I_1}\, \sigma_{I_2} \rho_{I_1 I_2}$$

with $\rho_{I_1 I_2}$ representing the covariance between I_1 and I_2.

8.3.2 Stationary random processes

A stationary random process is one where all statistical characteristics (i.e statistical averages) are invariant with time. A specific type of stationary random process is known as the 'wide sense stationary process' where the time invariant characteristics are satisfied for the mean and the autocorrelation functions, that is;

- The mean value is constant for all times:

$$\mathrm{E}\left[I(t)\right] = \mu \qquad (8.64)$$

- The autocorrelation function depends only on the time difference $t_2 - t_1$, that is

$$\begin{aligned} R_i(t_1, t_2) &= R_i(t_2 - t_1) \\ &= R_i(\tau) \end{aligned} \qquad (8.65)$$

with $\tau = t_2 - t_1$. The autocorrelation of a stationary random process is often written as;

$$R_i(\tau) = \mathrm{E}\left[I(t)\, I(t+\tau)\right] \qquad (8.66)$$

Setting $\tau = 0$ in eqn 8.66 we have $R_i(0) = \mathrm{E}\left[I(t)^2\right]$, that is, the mean square value and the variance of a stationary random process are constants.

8.3.3 Ergodic random processes

It is possible to take time averages of sample functions of a random process. For example, the mean value obtained by averaging the k-th sample function of a random process, $i_k(t)$, over time is given by the following expression;

$$\prec i_k(t) \succ \triangleq \lim_{T \to \infty} \frac{1}{T} \int_{-T/2}^{T/2} i_k(t)\, dt \qquad (8.67)$$

where $\prec \cdot \succ$ is a time averaging operator.

Similarly, the autocorrelation function obtained by averaging the k-th sample function of a random process over time is given by the following expression;

$$\prec i_k(t)\, i_k(t+\tau) \succ \triangleq \lim_{T \to \infty} \frac{1}{T} \int_{-T/2}^{T/2} i_k(t)\, i_k(t+\tau)\, dt \qquad (8.68)$$

When all time averages are equal to the corresponding ensemble averages the random process is said to be ergodic. Hence, for an ergodic random process we can write the following:

$$\prec i_k(t) \succ = \mathrm{E}\left[I(t)\right] \qquad (8.69)$$
$$\prec i_k^2(t) \succ = \mathrm{E}\left[I^2(t)\right] \qquad (8.70)$$
$$\prec i_k(t)\, i_k(t+\tau) \succ = \mathrm{E}\left[I(t)\, I(t+\tau)\right] \qquad (8.71)$$

8. Noise in electronic circuits

Ergodicity implies that a single sample function is representative of the entire random process since all statistics associated with this random process can be calculated from such a sample function.

In the context of ergodic random signals or noise we can state the following:

- The mean value of the process, μ_i, represents the DC value of the process $\prec i(t) \succ$;

- The square of the mean value, μ_i^2, represents the power of the DC component $\prec i(t) \succ^2$;

- The variance, σ_i^2, represents the average power associated with the AC components of the process, that is it represents the average power of the time varying component of the process;

- The mean square, $\mathrm{E}\left[i(t)^2\right]$, represents the total average power $\prec i^2(t) \succ$;

- The standard deviation, σ_i, represents the root-mean-square (RMS) value of the time varying component of the process.

As mentioned previously, the autocorrelation function of a random process can provide information about the bandwidth of this random process, at least in qualitative terms. To understand how this information is retrieved we refer now to figure 8.11 where we represent, in detail, the computation of $R_i(\tau_1)$ for a slowly varying zero mean random signal and a rapidly varying zero mean random signal[2]. Recall that slowly varying signals have an associated low bandwidth while rapidly varying signals exhibit high bandwidths. Figures 8.11 a) and 8.11 b) represent the sample waveforms of both random processes and also their replica delayed by τ_1. In figures 8.11 c) and 8.11 d) we represent the signals resulting from multiplying each sample waveform with its delayed replica. From figure 8.11 c) we observe that the percentage of positive area is significantly greater than the percentage of negative area of $i(t)\,i(t+\tau_1)$. Therefore the total area, that is $R_i(\tau_1)$, is non-zero indicating a strong correlation between $i(t)$ and $i(t+\tau_1)$. On the other hand, from figure 8.11 d) we observe that the percentages of positive and negative areas of $i(t)\,i(t+\tau_1)$ are approximately equal. Hence the total area, that is $R_i(\tau_1)$, tends to zero indicating that $i(t)$ is uncorrelated to $i(t+\tau_1)$. It is important to note that, in qualitative terms, the existence or the non-existence of correlation for a given time difference, τ_1, is a direct consequence of the random signal being slowly or rapidly varying. Figures 8.11 e) and 8.11 f) show the autocorrelation functions for the slowly varying and rapidly varying random signals as functions of time delay τ, respectively. From figure 8.11 e) we observe that the slowly varying, low frequency random signal, exhibits strong autocorrelation for values of τ well above τ_1. However, from figure 8.11 f) it can be seen that the rapidly varying, high bandwidth random signal only exhibits significant correlation for values of τ around zero.

---

[2]Signals are 'rapid' or 'slow' in relation to each other and to the observation period.

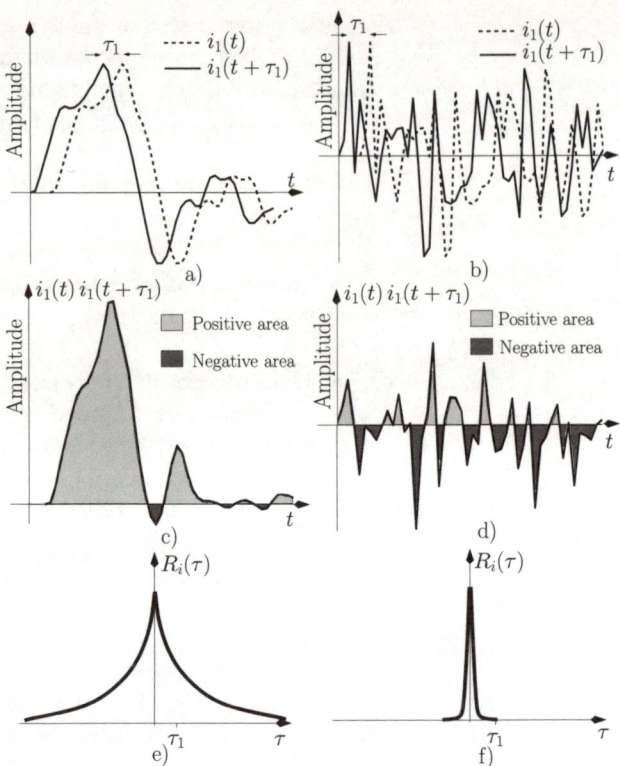

Figure 8.11: *a) Sample function of a slowly varying random process and its delayed replica. b) Sample function of a rapidly varying random process and its delayed replica. c) Product of the sample function of a slowly varying random process with its delayed replica. d) Product of the sample function of a rapidly varying random process with its delayed replica. e) Autocorrelation function of a slowly varying random process. f) Autocorrelation function of a rapidly varying random process.*

8.3.4 Power spectrum

We have just seen that the autocorrelation function can provide qualitative information about the bandwidth of a random process. In order to quantify this information we need to calculate the Fourier transform of the autocorrelation function. This is called the power spectral density (PSD)[3]

$$\begin{aligned} S_{ii^*}(f) &= \mathfrak{F}_\tau[R_i(\tau)] \\ &= \int_{-\infty}^{\infty} R_i(\tau) \, e^{-j\,2\,\pi\,f\,\tau} \, d\tau \end{aligned} \quad (8.72)$$

This eqn is the Wiener-Kinchine theorem and it applies to stationary random signals.

---

[3] The use of $_{ii^*}$ as an underscript will become clear with the discussion of eqn 8.80.

8. Noise in electronic circuits

For ergodic random processes the PSD can be calculated from the time averaging-based autocorrelation function, $\prec i_k(t)\, i_k(t+\tau) \succ$:

$$S_{ii^*}(f) = \int_{-\infty}^{\infty} \prec i_k(t)\, i_k(t+\tau) \succ e^{-j 2\pi f \tau}\, d\tau \qquad (8.73)$$

that is

$$S_{ii^*}(f) = \int_{-\infty}^{\infty} \lim_{T\to\infty} \frac{1}{T} \int_{-T/2}^{T/2} i_k(t)\, i_k(t+\tau)\, dt\, e^{-j 2\pi f \tau}\, d\tau \qquad (8.74)$$

Assuming that the random process sample function, $i_k(t)$ has a Fourier transform $\mathbf{i}_k(f)$ such that:

$$\mathbf{i}_k(f) = \int_{-\infty}^{\infty} i_k(t)\, e^{-j 2\pi f t}\, dt \qquad (8.75)$$

we can write eqn 8.74 as follows:

$$S_{ii^*}(f) = \lim_{T\to\infty} \frac{1}{T} \int_{-T/2}^{T/2} i_k(t)\, e^{+j 2\pi f t}\, dt\, \mathbf{i}_k(f) \qquad (8.76)$$

where we used the following result

$$\int_{-\infty}^{\infty} i_k(t+\tau)\, e^{-j 2\pi f \tau}\, d\tau = \mathbf{i}_k(f)\, e^{j 2\pi f t} \qquad (8.77)$$

Noting that

$$\lim_{T\to\infty} \int_{-T/2}^{T/2} i_k(t)\, e^{+j 2\pi f t}\, dt = \mathbf{i}_k^*(f) \qquad (8.78)$$

we get $S_{ii^*}(f)$ to be

$$S_{ii^*}(f) = \lim_{T\to\infty} \frac{\mathbf{i}_k(f)\mathbf{i}_k^*(f)}{T} \qquad (8.79)$$

that is

$$S_{ii^*}(f) = \lim_{T\to\infty} \frac{|\mathbf{i}_k(f)|^2}{T} \triangleq \langle \mathbf{i}\, \mathbf{i}^* \rangle \qquad (8.80)$$

The $\langle \cdot \rangle$ represents a mathematical operator which can be used to determine the power spectral density of an ergodic random process. We shall discuss some elementary properties of this operator in the next section (see also example 8.3.2).

The result expressed by eqn 8.80 is extremely important since it is the basis of the AC-based electronic noise analysis which is presented in section 8.4. In fact, this eqn tells us that if a *single* sample function $i_k(t)$ of an ergodic random process is observed for a long period of time, T, then the PSD of such a process can be estimated by

$$S_{ii^*}(f) = \langle \mathbf{i}\, \mathbf{i}^* \rangle \simeq \frac{|\mathbf{i}_k(f)|^2}{T} \qquad (8.81)$$

where $\mathbf{i}_k(f)$ is the Fourier transform of $i_k(t)$.

It should be noted that the total power, P_{ni} associated with a random signal, or noise modelled as a stationary random process, can be calculated as follows:

$$P_{ni} = \int_{-\infty}^{\infty} S_{ii^*}(f) \, df \tag{8.82}$$

This is equivalent to finding the autocorrelation function for $\tau = 0$, that is:

$$P_{ni} = R_i(0) \tag{8.83}$$

Example 8.3.1   1. Show that if a stationary random signal has a spectral density given by $S_{x\,x^*}(f) = \sigma^2/B \, \text{sinc}^2(f/B)$, where B is the bandwidth, then the correlation time is $\tau_T = 1/B$.

2. For the random process mentioned above consider $\tau_T = 1/B$ and $\tau_T = 1/B'$, $B' = 3B$. Compare the PSD and the autocorrelation function for each situation.

Solution:

1. The correlation time τ_T is defined as the maximum delay between a sample random waveform and its replica above which the autocorrelation function is zero. Taking the inverse Fourier transform of $S_x(f)$ we obtain

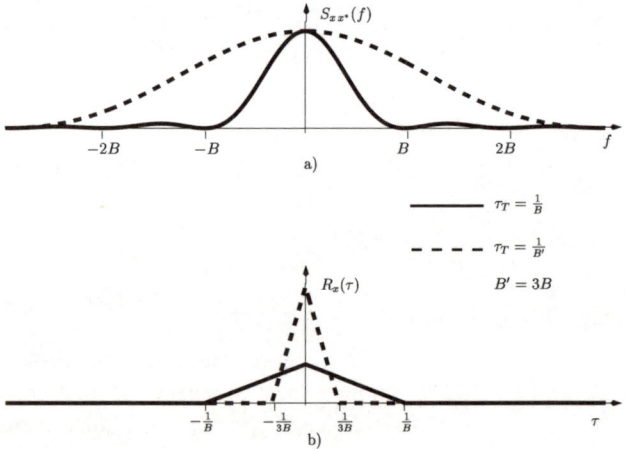

Figure 8.12: *a) Power spectral density. b) Autocorrelation function.*

the autocorrelation function (see appendix A)

$$R_x(\tau) = \sigma^2 \, \text{triang}(\tau B) \tag{8.84}$$

For $|\tau| \geq 1/B$ the autocorrelation function is zero. Hence $\tau_T = 1/B$.

8. Noise in electronic circuits

2. Figure 8.12 illustrates the PSD and the autocorrelation function of the random process $x(t)$ when $\tau_T = 1/B$ and when $\tau_T = 1/B'$, $B' = 3B$. As has been discussed previously (see also figure 8.11), as the bandwidth associated with the random process (or random signal) increases the correlation time decreases.

8.3.5 Cross-power spectrum

Quite often there is need to examine the joint statistics of two random processes, $u(t)$ and $w(t)$. This is especially relevant when a given random process $z(t)$ results from the sum of two random processes. The cross-correlation function of $u(t)$ and $w(t)$ is used to measure the relatedness of these random processes and it is defined as follows:

$$R_{uw}(t_1, t_2) \triangleq \mathrm{E}\left[u(t_1)\, w(t_2)\right] \tag{8.85}$$

If the two processes are uncorrelated then

$$R_{uw}(t_1, t_2) = \mathrm{E}\left[u(t_1)\right]\mathrm{E}\left[w(t_2)\right] \tag{8.86}$$

Moreover, if any of them has a zero mean value for all t then $R_{uw}(t_1, t_2) = 0$.

If $u(t)$ and $w(t)$ are jointly ergodic processes such that $R_{uw}(t_1, t_2) = R_{uw}(\tau)$, $\tau = t_2 - t_1$ then the cross spectral density is defined as the Fourier transform of $R_{uw}(\tau)$:

$$\begin{aligned}
S_{uw^*}(f) &\triangleq \int_{-\infty}^{\infty} R_{uw}(\tau)\, e^{-j2\pi f \tau} d\tau \\
&= \int_{-\infty}^{\infty} \prec u(t)\, w(t+\tau) \succ e^{-j2\pi f \tau} d\tau
\end{aligned} \tag{8.87}$$

Using the results of eqns 8.74–8.81 it is straightforward to show that

$$S_{uw^*}(f) = \lim_{T \to \infty} \frac{\mathbf{u}(f)\, \mathbf{w}^*(f)}{T} \triangleq \langle \mathbf{u}\, \mathbf{w}^* \rangle \tag{8.88}$$

where, as in eqn 8.80, $\langle \cdot \rangle$ is an operator which is now used to determine the cross-power spectral density of two or more jointly ergodic random processes. The next example illustrates some important properties of this operator.

It should be noted that $S_{uw^*}(f) = S^*_{wu^*}(f)$ that is $\langle \mathbf{u}\, \mathbf{w}^* \rangle = (\langle \mathbf{w}\, \mathbf{u}^* \rangle)^*$. If the two random processes are uncorrelated with zero mean then $\langle \mathbf{u}\, \mathbf{w}^* \rangle = 0$.

Example 8.3.2 Consider two jointly ergodic random processes $x(t)$ and $y(t)$. Consider the random process $z = x(t) + y(t)$. Show that the PSD of $z(t)$, $\langle \mathbf{z}\, \mathbf{z}^* \rangle$, can be determined as follows:

$$\begin{aligned}
\langle \mathbf{z}\, \mathbf{z}^* \rangle &= \langle (\mathbf{x}+\mathbf{y})\, (\mathbf{x}+\mathbf{y})^* \rangle \\
&= \langle \mathbf{x}\, \mathbf{x}^* \rangle + \langle \mathbf{x}\, \mathbf{y}^* \rangle + \langle \mathbf{y}\, \mathbf{x}^* \rangle + \langle \mathbf{y}\, \mathbf{y}^* \rangle
\end{aligned} \tag{8.89}$$

Solution: According to eqn 8.73, the PSD of $z(t)$ can be determined as follows:

$$
\begin{aligned}
\langle \mathbf{z\,z^*} \rangle &= \int_{-\infty}^{\infty} \prec z_k(t)\, z_k(t+\tau) \succ e^{-j2\pi f \tau}\, d\tau \\
&= \int_{-\infty}^{\infty} \lim_{T\to\infty} \frac{1}{T} \int_{-T/2}^{T/2} [x_k(t) + y_k(t)] \\
&\quad \times\ [x_k(t+\tau) + y_k(t+\tau)]\, dt\, e^{-j2\pi f \tau}\, d\tau
\end{aligned}
\qquad (8.90)
$$

where $z_k(t)$, $x_k(t)$ and $y_k(t)$ represent sample functions of $z(t)$, $x(t)$ and $y(t)$, respectively. The last eqn can be written as:

$$
\begin{aligned}
\langle \mathbf{z\,z^*} \rangle &= \int_{-\infty}^{\infty} \lim_{T\to\infty} \frac{1}{T} \int_{-T/2}^{T/2} x_k(t)\, x_k(t+\tau)\, dt\ e^{-j2\pi f\tau}\, d\tau \\
&+ \int_{-\infty}^{\infty} \lim_{T\to\infty} \frac{1}{T} \int_{-T/2}^{T/2} x_k(t)\, y_k(t+\tau)\, dt\ e^{-j2\pi f\tau}\, d\tau \\
&+ \int_{-\infty}^{\infty} \lim_{T\to\infty} \frac{1}{T} \int_{-T/2}^{T/2} y_k(t)\, x_k(t+\tau)\, dt\ e^{-j2\pi f\tau}\, d\tau \\
&+ \int_{-\infty}^{\infty} \lim_{T\to\infty} \frac{1}{T} \int_{-T/2}^{T/2} y_k(t)\, y_k(t+\tau)\, dt\ e^{-j2\pi f\tau}\, d\tau
\end{aligned}
\qquad (8.91)
$$

Using eqns 8.74–8.81 and eqn 8.88 this eqn can be written as

$$
\langle \mathbf{z\,z^*} \rangle = \langle \mathbf{x\,x^*} \rangle + \langle \mathbf{x\,y^*} \rangle + \langle \mathbf{y\,x^*} \rangle + \langle \mathbf{y\,y^*} \rangle
\qquad (8.92)
$$

8.3.6 Gaussian random processes

A random process $u(t)$ is called a Gaussian process if the PDF for any random variable considered at any arbitrary instant of time t is Gaussian. Also the joint probability function of any two random variables considered at two arbitrary instants of time t_1 and t_2 is a bivariate Gaussian PDF and likewise for every higher order joint PDF.

The following holds for a Gaussian process[4] $u(t)$:

- The process is completely described by $\mathrm{E}\,[u(t)]$ and by $R_u(t_1, t_2)$.
- If $R_u(t_1, t_2) = \mathrm{E}\,[u(t_1)]\mathrm{E}\,[u(t_2)]$ then $u(t_1)$ and $u(t_2)$ are uncorrelated and statistically independent.
- If $u(t)$ is wide-sense stationary then it is also ergodic.
- Any linear operation on $u(t)$ produces a Gaussian random process.

Gaussian processes are very important in electronics since the Gaussian model applies to the vast majority of random electrical phenomena, or noise, at least as a first approximation. All the electronic noise sources that are discussed later are considered as Gaussian and as ergodic random processes.

---

[4]For detailed discussion of properties of random processes see [1].

8.3.7 Filtered random signals

The output signal, $v(t)$, resulting from applying a random signal, $u(t)$ to a linear system is given by the convolution operation

$$v(t) = \int_{-\infty}^{\infty} h(\lambda)\, u(t-\lambda)\, d\lambda \tag{8.93}$$

where $h(t)$ is the linear system impulse response (see figure 8.13).

The main statistics of $v(t)$, namely the mean and the autocorrelation function, can be determined as follows:

$$\mathrm{E}\left[v(t)\right] = \int_{-\infty}^{\infty} \mathrm{E}\left[u(t-\lambda)\right] h(\lambda)\, d\lambda \tag{8.94}$$

$$R_v(t_1, t_2) = \int_{-\infty}^{\infty}\int_{-\infty}^{\infty} R_u(t_1-\lambda_1, t_2-\lambda_2)\, h(\lambda_1)\, h(\lambda_2)\, d\lambda_1\, d\lambda_2 \tag{8.95}$$

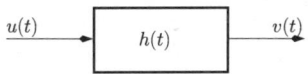

Figure 8.13: *Linear system with impulse response $h(t)$.*

If $u(t)$ is a stationary random process then $v(t)$ is also a stationary random process, and for this situation we have:

$$\mathrm{E}\left[v(t)\right] = \mu_u \int_{-\infty}^{\infty} h(\lambda)\, d\lambda \tag{8.96}$$

$$= \mu_u\, H(0) \tag{8.97}$$

$$R_v(\tau) = R_u(\tau) * h(-\tau) * h(\tau) \tag{8.98}$$

with $H(0)$ representing the transfer function of the linear system at zero frequency (DC).

The power spectrum density of $v(t)$ can be determined by the Fourier transform of eqn 8.98, that is:

$$S_{vv^*}(f) = |H(f)|^2\, S_{uu^*}(f) \tag{8.99}$$

This eqn indicates that the system transfer function shapes the power spectrum density as illustrated by figure 8.14.

The power of $v(t)$ can be determined as follows;

$$\mathrm{E}\left[v(t)^2\right] = R_v(0) = \int_{-\infty}^{\infty} |H(f)|^2\, S_{uu^*}(f)\, df \tag{8.100}$$

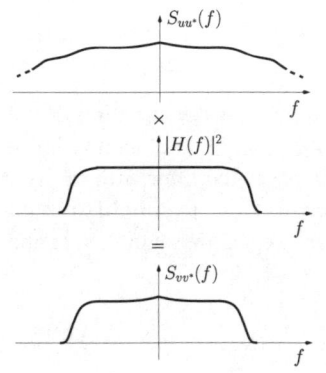

Figure 8.14: *Illustration of eqn 8.99.*

White noise and equivalent noise bandwidth

As will be discussed in section 8.4.1 thermal noise, and many other types of noise sources which can be modelled as random processes, have a flat spectral density over a very wide range of frequencies. This type of spectrum is called *white* by analogy to the spectrum of white light which also has a constant (or flat) spectrum.

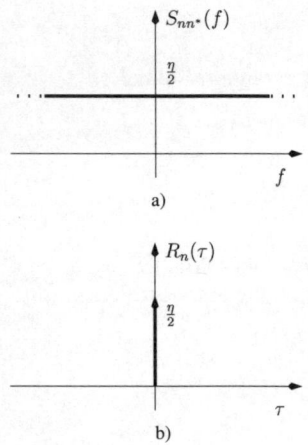

The power spectral density of a white noise can be written as:

$$S_{nn^*}(f) = \frac{\eta}{2} \tag{8.101}$$

where the $1/2$ factor is included to indicate that the PSD is considered to be bilateral, that is, half of the noise power is associated with positive frequencies and the other half is associated with the corresponding negative frequencies.

The autocorrelation function of the white noise is given by the inverse Fourier transform of $S_{nn^*}(f)$:

$$R_n(\tau) = \frac{\eta}{2} \delta(t) \tag{8.102}$$

Figure 8.15 shows the PSD and the autocorrelation function of white noise. It should be noted that, according to eqn 8.82 (or eqn 8.83), white noise must have infinite power. Since it is well known that there is no practical random signal or random noise source with infinite power, we emphasise that, although the theoretical concept of white noise is very useful to model flat and broad bandwidth noise sources, we must bear in mind that white noise sources are always filtered by finite bandwidth systems. It is also important to note that the spectral density of filtered white noise takes the shape of the system transfer function according to:

$$\frac{\eta}{2} |H(f)|^2 \tag{8.103}$$

Figure 8.15: *White noise. a) Power Spectral density. b) Autocorrelation function.*

The noise power associated with the filtered white noise is given by:

$$\begin{aligned} P_n &= \int_{-\infty}^{\infty} \frac{\eta}{2} |H(f)|^2 \, df \\ &= \frac{\eta}{2} \int_{-\infty}^{\infty} |H(f)|^2 \, df \end{aligned} \tag{8.104}$$

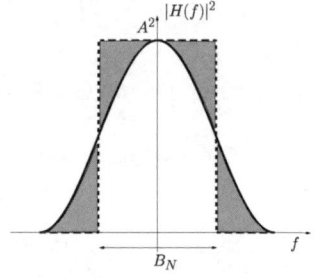

Figure 8.16: *The equivalent noise bandwidth.*

Since the integral of eqn 8.104 depends only on the transfer function $H(f)$ we can define the equivalent noise bandwidth, B_N. This is the bandwidth of an ideal low-pass filter that would pass as much noise as the filter with transfer function $H(f)$ (see also figure 8.16). This is normally done to simplify circuit and system noise calculations by getting rid of the integral (see eqn 8.104) and replacing it by a simple product as shown below:

$$P_n = \frac{\eta}{2} A^2 B_N \tag{8.105}$$

with

$$B_N \triangleq \frac{1}{A^2} \int_{-\infty}^{\infty} |H(f)|^2 \, df \tag{8.106}$$

where $A = |H(f)|_{max}$ is the maximum amplitude of $|H(f)|$. Usually, for a low-pass filter $|H(f)|_{max}$ coincides with the DC gain.

Example 8.3.3 Determine the equivalent noise bandwidth of an RC low-pass filter with a time constant τ.

<u>Solution</u>: The transfer function for the RC low-pass filter is given by:

$$H(f) = \frac{1}{1 + j\, 2\pi f \tau}$$

therefore $A = |H(f)|_{max} = 1$.

$$\begin{aligned}
B_N &= \int_{-\infty}^{\infty} |H(f)|^2\, df \\
&= \int_{-\infty}^{\infty} \frac{1}{1 + (2\pi f \tau)^2} \\
&= \left.\frac{1}{2\pi\tau} \tan^{-1}(2\pi f \tau)\right|_{-\infty}^{\infty} \\
&= \frac{1}{2\tau}
\end{aligned}$$

We note that the equivalent noise bandwidth for a filter of order greater than or equal to three is approximately equal to the 3 dB bandwidth.

8.4 Noise in electronic circuits

Knowledge of noise statistics is necessary for estimating the noise behaviour of electronic circuits. This, in turn, is crucial for the estimation of the circuits' behaviour and their figures of merit. In this section the basic sources of noise in electronic circuits are introduced and methods of analysis of such sources and their interactions are presented.

8.4.1 Thermal noise

Thermal noise or Johnson noise consists of thermal induced random fluctuations in the charge carriers of any material with a finite resistivity. Although the average motion of the charge carriers is zero, the instantaneous random motion of such free carriers generates instantaneous charge gradients which, in turn, produce wide-band random voltage fluctuations. These fluctuations are characterised by an ergodic Gaussian random process and they can be modelled either by an equivalent voltage noise source, $\mathbf{u_n}$, in series with a noise-free resistor or an equivalent current noise source in parallel with a noise-free resistor as shown in figure 8.17. The power spectral density for thermal noise is constant from DC to frequencies up to near infrared and can be considered as white noise. The PSD associated with the Thévenin equivalent model for this type of noise is given by (see also eqn 8.80):

$$S_{u_n u_n^*}(f) = \langle \mathbf{u_n\, u_n^*}\rangle = 2R\mathcal{KT} \quad (\text{V}^2/\text{Hz}) \qquad (8.107)$$

and the PSD associated with the Norton equivalent model is given by

$$S_{i_n i_n^*}(f) = \langle \mathbf{i_n\, i_n^*}\rangle = \frac{2\mathcal{KT}}{R} \quad (\text{A}^2/\text{Hz}) \qquad (8.108)$$

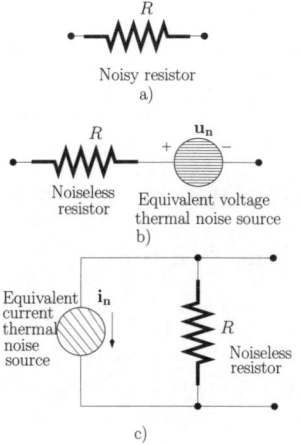

Figure 8.17: *a) Noisy resistor. b) Thévenin equivalent. c) Norton equivalent.*

where $\mathcal{K} = 1.38 \times 10^{-23}$ joule/kelvin is the Boltzmann constant and $\mathcal{T}$ is the temperature in kelvin. At room temperature $\mathcal{T} = 290$ kelvin. R is the resistance. We emphasise that eqns 8.107 and 8.108 are valid for frequencies up to about 10^{12} Hz. Also, these eqns represent bilateral PSDs, that is, they account for positive and negative frequencies. It is worthwhile noting that these eqns may appear contradictory when it comes to the level of noise associated with the resistor. Does the noise increase (eqn 8.107) or decrease (eqn 8.108) with the value of the resistance? The answer to this is simply dependent on the location of the resistance in a circuit and whether it is best to deal with the resistance as a voltage or as a current noise source. Essentially, the noise source is a power source. From eqn 8.107 we can calculate the open circuit RMS voltage, σ_{u_n} produced by the resistance R for a bandwidth B as follows:

$$\sigma_{u_n}^2 = \int_{-B}^{B} S_{u_n u_n^*}(f)\, df \qquad (8.109)$$

$$= 4 R \mathcal{K} \mathcal{T} B \quad (\text{V}^2) \qquad (8.110)$$

$$\sigma_{u_n} = \sqrt{4 R \mathcal{K} \mathcal{T} B} \quad (\text{volts RMS}) \qquad (8.111)$$

8.4.2 Electronic shot-noise

Electronic shot-noise is associated with the passage of carriers across a potential barrier such as those encountered in p–n junctions of semiconductor diodes and transistors. The statistics that describe charge motion determine the noise characteristics. The number of carriers that cross the barrier is random and is characterised by a Poisson distribution. However, when the number of events that occur per unit observation time is large then the Poisson distribution can be replaced by the distribution of a zero mean and ergodic Gaussian process with a white power spectral density. In terms of an equivalent electronic model this noise is modelled as a current noise source in parallel with the p–n junction. Figure 8.18 shows the noise equivalent model for a diode.

The flat PSD of the electronic shot noise associated with a DC current, I_{DC}, which crosses a potential barrier is given by

$$S_{i_n i_n^*}(f) = \langle \mathbf{i_n}\, \mathbf{i_n^*} \rangle = q\, I_{DC} \quad (\text{A}^2/\text{Hz}) \qquad (8.112)$$

where $q = 1.6 \times 10^{-19}$ coulomb is the electronic charge.

8.4.3 $1/f$ noise

$1/f$ noise is observed in some resistors and semiconductor devices. The origins of this type of noise are usually associated with imperfections in the material from which the devices are made.

$1/f$ noise is considered as a zero mean ergodic Gaussian process. The Norton equivalent model for this type of noise has a general spectral density given by:

$$S_{i_n i_n^*}(f) = \langle \mathbf{i_n}\, \mathbf{i_n^*} \rangle = \frac{K_f}{f^\alpha} \quad (\text{A}^2/\text{Hz}) \qquad (8.113)$$

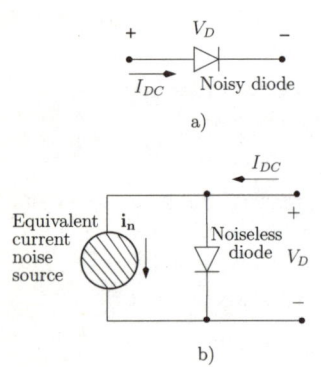

Figure 8.18: *Noisy p–n junction. b) Norton equivalent model for shot noise.*

8. Noise in electronic circuits

where K_f is a factor which depends on the current passing on the device and α is a coefficient with a range of 0.8 to 1.4. For calculation purposes α is taken as unity and the noise power varies as the inverse of the frequency. The total noise power considered between f_a and f_b is:

$$P_n = \int_{f_a}^{f_b} \frac{K_f}{f} df \qquad (8.114)$$

$$= K_f \ln \frac{f_b}{f_a} \quad (A^2) \qquad (8.115)$$

Note that over any decade in frequency, that is, for any $f_b = 10\, f_a$, the noise power is the same and given by $K_f \ln(10)$.

In terms of a noise model the $1/f$ noise component can be accounted for by adding an additional noise source to the white noise source as illustrated in figure 8.19.

Since the two noise sources are uncorrelated, the total noise power spectral density is just the sum of each power spectral density:

$$S_{i_n i_n^*}(f) = \langle \mathbf{i_n}\, \mathbf{i_n^*} \rangle = \frac{\eta}{2}\left(1 + \frac{f_c}{f}\right) \quad (A^2/Hz) \qquad (8.116)$$

where $\eta/2$ is the PSD of the white noise source component and f_c is the 'corner frequency' of the $1/f$ noise.

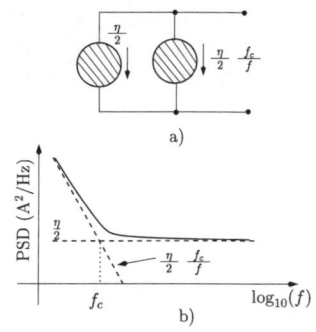

Figure 8.19: *$1/f$ noise and white noise. a) Noise sources. b) Equivalent noise power spectral density.*

8.4.4 Noise models for passive devices

The ideal passive devices are the resistor, the capacitor and the inductor. Among all these the only one which is noisy is the resistor since, as discussed above, it produces thermal noise. Both the ideal inductor and the ideal capacitor are energy storage elements, do not dissipate energy and do not induce thermal noise. However, practical devices always have parasitic resistances and these components will produce thermal noise and $1/f$ noise.

Resistor

The practical resistor generates thermal noise and $1/f$ noise. The noise model for the resistor is presented in figure 8.20. For the series voltage sources model the PSD is given by:

$$\langle \mathbf{u_{nt}}\, \mathbf{u_{nt}^*} \rangle = 2\mathcal{K}TR \qquad (8.117)$$

$$\langle \mathbf{u_{nf}}\, \mathbf{u_{nf}^*} \rangle = 2\mathcal{K}TR\frac{f_c}{f} \qquad (8.118)$$

where $\mathbf{u_{nt}}$ accounts for the thermal noise contribution and $\mathbf{u_{nf}}$ accounts for the $1/f$ noise contribution. For the parallel current sources model the PSD is given by:

$$\langle \mathbf{i_{nt}}\, \mathbf{i_{nt}^*} \rangle = \frac{2\mathcal{K}T}{R} \qquad (8.119)$$

$$\langle \mathbf{i_{nf}}\, \mathbf{i_{nf}^*} \rangle = \frac{2\mathcal{K}T}{R}\frac{f_c}{f} \qquad (8.120)$$

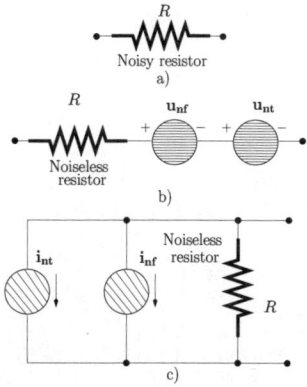

Figure 8.20: *a) Noisy resistor. b) Series voltage sources. c) Parallel current sources.*

Capacitor

The real capacitor has two resistive components, as shown in figure 8.21. R_d is associated with dielectric losses and R_s includes the lead resistances. While R_d is greater than 10^9 Ω, R_s is, typically, less than 1 Ω. There is thermal and $1/f$ noise associated with R_s represented by $\mathbf{u_{nt1}}$ and by $\mathbf{u_{nf}}$, respectively. R_d generates only thermal noise represented by $\mathbf{u_{nt2}}$. In practice their noise is negligible compared to other circuit noise generators.

Inductor

The real inductor has a main resistive component R_s, as shown in figure 8.22, from the wire resistance. R_s generates both thermal and $1/f$ noise represented by $\mathbf{u_{nt}}$ and by $\mathbf{u_{nf}}$, respectively. Like the capacitor the noise sources associated with the practical inductor are usually neglected when compared with other circuit noise generators like resistors.

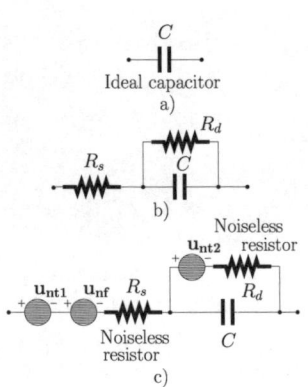

Figure 8.21: *a) Ideal capacitor. b) Electrical model for a practical capacitor. c) Noise model for a practical capacitor.*

8.4.5 Noise models for active devices

Noise model for field effect transistors

Figure 8.23 shows the high frequency small-signal model for the FET including the main noise source contributions. $\mathbf{i_{nd}}$ is the thermal noise associated with the Ohmic resistance of the channel, R_{ch}, which is related to the transconductance as follows:

$$R_{ch} = (g_m K_d)^{-1} \qquad (8.121)$$

where K_d is a constant associated with the physical dimensions of the FET and is normally taken as $2/3$. The power spectral density of $\mathbf{i_{nd}}$ is given by:

$$\langle \mathbf{i_{nd}} \mathbf{i_{nd}^*} \rangle = 2 \mathcal{K} \mathcal{T} \frac{2 g_m}{3} \qquad (8.122)$$

Since r_o is a dynamic resistance it does not dissipate power and, therefore, there is no noise source associated with it.

The $1/f$ noise power spectral density can be written as:

$$\langle \mathbf{i_{nf}} \mathbf{i_{nf}^*} \rangle = 2 \mathcal{K} \mathcal{T} \frac{2 g_m}{3} \frac{f_c}{f} \qquad (8.123)$$

where f_c is the corner frequency. $\mathbf{i_{ng}}$ accounts for shot noise resulting from the DC gate leakage current, I_G. The PSD for $\mathbf{i_{ng}}$ can be written as:

$$\langle \mathbf{i_{ng}} \mathbf{i_{ng}^*} \rangle = q\, I_G \qquad (8.124)$$

In some FETs there is correlation between the noise at the gate and the noise at the drain especially at high frequencies. However, at mid-range frequency this correlation is usually neglected.

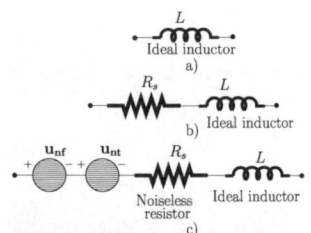

Figure 8.22: *a) Ideal inductor. b) Electrical model for the practical inductor. c) Noise model.*

8. Noise in electronic circuits

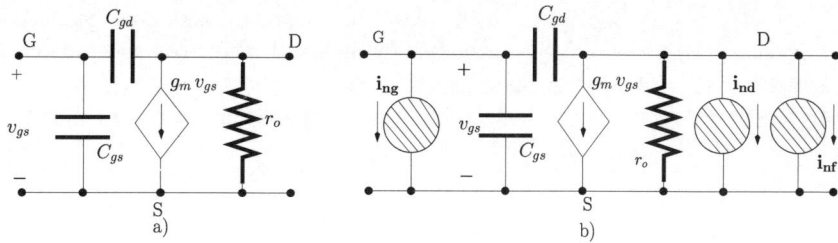

Figure 8.23: *a) Noiseless small-signal model for the FET. b) Small-signal model including noise sources.*

Noise model for bipolar junction transistors

The noise model for the silicon bipolar junction transistor (BJT) is shown in figure 8.24. u_{nB} accounts for the thermal noise associated with the base spreading resistance r_x. It should be noted that this is the only dissipative resistance in the hybrid-π model. All the remaining resistances in the model are dynamic resistances and are therefore noiseless. i_{nb} and i_{nc} are the shot noise current sources associated with the base current and the collector current, respectively. i_{nf} is the $1/f$ current noise source in the base current.

Each noise source is characterised by its power spectral density as follows:

$$\langle u_{nB} \, u_{nB}^* \rangle = 2\mathcal{K}\mathcal{T}\, r_x \tag{8.125}$$

$$\langle i_{nb} \, i_{nb}^* \rangle = q\, I_B \tag{8.126}$$

$$\langle i_{nc} \, i_{nc}^* \rangle = q\, I_C \tag{8.127}$$

$$\langle i_{nf} \, i_{nf}^* \rangle = q\, I_B \frac{f_c}{f} \tag{8.128}$$

where I_B and I_C are the base and the collector bias currents, respectively.

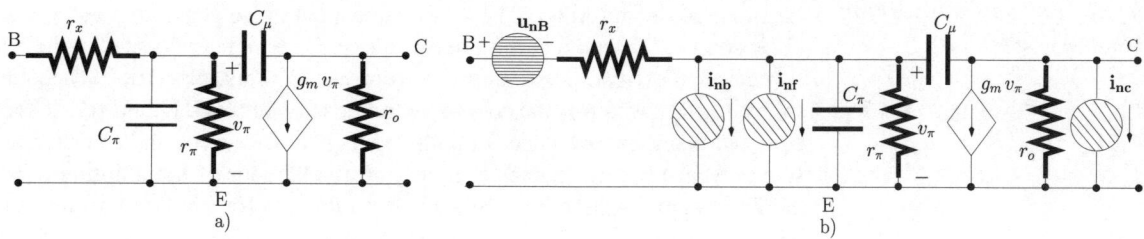

Figure 8.24: *a) Noiseless hybrid-π model for the bipolar transistor. b) Hybrid-π model including noise sources.*

All the noise sources described above are assumed to be uncorrelated. However, correlation between i_{nb} and i_{nc} occurs at high frequencies. For example, in Hetero-junction Bipolar Transistors (HBTs)[5] the shot noise induced

---

[5] HBTs are bipolar transistors with f_T of the order of tens of GHz which are suitable for Radio-Frequency/Microwave applications.

by the base current and the shot noise induced by the collector current are characterised by their self- and cross-spectral densities as follows:

$$\langle i_{nb} \, i_{nb}^* \rangle = q \left(I_B + \left|1 - e^{-j\omega\tau_b}\right|^2 I_C \right) \quad (8.129)$$

$$\langle i_{nc} \, i_{nc}^* \rangle = q \, I_C \quad (8.130)$$

$$\langle i_{nb} \, i_{nc}^* \rangle = q \left(e^{j\omega\tau_b} - 1 \right) I_C \quad (8.131)$$

where τ_b represents the finite base transit time for the charge carriers. It should be noted that at low-frequencies $\langle i_{nb} \, i_{nb}^* \rangle$, given by eqn 8.129, tends to $q \, I_B$ as in eqn 8.126. Also, at low frequencies the cross-spectral density, given by eqn 8.131, tends to zero.

It is interesting to note that the dominant noise process in FET devices is thermal while bipolar devices have dominant shot-noise components. This is understandable given the nature of FETs and BJTs. A FET can be viewed as a variable semiconductor resistor generating the thermal noise components of eqns 8.122 and 8.124. On the other hand a BJT is effectively three semiconductor regions forming two p–n junctions giving rise to the shot noise components of eqns 8.126 and 8.127.

8.4.6 The equivalent input noise sources

Now that the noise models for the main circuit elements have been presented it is clear that even simple circuits can have a significant number of noise sources contributing to the overall noise of the circuit. Therefore, it is necessary to assess the impact of each noise source on the circuit performance.

A noise assessment method used to quantify the noise performance of an amplifier, or of any linear two-port circuit, is based on modelling the noisy amplifier using equivalent input noise sources, as shown in figure 8.25. The noisy amplifier is replaced by a noise-free amplifier, which accounts for the electrical response, and by two equivalent noise generators which characterise the noise of the amplifier. These two equivalent noise generators are a voltage noise source, u_n, and a current noise source, i_n, which are located at the input. In general, these two noise sources are correlated. This correlation is accounted for by $\langle i_n \, u_n^* \rangle$ which is the cross-spectral density between i_n and u_n^*. It should be noted that there is a need for both a voltage noise source and a current noise source to accurately characterise the noise behaviour of the amplifier. When the input signal source is a voltage source its zero (or low) output impedance absorbs the current noise i_n. If there was not a voltage noise source, the noisy behaviour of the amplifier would not be taken into account. Similar reasoning accounts for the need of a current noise source when the input is a current source.

This noise representation technique is very useful since it allows amplifiers to be compared in terms of noise performance, regardless of gain, impedance, or transfer function of the amplifiers. In order to characterise an amplifier according to the model shown in figure 8.25 b) it is necessary to relate all the individual noise sources of the noisy amplifier with the equivalent noise sources u_n and i_n.

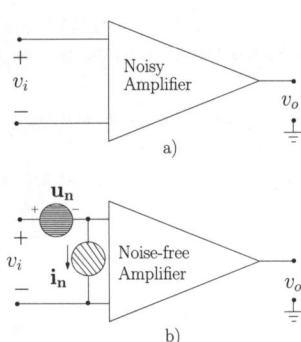

Figure 8.25: *a) Noisy amplifier. b) Equivalent model for the noisy amplifier.*

8. Noise in electronic circuits

We describe a circuit analysis method to calculate the equivalent input voltage and equivalent input current noise spectral densities for a general amplifier, or any linear noisy circuit. As an example, we consider a common-emitter HBT based amplifier shown in figure 8.26 a). The circuit does not include the input bias circuit. The HBT is chosen for this example due to its correlated noise

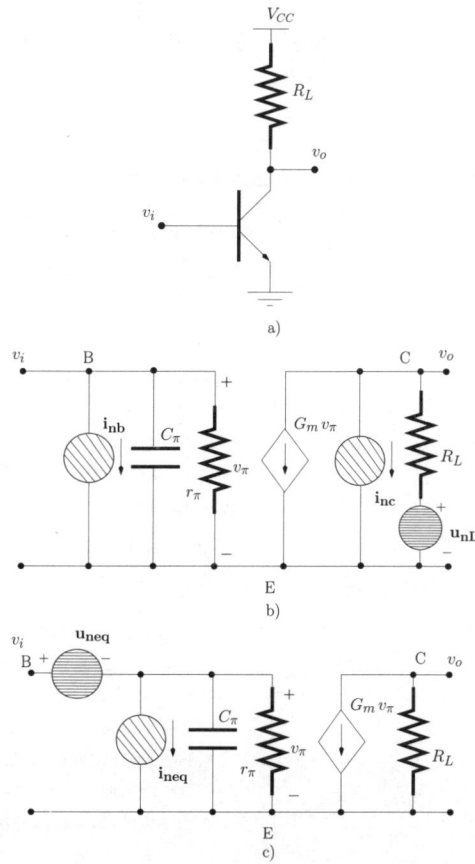

Figure 8.26: *HBT common-emitter amplifier. a) Simplified schematic. b) Small-signal equivalent circuit including intrinsic noise sources. c) Small-signal equivalent circuit with equivalent noise sources referred to the input of the amplifier.*

sources discussed above, making the analysis most general. Figure 8.26 b) shows the small-signal model of the amplifier including the intrinsic noise sources of the HBT. i_{nb} and i_{nc} are characterised by their self- and cross-spectral densities given eqns 8.129–8.131. Figure 8.26 b) also shows the noise source of the load resistance R_L, u_{nL}, with power spectral density given by:

$$\langle u_{nL}\, u_{nL}^* \rangle = 2\mathcal{K}\mathcal{T} R_L \qquad (8.132)$$

For reasons of simplicity the model of the circuit described by figure 8.26 b) does not include the effect of the base resistance r_x and C_μ. Also, it assumes that the Early effect can be neglected, that is, r_o is very large compared to R_L. G_m is the transistor transconductance given by:

$$G_m = g_m e^{-j\omega\tau} \tag{8.133}$$

$$g_m = \frac{I_C\, q}{\mathcal{K}\mathcal{T}} \tag{8.134}$$

where the term $e^{-j\omega\tau}$ accounts for the delay, τ, associated with the transistor transconductance. In the following analysis, we shall assume $\tau \simeq 0$, that is, $G_m \simeq g_m$.

The analysis method

1. **Each noise source is considered as an ergodic random process and, therefore, can be characterised by a single sample function which is considered in the frequency domain (see eqn 8.75).**

2. **The total open-circuit noise voltage at the output of the linear circuit is calculated assuming an input voltage source. Applying the superposition theorem, the contribution of each voltage noise source and each current noise source is obtained by replacing all the other noise sources and the input signal source by their corresponding output impedances.**

We calculate now the contributions of the various noise sources to the output voltage.

- **$\mathbf{u_{nL}}$**: Figure 8.27 a) shows the equivalent circuit for the calculation of the contribution of this noise source to the output voltage. It can be seen that $v_\pi = 0$ since the zero impedance of the input voltage signal source short-circuits the base-emitter terminal of the transistor. Therefore, the voltage-controlled current source can be replaced by an open-circuit. Since there is no current flowing across R_L the output voltage is

$$v_o = \mathbf{u_{nL}} \tag{8.135}$$

- **$\mathbf{i_{nc}}$**: Figure 8.27 b) shows the relevant equivalent circuit. Again $v_\pi = 0$ and the current flowing through R_L is just $-\mathbf{i_{nc}}$. Hence, the output voltage is:

$$v_o = -\mathbf{i_{nc}}\, R_L \tag{8.136}$$

- **$\mathbf{i_{nb}}$**: See figure 8.27 c). Again, we have $v_\pi = 0$. Hence, the voltage-controlled current source does not produce any current and, therefore, the output voltage is zero.

- **Total noise voltage**: The sum of all noise contributions to the noise voltage at the output of the amplifier is given by:

$$v_o = \mathbf{u_{nL}} - \mathbf{i_{nc}}\, R_L \tag{8.137}$$

8. Noise in electronic circuits

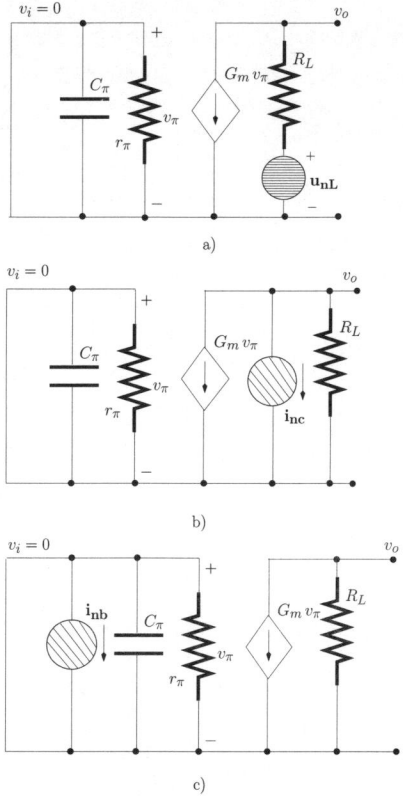

Figure 8.27: *Equivalent circuit for the calculation of the various contributions to the output noise voltage. a) Contribution from* u_{nL}. *b) Contribution from* i_{nc}. *c) Contribution from* i_{nb}.

3. **The total short-circuit noise current at the output is calculated by assuming an input current signal source. The contribution of each voltage and current noise source to the total output current is obtained applying the superposition theorem. We calculate the individual contribution of each noise source to the output short-circuit current replacing all the other noise sources by their output impedances and by replacing the input signal source by an open-circuit.**

   We now calculate the contributions of the various noise sources to the output short-circuit current.

   - u_{nL}: See figure 8.28 a). Since there is no signal applied to r_π and C_π, $v_\pi = 0$. Therefore, the voltage controlled current source can be replaced by an open circuit. From this figure we also observe that the voltage across across R_L is u_{nL} and therefore the output current is

$$i_o = -\frac{u_{nL}}{R_L} \qquad (8.138)$$

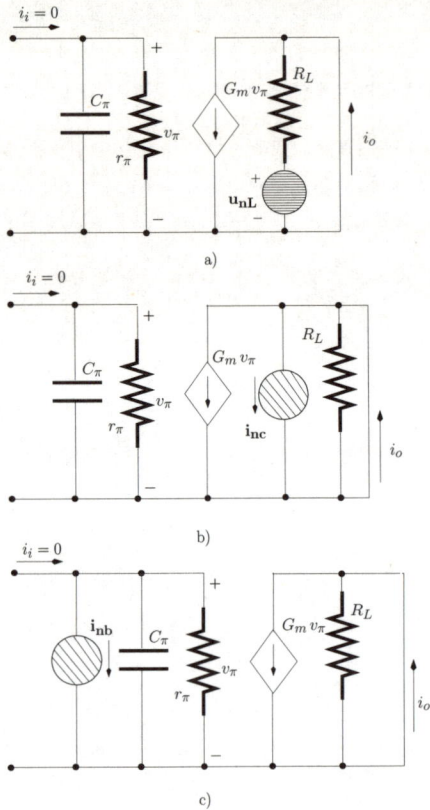

Figure 8.28: *Equivalent circuit for the calculation of the various contributions to the output noise current. a) Contribution from $\mathbf{u_{nL}}$. b) Contribution from $\mathbf{i_{nc}}$. c) Contribution from $\mathbf{i_{nb}}$.*

- $\underline{\mathbf{i_{nc}}}$: In figure 8.28 b) we see that $v_\pi = 0$ and the voltage controlled current source can be replaced by an open circuit. Since R_L is short-circuited, there is no current flowing through its terminals and the output current is

$$i_o = \mathbf{i_{nc}} \tag{8.139}$$

- $\underline{\mathbf{i_{nb}}}$: Figure 8.28 c) shows the equivalent circuit for this calculation. Now we have:

$$v_\pi = -\mathbf{i_{nb}} \frac{r_\pi}{1 + j\, 2\pi f\, C_\pi\, r_\pi} \tag{8.140}$$

$$i_o = g_m v_\pi$$
$$= -\mathbf{i_{nb}} \frac{g_m\, r_\pi}{1 + j\, 2\pi f\, C_\pi\, r_\pi} \tag{8.141}$$

8. Noise in electronic circuits 315

- **Total noise current**: The sum of all noise contributions to the noise current at the output of the amplifier is given by:

$$i_o = -\frac{u_{nL}}{R_L} + i_{nc} - i_{nb}\frac{g_m\, r_\pi}{1 + j\,2\,\pi\,f\,C_\pi\,r_\pi} \quad (8.142)$$

4. **The equivalent input voltage noise source is obtained by dividing the total open-circuit noise voltage at the output by the circuit voltage gain.**

The circuit voltage gain, A_v, is:

$$A_v = -g_m\, R_L \quad (8.143)$$

and, therefore, the equivalent input voltage noise source is given by (see eqn 8.137):

$$u_{neq} = -\frac{u_{nL}}{g_m\, R_L} + \frac{i_{nc}}{g_m} \quad (8.144)$$

5. **The equivalent input current noise source is obtained by dividing the total short-circuit noise current at the output by the circuit current gain.**

The circuit current gain, A_i, is:

$$A_i = \frac{g_m\, r_\pi}{1 + j\,2\,\pi\,f\,C_\pi\,r_\pi} \quad (8.145)$$

and, therefore, the equivalent input current noise source is given by (see eqn 8.142):

$$i_{neq} = -u_{nL}\frac{1 + j\,2\,\pi\,f\,C_\pi\,r_\pi}{g_m\, r_\pi\, R_L} + i_{nc}\frac{1 + j\,2\,\pi\,f\,C_\pi\,r_\pi}{g_m\, r_\pi} - i_{nb}$$

6. **Finally, the self- and the cross-spectral densities of these two equivalent noise sources can be calculated by simplifying** $\langle u_{neq}\, u_{neq}^*\rangle$, $\langle i_{neq}\, i_{neq}^*\rangle$ **and** $\langle u_{neq}\, i_{neq}^*\rangle$ **as defined by eqn 8.80 and eqn 8.88.**

- The power spectral density of u_{neq} is calculated as follows:

$$\begin{aligned}\langle u_{neq}\, u_{neq}^*\rangle &= \left\langle \left(-\frac{u_{nL}}{g_m\, R_L} + \frac{i_{nc}}{g_m}\right)\left(-\frac{u_{nL}}{g_m\, R_L} + \frac{i_{nc}}{g_m}\right)^*\right\rangle\\ &= \langle u_{nL}\, u_{nL}^*\rangle\frac{1}{(g_m\, R_L)^2} + \langle i_{nc}\, i_{nc}^*\rangle\frac{1}{g_m^2}\\ &= \frac{2\mathcal{K}\mathcal{T}}{g_m^2\, R_L} + \frac{q\, I_C}{g_m^2}\\ &= \frac{2\,(\mathcal{K}\mathcal{T})^3}{q^2\, R_L\, I_C^2} + \frac{(\mathcal{K}\mathcal{T})^2}{q\, I_C} \quad (8.146)\end{aligned}$$

where we have used the definition of transconductance for a BJT, given by eqn 8.134. It should be noted that, since u_{nL} and i_{nc} are uncorrelated and zero mean random processes, we have:

$$\langle u_{nL}\, i_{nc}^*\rangle = \langle i_{nc}\, u_{nL}^*\rangle = 0 \quad (8.147)$$

- The power spectral density of $\mathbf{i_{neq}}$ is calculated as follows:

$$\begin{aligned}
\langle \mathbf{i_{neq}\, i_{neq}^*} \rangle &= \left\langle \left(-\mathbf{u_{nL}} \frac{1 + j\,2\pi f\,C_\pi\,r_\pi}{g_m\,r_\pi\,R_L} \right.\right. \\
&\quad + \left. \mathbf{i_{nc}} \frac{1 + j\,2\pi f\,C_\pi\,r_\pi}{g_m\,r_\pi} - \mathbf{i_{nb}} \right) \\
&\quad \times \left(-\mathbf{u_{nL}} \frac{1 + j\,2\pi f\,C_\pi\,r_\pi}{g_m\,r_\pi\,R_L} \right. \\
&\quad + \left.\left. \mathbf{i_{nc}} \frac{1 + j\,2\pi f\,C_\pi\,r_\pi}{g_m\,r_\pi} - \mathbf{i_{nb}} \right)^* \right\rangle \\
&= \left(\langle \mathbf{u_{nL}\, u_{nL}^*} \rangle \frac{1}{R_L^2} + \langle \mathbf{i_{nc}\, i_{nc}^*} \rangle \right) \frac{1 + (2\pi f\,C_\pi\,r_\pi)^2}{(g_m\,r_\pi)^2} \\
&\quad + \langle \mathbf{i_{nb}\, i_{nb}^*} \rangle - 2\,\mathrm{Real}\left[\langle \mathbf{i_{nc}\, i_{nb}^*} \rangle \frac{1 + j\,2\pi f\,C_\pi\,r_\pi}{g_m\,r_\pi} \right]
\end{aligned} \tag{8.148}$$

This can be written as:

$$\begin{aligned}
\langle \mathbf{i_{neq}\, i_{neq}^*} \rangle &= \left(\frac{2\mathcal{K}\mathcal{T}}{R_L} + q\,I_C \right) \frac{1 + (2\pi f\,C_\pi\,r_\pi)^2}{(g_m\,r_\pi)^2} \\
&\quad + q\left(I_B + \left| 1 - e^{-j\omega\tau_b} \right|^2 I_C \right) \\
&\quad - 2\,\mathrm{Real}\left[q\left(e^{-j\omega\tau_b} - 1 \right) I_C \frac{1 + j\,2\pi f\,C_\pi\,r_\pi}{g_m\,r_\pi} \right]
\end{aligned} \tag{8.149}$$

- The cross-power spectral density of $\mathbf{i_{neq}}$ and $\mathbf{u_{neq}}$ is calculated as follows:

$$\begin{aligned}
\langle \mathbf{i_{neq}\, u_{neq}^*} \rangle &= \left\langle \left(-\mathbf{u_{nL}} \frac{1 + j\,2\pi f\,C_\pi\,r_\pi}{g_m\,r_\pi\,R_L} + \mathbf{i_{nc}} \frac{1 + j\,2\pi f\,C_\pi\,r_\pi}{g_m\,r_\pi} \right.\right. \\
&\quad - \left.\left. \mathbf{i_{nb}} \right) \left(-\frac{\mathbf{u_{nL}}}{g_m\,R_L} + \frac{\mathbf{i_{nc}}}{g_m} \right)^* \right\rangle \\
&= \langle \mathbf{u_{nL}\, u_{nL}^*} \rangle \frac{1 + j\,2\pi f\,C_\pi\,r_\pi}{g_m^2\,r_\pi\,R_L^2} \\
&\quad + \langle \mathbf{i_{nc}\, i_{nc}^*} \rangle \frac{1 + j\,2\pi f\,C_\pi\,r_\pi}{g_m^2\,r_\pi} - \langle \mathbf{i_{nb}\, i_{nc}^*} \rangle \frac{1}{g_m} \\
&= 2\mathcal{K}\mathcal{T} \frac{1 + j\,2\pi f\,C_\pi\,r_\pi}{g_m^2\,r_\pi\,R_L} + q\,I_C \frac{1 + j\,2\pi f\,C_\pi\,r_\pi}{g_m^2\,r_\pi} \\
&\quad - q\left(e^{j\omega\tau_b} - 1 \right) \frac{I_C}{g_m}
\end{aligned} \tag{8.150}$$

and $\langle \mathbf{u_{neq}\, i_{neq}^*} \rangle = \left(\langle \mathbf{i_{neq}\, u_{neq}^*} \rangle \right)^*$.

8. Noise in electronic circuits

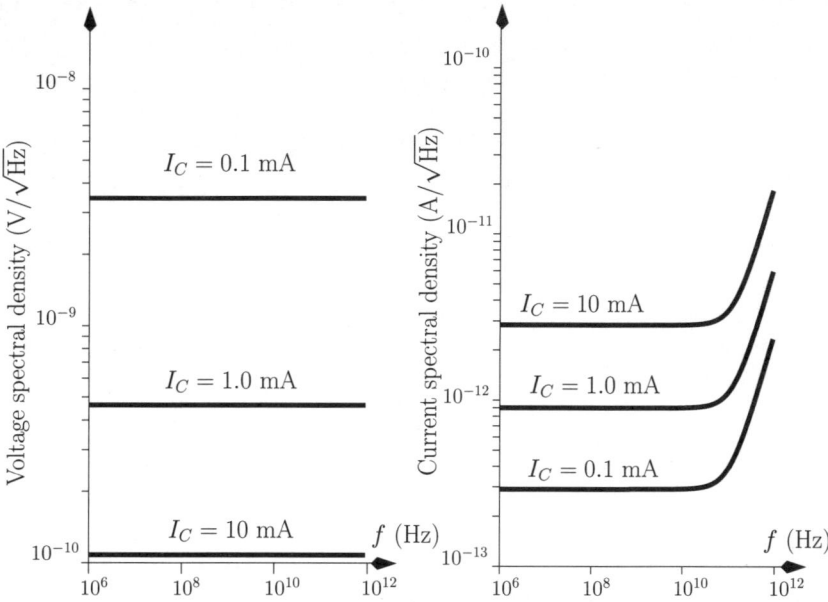

Figure 8.29: *Equivalent input noise spectral densities for the common-emitter amplifier. Note that both axes are logarithmic.*

Figure 8.29 shows the RMS voltage spectral density, $(\langle \mathbf{u_{neq}} \mathbf{u_{neq}^*} \rangle)^{1/2}$, and the RMS current spectral density, $(\langle \mathbf{i_{neq}} \mathbf{i_{neq}^*} \rangle)^{1/2}$, obtained from eqns 8.146 and 8.149, respectively, for three collector bias currents: $I_C = 0.1$ mA, $I_C = 1.0$ mA and $I_C = 10$ mA. We have also assumed that the HBT is characterised by $\beta = 200$, $\tau_\pi = r_\pi C_\pi = 4$ ps and $\tau_b = 0.1$ ps. $R_L = 50$ Ω.

From this figure it can be observed that while the spectral density associated with $\langle \mathbf{u_{neq}} \mathbf{u_{neq}^*} \rangle$ decreases with increasing I_C, the PSD associated with $\langle \mathbf{i_{neq}} \mathbf{i_{neq}^*} \rangle$ increases. These opposite trends suggest that there is an optimum bias collector current for which the overall noise PSD can be minimised. It is also interesting to examine what are the most significant noise contributions for $\langle \mathbf{i_{neq}} \mathbf{i_{neq}^*} \rangle$ and $\langle \mathbf{u_{neq}} \mathbf{u_{neq}^*} \rangle$. Figure 8.30 shows the various contributions to the equivalent input current noise spectral density as given by eqn 8.149, for a bias collector current $I_C = 0.1$ mA. It can be seen that the dominant contribution is the current shot noise induced by the HBT base current (see also eqn 8.129). It is left as an exercise for the reader to determine which contributions dominate the equivalent input voltage noise spectral density of the common-emitter amplifier.

Equivalent current noise spectral density

Figure 8.31 shows a noisy amplifier connected to a current source with an output impedance $Z_s(\omega)$. The resistive part of $Z_s(\omega)$, $R_s = \text{Real}\,[Z_s(\omega)]$, produces thermal noise $\mathbf{i_{ns}}$. It is very useful to be able to represent such circuits as having a single 'equivalent' noise source at the input. The spectral density of

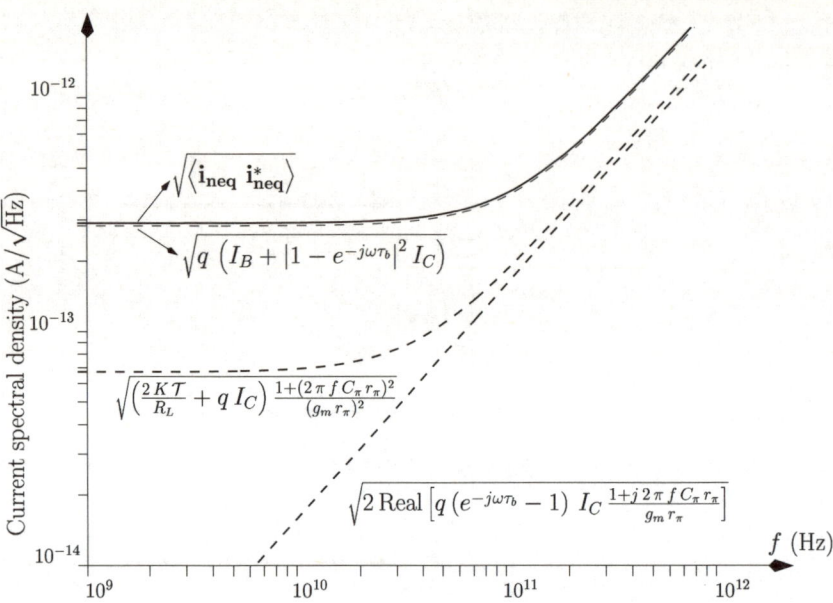

Figure 8.30: *Contributions to the equivalent input current noise spectral densities for the common-emitter amplifier ($I_C = 0.1$ mA).*

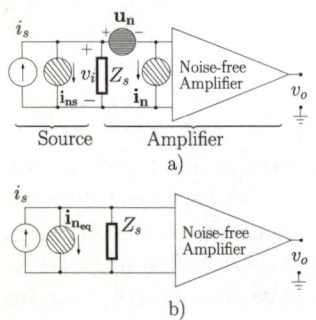

Figure 8.31: *a) Noisy amplifier connected to a current signal source. b) Equivalent noise model for the signal source and amplifier.*

such a source can be calculated by applying simple Thévenin or Norton equivalent models. For this example the circuit can be represented by applying the Norton equivalent model to the circuit in figure 8.31 a). This results in the simplified representation of figure 8.31 b) where the equivalent current noise $i_{n_{eq}}$ describes the overall noise performance of the current source-amplifier combination. Using the noise analysis method explained previously it can be shown (see problem 8.8) that the equivalent noise current source can be expressed by:

$$i_{n_{eq}} = \frac{u_n}{Z_s(\omega)} + i_{ns} + i_n \qquad (8.151)$$

and the PSD of $i_{n_{eq}}$ can be obtained by simplifying the following equation:

$$\langle i_{n_{eq}} i_{n_{eq}}^* \rangle = \left\langle \left(\frac{u_n}{Z_s(\omega)} + i_{ns} + i_n\right)\left(\frac{u_n}{Z_s(\omega)} + i_{ns} + i_n\right)^* \right\rangle$$

that is

$$\langle i_{n_{eq}} i_{n_{eq}}^* \rangle = \frac{\langle u_n u_n^* \rangle}{|Z_s(\omega)|^2} + \langle i_n i_n^* \rangle + \langle i_{ns} i_{ns}^* \rangle + 2\,\mathrm{Real}\left[\frac{\langle u_n i_n^* \rangle}{Z_s(\omega)}\right]$$

$$= \frac{\langle u_n u_n^* \rangle}{|Z_s(\omega)|^2} + \langle i_n i_n^* \rangle + \frac{2KT}{R_s} + 2\,\mathrm{Real}\left[\frac{\langle u_n i_n^* \rangle}{Z_s(\omega)}\right]$$

(8.152)

As an example we consider the common-emitter amplifier of figure 8.26 driven by a current source with an output impedance $Z_s(\omega) = R_s = 50\ \Omega$.

8. Noise in electronic circuits

$\langle \mathbf{u_n}\, \mathbf{u_n^*} \rangle$, $\langle \mathbf{i_n}\, \mathbf{i_n^*} \rangle$ and $\langle \mathbf{u_n}\, \mathbf{i_n^*} \rangle$ are given by eqns 8.146, 8.149 and 8.150, respectively.

Figure 8.32 shows the variation of the current spectral density of $\mathbf{i_{n_{eq}}}$ with bias collector current I_C at 800 MHz. It can be seen that there is an optimum bias value for I_C (about 8 mA) which minimises $\left(\left\langle \mathbf{i_{n_{eq}}}\, \mathbf{i_{n_{eq}}^*} \right\rangle\right)^{1/2}$.

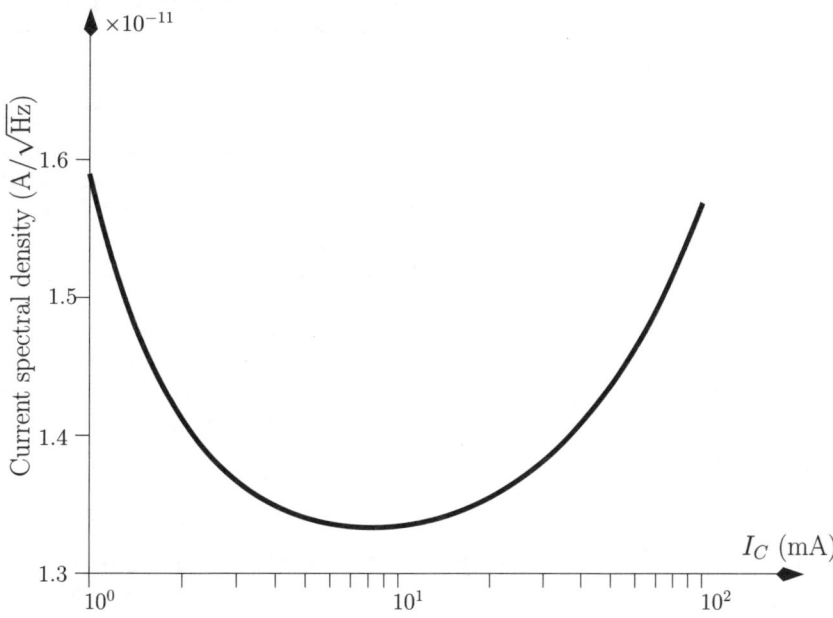

Figure 8.32: *Current spectral density of $\mathbf{i_{n_{eq}}}$ versus collector bias current I_C ($f = 800$ MHz).*

Equivalent voltage noise spectral density

If the noisy amplifier of figure 8.31 a) is now driven by a voltage source with output impedance $Z_s(\omega)$, as illustrated in figure 8.33 a), then the noise behaviour of the overall configuration (voltage source and amplifier) can be best represented (see figure 8.33 b)) by an equivalent voltage noise source, $\mathbf{u_{n_{eq}}}$ with a PSD given by (see exercise 8.9):

$$\begin{aligned}\langle \mathbf{u_{n_{eq}}}\, \mathbf{u_{n_{eq}}}^* \rangle &= \langle \mathbf{u_{ns}}\, \mathbf{u_{ns}^*} \rangle + \langle \mathbf{u_n}\, \mathbf{u_n^*} \rangle + 2\,\text{Real}[\langle \mathbf{u_n}\, \mathbf{i_n^*} \rangle\, Z_s^*(\omega)] \\ &+ \langle \mathbf{i_n}\, \mathbf{i_n^*} \rangle\, |Z_s(\omega)|^2\end{aligned} \quad (8.153)$$

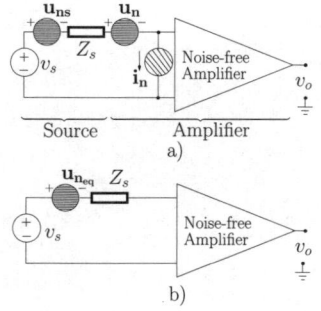

Figure 8.33: *a) Noisy amplifier connected to a voltage signal source. b) Equivalent noise model for the signal source and amplifier.*

8.4.7 The noise figure

The noise figure is widely used to quantify the noise performance of active devices and amplifiers specially when both the signal source impedance and the output load impedances are resistive.

The noise factor, F, can be defined as:

$$F = \frac{\text{SNR}_{in}}{\text{SNR}_{out}} \tag{8.154}$$

where SNR_i and SNR_o are the signal-to-noise ratios at the input and at the output of the amplifier, respectively. The signal-to-noise ratio is defined as the ratio of the power of the signal to the power of the noise. The noise figure, NF, is the noise factor expressed in dB, that is:

$$NF = 10\log_{10}\left[\frac{\text{SNR}_{in}}{\text{SNR}_{out}}\right] \tag{8.155}$$

For example, a noise figure of 0 dB would describe a noiseless amplifier and a noise figure of 3 dB would indicate that the amplifier contributes as much noise as the signal source.

The noise factor can be expressed as follows:

$$F = \frac{\frac{P_i}{P_{n_i}}}{\frac{P_o}{P_{n_o}}} = \frac{P_i}{P_o}\frac{P_{n_o}}{P_{n_i}} \tag{8.156}$$

$$= \frac{1}{G_p}\frac{P_{n_o}}{P_{n_i}} \tag{8.157}$$

where P_i and P_o are the input signal power and the output signal power, respectively. P_{n_i} is the noise power at the input of the amplifier and P_{n_o} is the noise power at the output of the amplifier. G_p is the power gain of the amplifier. For the amplifier of figure 8.31 we can write:

$$P_{n_o} = G_p \left\langle \mathbf{i_{n_{eq}}} \mathbf{i^*_{n_{eq}}} \right\rangle \tag{8.158}$$

$$P_{n_i} = \left\langle \mathbf{i_{ns}} \mathbf{i^*_{ns}} \right\rangle \tag{8.159}$$

and the noise factor can be written as

$$F = \frac{\left\langle \mathbf{i_{n_{eq}}} \mathbf{i^*_{n_{eq}}} \right\rangle}{\left\langle \mathbf{i_{ns}} \mathbf{i^*_{ns}} \right\rangle} \tag{8.160}$$

where

$$\left\langle \mathbf{i_{ns}} \mathbf{i^*_{ns}} \right\rangle = 2\mathcal{K}\mathcal{T}\,\text{Real}[Y_s(\omega)] \tag{8.161}$$

$$Y_s(\omega) = \frac{1}{Z_s(\omega)} \tag{8.162}$$

$$\text{Real}[Y_s(\omega)] = G_s \tag{8.163}$$

Using eqn 8.152, F is given by:

$$F = 1 + \frac{\langle \mathbf{i_n}\mathbf{i^*_n}\rangle + 2\,\text{Real}[\langle \mathbf{u_n}\mathbf{i^*_n}\rangle Y_s(\omega)] + \langle \mathbf{u_n}\mathbf{u^*_n}\rangle|Y_s(\omega)|^2}{2\mathcal{K}\mathcal{T}\,\text{Real}[Y_s(\omega)]} \tag{8.164}$$

From eqn 8.164 we can conclude that when $Y_s(\omega)$ tends to zero the terms $\langle \mathbf{i_n} \mathbf{i_n^*} \rangle (2\mathcal{K}\mathcal{T}\,\text{Real}[Y_s(\omega)])^{-1}$ and the noise factor tend to infinity. Also, when $Y_s(\omega)$ tends to infinity both the term $\langle \mathbf{u_n} \mathbf{u_n^*} \rangle |Y_s(\omega)|^2$ and F tend to infinity. Therefore, there must be an optimum value of the signal source output admittance, $Y_{s_{opt}}(\omega)$, which minimises the noise factor. In order to simplify this calculation we consider a pure resistive output admittance $Y_s(\omega) = G_s$, for which we can write

$$F = 1 + \frac{\langle \mathbf{i_n} \mathbf{i_n^*} \rangle + 2G_s\,\text{Real}[\langle \mathbf{u_n} \mathbf{i_n^*} \rangle] + \langle \mathbf{u_n} \mathbf{u_n^*} \rangle G_s^2}{2\mathcal{K}\mathcal{T}G_s} \quad (8.165)$$

The optimum G_s can be obtained by differentiating eqn 8.165 and setting the differential to zero, that is

$$\frac{dF}{dG_s} = 0$$

$$\frac{2\mathcal{K}\mathcal{T}\left[G_s^2 \langle \mathbf{u_n} \mathbf{u_n^*} \rangle - \langle \mathbf{i_n} \mathbf{i_n^*} \rangle\right]}{(2\mathcal{K}\mathcal{T}G_s)^2} = 0$$

and

$$G_{s_{opt}}^2 = \frac{\langle \mathbf{i_n} \mathbf{i_n^*} \rangle}{\langle \mathbf{u_n} \mathbf{u_n^*} \rangle} \quad (8.166)$$

Hence the minimum noise factor, F_{min}, can be written as

$$F_{min} = 1 + \frac{2G_{s_{opt}}\,\text{Real}[\langle \mathbf{u_n} \mathbf{i_n^*} \rangle] + 2\langle \mathbf{u_n} \mathbf{u_n^*} \rangle G_{s_{opt}}^2}{2\mathcal{K}\mathcal{T}G_{s_{opt}}} \quad (8.167)$$

It should be noted that usually F_{min} varies with the frequency.

The noise factor also provides insight about the relative importance of noise sources along a chain of amplifiers. Figure 8.34 shows a chain of three amplifiers each one with a noise factor F_k and a power gain G_{p_k}. It can be shown that the noise factor of the amplifier chain is given by:

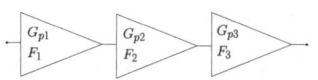

Figure 8.34: *Chain of three noisy amplifiers.*

$$F = F_1 + \frac{F_2 - 1}{G_{p1}} + \frac{F_3 - 1}{G_{p1} G_{p2}} \quad (8.168)$$

From this we conclude that the noise performance of the amplifier is dominated by the noise performance of the first amplifier stage as long as the gain of this first stage, G_{p_1}, is large.

Clearly, equation 8.168 can be generalised for a chain of N amplifiers each one with noise factor F_k and a power gain G_{p_k}:

$$F = F_1 + \sum_{k=2}^{N} \frac{F_k - 1}{\prod_{n=1}^{k-1} G_{p_n}} \quad (8.169)$$

Example 8.4.1 An amplifier with a power gain of 24 dB is to be built using the cascade of two power amplifiers. Amplifier A has a power gain of 15 dB and a noise figure of 8 dB. Amplifier B has a gain of 9 dB and a noise figure of 5 dB. Determine which amplifier is best suited as the first stage of amplification in order to have a final amplifier with the lowest possible noise figure.

Solution: The gain and the noise factor of each amplifier are given by:

$$G_{pA} = 31.6$$
$$G_{pB} = 7.9$$
$$F_A = 6.31$$
$$F_B = 3.16$$

Assuming that amplifier A is the first stage of amplification the noise factor of the final amplifier is, according to eqn 8.169, equal to:

$$F_A = 6.38 \qquad (8.170)$$

If amplifier B is the first stage of amplification then the noise factor of the final amplifier is

$$F_B = 3.83 \qquad (8.171)$$

From the above we conclude that amplifier B should be the first stage of amplification followed by amplifier A. For this situation the noise figure of the final amplifier is 5.8 dB.

The equivalent amplifier noise resistance and conductance

The equivalent noise resistance, R_n, of an amplifier characterised by two equivalent input noise sources $\mathbf{u_n}$ and $\mathbf{i_n}$ (see figure 8.25) is defined as the value of a resistance having a thermal noise PSD equal to the PSD of $\mathbf{u_n}$ at a standard temperature, usually 290 kelvin. Hence,

$$2\mathcal{K}T R_n = \langle \mathbf{u_n} \mathbf{u_n^*} \rangle \qquad (8.172)$$
$$R_n = \frac{\langle \mathbf{u_n} \mathbf{u_n^*} \rangle}{2\mathcal{K}T} \qquad (8.173)$$

Similarly, the equivalent noise conductance of an amplifier, g_n, is defined as the value of a conductance having a thermal noise PSD equal to the PSD of $\mathbf{i_n}$ at 290 kelvin

$$2\mathcal{K}T g_n = \langle \mathbf{i_n} \mathbf{i_n^*} \rangle \qquad (8.174)$$
$$g_n = \frac{\langle \mathbf{i_n} \mathbf{i_n^*} \rangle}{2\mathcal{K}T} \qquad (8.175)$$

It can be shown (see problem 8.11) that g_n and R_n can be related by the following equation:

$$g_n = R_n |Y_{s_{opt}}|^2 \qquad (8.176)$$

8. Noise in electronic circuits

where $Y_{s_{opt}}$ is the optimum source admittance which will minimise the noise figure.

The equivalent amplifier noise temperature

The equivalent noise temperature, $\mathcal{T}_n$, of an amplifier characterised by two equivalent input noise sources $\mathbf{u_n}$ and $\mathbf{i_n}$ (see figure 8.25), and driven by a signal source with an output resistance R_s is defined as the temperature at which the thermal noise contribution from R_s is equal to the overall noise of the amplifier. $\mathcal{T}_n$ satisfies the following eqn:

$$2\mathcal{K}\mathcal{T}_n R_s = \langle \mathbf{u_n}\,\mathbf{u_n^*}\rangle + R_s^2\,\langle \mathbf{i_n}\,\mathbf{i_n^*}\rangle + 2\,R_s\,\text{Real}[\langle \mathbf{u_n}\,\mathbf{i_n^*}\rangle] \quad (8.177)$$

or

$$\mathcal{T}_n = \frac{\langle \mathbf{u_n}\,\mathbf{u_n^*}\rangle + R_s^2\,\langle \mathbf{i_n}\,\mathbf{i_n^*}\rangle + 2\,R_s\,\text{Real}[\langle \mathbf{u_n}\,\mathbf{i_n^*}\rangle]}{2\mathcal{K} R_s} \quad (8.178)$$

Note that $\mathcal{T}_n$ is, in general, frequency dependent.

8.5 Computer-aided noise modelling and analysis

The combination of matrix algebra and circuit analysis can be used to simplify the computation of noise parameters in complex circuits. This can be achieved by following a method based on dividing a complex circuit into its elementary constituents. This powerful method was originally proposed by Hillbrand and Russer [7] and it has many similarities with the method described in Chapter 5 for the computation of the electrical response of two-port circuits. In fact, Hillbrand and Russer's method is also based on the analysis of a two-port circuit as an interconnection of basic two-port sub-circuits. These basic two-port sub-circuits are elementary passive impedances, and elementary active gain elements, such as transistors. These, in turn, can be modelled as the interconnection of passive devices with voltage or current-controlled sources. The noise behaviour of these elementary sub-circuits has been discussed previously. Starting from these basic two-port circuits, the analysis is performed by interconnecting these basic circuits in order to obtain the noise performance of the whole circuit. Figure 8.35 a) shows a common-emitter amplifier and figure 8.35 b) shows the AC equivalent circuit of the amplifier as an interconnection of elementary two-port circuits. From figure 8.35 b) it can be seen that the two-port network which represents the transistor is in series with the two-port network which represents R_E. The resulting two-port circuit is in parallel with the two-port network representing R_B and, so on.

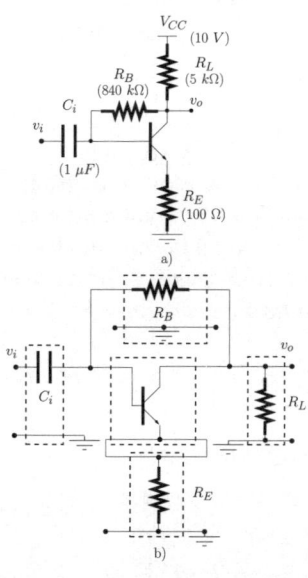

Figure 8.35: *a) Common-emitter amplifier. b) AC equivalent circuit.*

8.5.1 Noise representations

All noisy two-port circuits can be replaced by equivalent noiseless two-port circuits which characterise the noise-free electrical response and by two noise sources which account for the noisy behaviour. The natures of these noise sources (current or voltage) and their position relative to the input or output of the noiseless two-port network define what is called a noise representation. Here, we consider the representations shown in figure 8.36, that is, the admit-

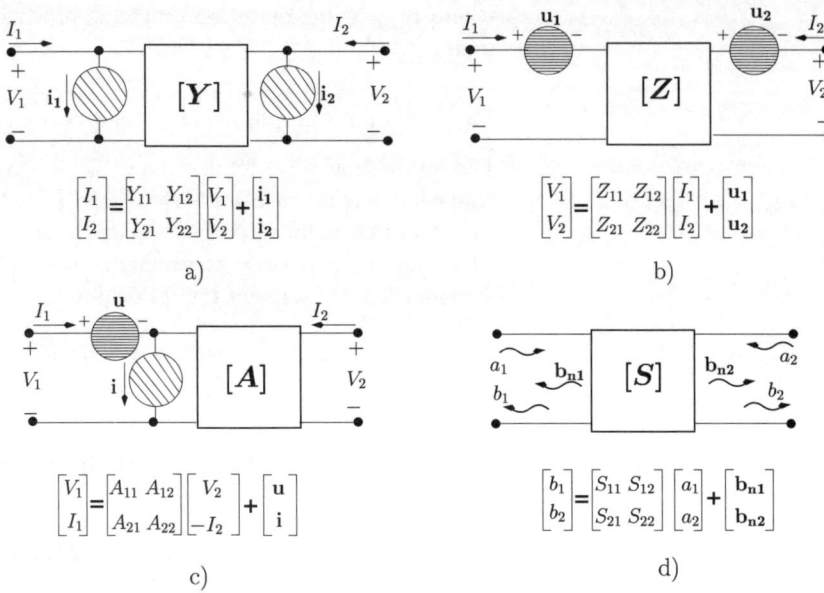

Figure 8.36: *Two-port noise representations: a) Admittance representation; b) Impedance representation; c) Chain representation; d) Scattering representation.*

tance representation, the impedance representation, the cascade/chain representation and the scattering representation. The two-port electrical response associated with each representation has been studied in detail in Chapter 5 (see section 5.3).

The admittance representation

The admittance representation characterises the electrical circuit response using $[Y]$ parameters. The noise of the circuit is modelled by two current noise sources, one located at the input and the other at the output of the circuit. These two noise sources are characterised by their self- and cross-power spectral densities arranged in a matrix form defined as the *admittance correlation matrix*:

$$[C_Y] = \begin{bmatrix} \langle i_1 i_1^* \rangle & \langle i_1 i_2^* \rangle \\ \langle i_2 i_1^* \rangle & \langle i_2 i_2^* \rangle \end{bmatrix} \tag{8.179}$$

The admittance representation is a useful way to deal with two-port sub-circuits which are in parallel.

The impedance representation

For the impedance representation the characterisation is effected by using the $[Z]$ parameters while the noise is characterised by two voltage noise sources located at the input and output of the circuit. This representation is the electrical dual of the admittance representation. The two noise sources are characterised by their self- and cross-power spectral densities arranged in a matrix form defined as the *impedance correlation matrix*:

$$[C_Z] = \begin{bmatrix} \langle u_1 u_1^* \rangle & \langle u_1 u_2^* \rangle \\ \langle u_2 u_1^* \rangle & \langle u_2 u_2^* \rangle \end{bmatrix} \quad (8.180)$$

The impedance representation is useful in dealing with two-port sub-circuits which are in series.

The chain representation

For this representation the characterisation is described by $[A]$ parameters. The noise is characterised by a voltage noise source and a current noise source, both at the input of the circuit. The two noise sources are characterised by their self- and cross-power spectral densities arranged in a matrix form defined as the *chain correlation matrix*:

$$[C_A] = \begin{bmatrix} \langle u u^* \rangle & \langle u i^* \rangle \\ \langle i u^* \rangle & \langle i i^* \rangle \end{bmatrix} \quad (8.181)$$

The cascade (or chain) representation, which has been discussed in detail in section 8.4.6 is also useful when analysing chains of two-port sub-circuits.

The scattering representation

The characterisation of the electrical circuit response is performed by using the $[S]$ parameters, studied in Chapter 7. There are two electromagnetic wave noise sources, one located at the input and one at the output of the circuit. These are characterised by their self- and cross-power spectral densities also arranged in a matrix form defined as the *scattering correlation matrix*:

$$[C_S] = \begin{bmatrix} \langle b_1 b_1^* \rangle & \langle b_1 b_2^* \rangle \\ \langle b_2 b_1^* \rangle & \langle b_2 b_2^* \rangle \end{bmatrix} \quad (8.182)$$

The scattering representation can be useful specially when data on high-frequency (RF) circuits and devices are provided by the manufacturers as S-parameters.

8.5.2 Calculation of the correlation matrices

For passive networks, and neglecting the $1/f$ noise, the correlation matrices can be obtained from $[Z]$ and $[Y]$ as follows:

$$[C_Z] = 2\mathcal{K}\mathcal{T}\,\text{Real}\,[Z] \quad (8.183)$$
$$[C_Y] = 2\mathcal{K}\mathcal{T}\,\text{Real}\,[Y] \quad (8.184)$$

If the $1/f$ noise cannot be neglected then the correlation matrices for passive devices can be obtained after calculating $[C_A]$ as described in section 8.4.6.

For active devices $[C_A]$ can also be derived using the noise analysis method described in section 8.4.6. In situations where the correlation matrix cannot be derived from theory, measurements of the noise performance can provide the required information. Such measurements are often performed by determining the equivalent noise resistance, R_n, and the optimum source admittance, Y_{opt}, for which the noise factor, is a minimum, F_{min}. Knowing these quantities it is possible to derive (see problem 8.12) the correlation matrix as follows:

$$[\mathbf{C_A}] = 2\mathcal{K}\mathcal{T} \begin{bmatrix} R_n & \frac{F_{min}-1}{2} - R_n Y^*_{s_{opt}} \\ \frac{F_{min}-1}{2} - R_n Y_{s_{opt}} & R_n |Y_{s_{opt}}|^2 \end{bmatrix} \quad (8.185)$$

Example 8.5.1 Determine the chain and the impedance representations for the noisy shunt admittance, Y, represented in figure 8.37 a).

Solution: The admittance Y can be represented by its corresponding impedance Z composed by a resistance R and a reactance jX:

$$Z = R + jX \quad (8.186)$$

Therefore, neglecting the $1/f$ noise, the thermal noise associated with Z can be modelled by its Thévenin model (see figure 8.37 b)). The PSD associated with **u** is given by:

$$\langle \mathbf{u}\, \mathbf{u}^* \rangle = 2\mathcal{K}\mathcal{T} R \quad (8.187)$$

In order to find the self- and cross-power spectral density of the two equivalent voltage noise sources, corresponding to the chain representation, we perform an analysis similar to that presented in section 8.4.6. Figure 8.37 c) shows the circuit to determine $\mathbf{u_n}$. We assume an input voltage source and we apply the superposition theorem to determine the open-circuit output voltage caused by **u**. Since the output is short-circuited we have:

$$v_o = 0$$

and therefore $\langle \mathbf{u_n}\, \mathbf{u_n}^* \rangle = 0$. The procedure to determine $\mathbf{i_n}$ is illustrated in figure 8.37 d). From this figure we observe that

$$i_o = \mathbf{u_n} Y$$

Since the current gain for this simple two-port circuit is unity then $\mathbf{i_n} = i_o$, that is

$$\begin{aligned} \mathbf{i_n} &= \mathbf{u_n} Y \\ \langle \mathbf{i_n}\, \mathbf{i_n}^* \rangle &= \langle \mathbf{u}\, \mathbf{u}^* \rangle |Y|^2 \\ &= 2\mathcal{K}\mathcal{T} R |Y|^2 \end{aligned} \quad (8.188)$$

and $\langle \mathbf{u_n}\, \mathbf{i_n}^* \rangle = \langle \mathbf{i_n}\, \mathbf{u_n}^* \rangle = 0$. Hence we can write:

$$[\mathbf{C_A}] = \begin{bmatrix} 0 & 0 \\ 0 & 2\mathcal{K}\mathcal{T} R |Y|^2 \end{bmatrix} \quad (8.189)$$

The impedance characterisation for Y is given by (see also appendix C)

$$[\mathbf{Z}] = \begin{bmatrix} Z & Z \\ Z & Z \end{bmatrix} \quad (8.190)$$

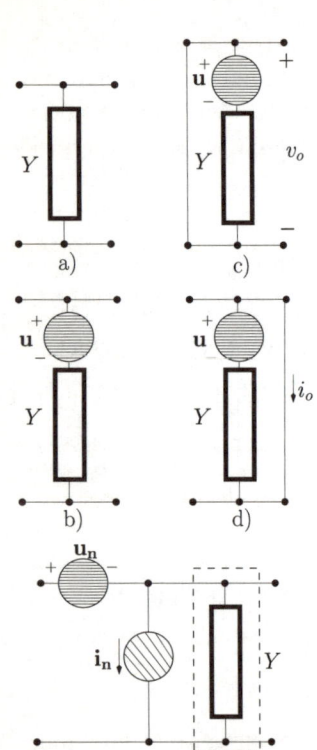

Figure 8.37: *a) Noisy shunt admittance. b) Equivalent Thévenin noise model. c) Circuit to determine $\mathbf{u_n}$. d) Circuit to determine $\mathbf{i_n}$. e) Two-port chain representation of Y.*

8. Noise in electronic circuits

and, from eqn 8.183, $[\mathbf{C_Z}]$ can be written as:

$$[\mathbf{C_Z}] = 2\mathcal{K}\mathcal{T} \begin{bmatrix} R & R \\ R & R \end{bmatrix} \quad (8.191)$$

8.5.3 Elementary two-port interconnections

As discussed in Chapter 5 there are three fundamental types of interconnections for two-port networks: parallel, series and chain interconnections.

Parallel interconnection

If two elementary noisy two-port networks are in parallel, as illustrated in figure 8.38, it is desirable that both such networks are described according to admittance representations. This is because the equivalent noisy network can be described as an admittance representation for which the equivalent admittance correlation matrix is the sum of the individual admittance correlation matrices:

$$[\mathbf{C_{Y_{total}}}] = [\mathbf{C_{Y_1}}] + [\mathbf{C_{Y_2}}] \quad (8.192)$$

and the equivalent electrical response can be represented by (see also section 5.2.2):

$$[\mathbf{Y_{total}}] = [\mathbf{Y_1}] + [\mathbf{Y_2}] \quad (8.193)$$

For the parallel connection of N elementary two-port circuits the noise characterisation can be described by an equivalent admittance correlation matrix:

$$[\mathbf{C_{Y_{total}}}] = \sum_{j=1}^{N}[\mathbf{C_{Y\,j}}] \quad (8.194)$$

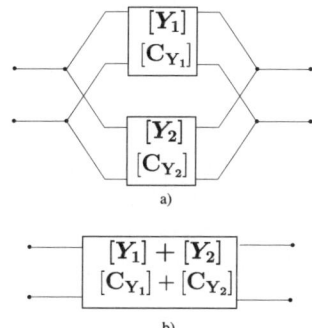

Figure 8.38: *a) Noisy two-port circuits in parallel. b) Equivalent two-port circuit.*

Series interconnection

If two elementary noisy networks are in series then both networks are best described according to impedance representations. Now, the equivalent noisy network can be described as an impedance representation for which the equivalent impedance correlation matrix is the sum of the individual impedance correlation matrices:

$$[\mathbf{C_{Z_{total}}}] = [\mathbf{C_{Z_1}}] + [\mathbf{C_{Z_2}}] \quad (8.195)$$

and the equivalent electrical response can be represented by (see also section 5.2.1):

$$[\mathbf{Z_{total}}] = [\mathbf{Z_1}] + [\mathbf{Z_2}] \quad (8.196)$$

For the series connection of N elementary two-port circuits the noise characterisation can be described by an equivalent impedance correlation matrix:

$$[\mathbf{C_{Z_{total}}}] = \sum_{j=1}^{N}[\mathbf{C_{Z_j}}] \quad (8.197)$$

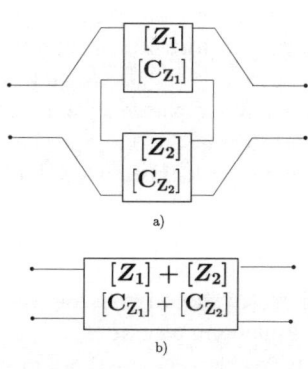

Figure 8.39: *Two-port circuits in series. b) Equivalent two-port circuit.*

Chain/cascade interconnection

When two elementary noisy networks are in a cascaded connection both networks are best described according to chain representations. Now, the equivalent noisy network can be described as a chain representation which is given by the following expression:

$$[C_{A_{total}}] = [A_1] [C_{A_2}] [A_1]^+ + [C_{A_1}] \quad (8.198)$$

where $[A_1]$ represents the electrical chain matrix of the first network of the cascade. $[A_1]^+$ is the Hermitian conjugate of $[A_1]$:

$$[A_1]^+ = \begin{bmatrix} A_{11}^* & A_{21}^* \\ A_{12}^* & A_{22}^* \end{bmatrix} \quad (8.199)$$

The equivalent electrical response can be represented by (see also section 5.2.3):

$$[A_{total}] = [A_1] [A_2] \quad (8.200)$$

The noise characterisation of a chain (or cascade) of N two-port sub-circuits can then be described by generalising eqn 8.198 as follows;

$$[C_{A_{total}}] = \sum_{i=0}^{N-1} [A_{1\to i}] [C_{A_{i+1}}] [A_{1\to i}]^+ \quad (8.201)$$

$$[A_{1\to i}] \triangleq \begin{cases} \prod_{k=1}^{i} [A_k] & \text{for } i \geq 1 \\ [1] & \text{for } i < 1 \end{cases} \quad (8.202)$$

where $[1]$ represents the two-by-two identity matrix (see appendix B).

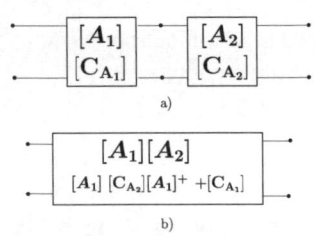

Figure 8.40: *a) Chain of two-port circuits. b) Equivalent two-port circuit.*

8.5.4 Transformation matrices

According to the type of interconnection (parallel, series or chain) it is often desirable to transform between representations. These transformations are provided by the transfer function, in matrix form represented by $[T]$, between the two positions in the electrical network where the equivalent noise generators of the new representation and the equivalent noise generators of the old representation are considered. The new correlation matrix can be calculated according to the transformation formula:

$$[C'] = [T] [C] [T]^+ \quad (8.203)$$

where $[C]$ and $[C']$ represent the correlation matrices of the original and the new representation, respectively. $[T]^+$ is the Hermitian conjugate of $[T]$.

Tables 8.1 and 8.2 show the transformation matrices between the representations considered here.

8. Noise in electronic circuits

		Original representation	
		Admittance	Impedance
Resulting representation	Admittance	$\begin{bmatrix} 1 & 0 \\ 0 & 1 \end{bmatrix}$	$\begin{bmatrix} Y_{11} & Y_{12} \\ Y_{21} & Y_{22} \end{bmatrix}$
	Impedance	$\begin{bmatrix} Z_{11} & Z_{12} \\ Z_{21} & Z_{22} \end{bmatrix}$	$\begin{bmatrix} 1 & 0 \\ 0 & 1 \end{bmatrix}$
	Chain	$\begin{bmatrix} 0 & A_{12} \\ 1 & A_{22} \end{bmatrix}$	$\begin{bmatrix} 1 & -A_{11} \\ 0 & -A_{21} \end{bmatrix}$
	Scattering	$\begin{bmatrix} \frac{1+S_{11}}{2\sqrt{Y_o}} & \frac{S_{12}}{2\sqrt{Y_o}} \\ \frac{S_{21}}{2\sqrt{Y_o}} & \frac{1+S_{22}}{2\sqrt{Y_o}} \end{bmatrix}$	$\begin{bmatrix} \frac{1-S_{11}}{2\sqrt{Z_o}} & \frac{-S_{12}}{2\sqrt{Z_o}} \\ \frac{-S_{21}}{2\sqrt{Z_o}} & \frac{1-S_{22}}{2\sqrt{Z_o}} \end{bmatrix}$

Table 8.1: Transformations between noise representations.

		Original representation	
		Chain	Scattering
Resulting representation	Admittance	$\begin{bmatrix} -Y_{11} & 1 \\ -Y_{21} & 0 \end{bmatrix}$	$\begin{bmatrix} \frac{Y_o+Y_{11}}{\sqrt{Y_o}} & \frac{Y_{12}}{\sqrt{Y_o}} \\ \frac{Y_{21}}{\sqrt{Y_o}} & \frac{Y_o+Y_{22}}{\sqrt{Y_o}} \end{bmatrix}$
	Impedance	$\begin{bmatrix} 1 & -Z_{11} \\ 0 & -Z_{21} \end{bmatrix}$	$\begin{bmatrix} \frac{Z_o+Z_{11}}{\sqrt{Z_o}} & \frac{Z_{12}}{\sqrt{Z_o}} \\ \frac{Z_{21}}{\sqrt{Z_o}} & \frac{Z_o+Z_{22}}{\sqrt{Z_o}} \end{bmatrix}$
	Chain	$\begin{bmatrix} 1 & 0 \\ 0 & 1 \end{bmatrix}$	$\begin{bmatrix} \sqrt{Z_o} & \frac{-(A_{12}+A_{11}Z_o)}{\sqrt{Z_o}} \\ \frac{-1}{\sqrt{Z_o}} & \frac{-(A_{22}+A_{21}Z_o)}{\sqrt{Z_o}} \end{bmatrix}$
	Scattering	$\begin{bmatrix} \frac{1-S_{11}}{2\sqrt{Z_o}} & \frac{-(1+S_{11})\sqrt{Z_o}}{2} \\ \frac{-S_{21}}{2\sqrt{Z_o}} & \frac{-S_{21}\sqrt{Z_o}}{2} \end{bmatrix}$	$\begin{bmatrix} 1 & 0 \\ 0 & 1 \end{bmatrix}$

Table 8.2: Transformations between noise representations. *(Cont.)*

Example 8.5.2 Determine the transformation matrix $[\mathbf{T}]$ to transform from the scattering representation to the impedance representation.

Solution: The impedance representation (see also figure 8.36 b)) can be expressed as follows:

$$V_1 = Z_{11} I_1 + Z_{12} I_2 + \mathbf{u_1} \tag{8.204}$$
$$V_2 = Z_{21} I_1 + Z_{22} I_2 + \mathbf{u_2} \tag{8.205}$$

and the scattering representation (see section 7.4) can be expressed as:

$$b_1 = S_{11} a_1 + S_{12} a_2 + \mathbf{b_{n1}} \tag{8.206}$$
$$b_2 = S_{21} a_1 + S_{22} a_2 + \mathbf{b_{n2}} \tag{8.207}$$

with

$$b_i = \frac{1}{2\sqrt{Z_o}}(V_i - Z_o I_i), \quad i = 1,2 \tag{8.208}$$

$$a_i = \frac{1}{2\sqrt{Z_o}}(V_i + Z_o I_i), \quad i = 1,2 \tag{8.209}$$

Equations 8.206 and 8.207 can be written as follows:

$$V_1(1 - S_{11}) = Z_o(1 + S_{11}) I_1 + S_{12} Z_o I_2 + S_{12} V_2 + 2\sqrt{Z_o}\,\mathbf{b_{n1}} \tag{8.210}$$

$$V_2(1 - S_{22}) = Z_o(1 + S_{22}) I_2 + S_{21} Z_o I_1 + S_{21} V_1 + 2\sqrt{Z_o}\,\mathbf{b_{n2}} \tag{8.211}$$

Solving these last two eqns in order to obtain V_1 and V_2 we get;

$$\begin{aligned}
V_1 = & \; Z_o \frac{(1-S_{22})(1+S_{11}) + S_{12}S_{21}}{(1-S_{11})(1-S_{22}) - S_{12}S_{21}} I_1 \\
& + Z_o \frac{2S_{12}}{(1-S_{11})(1-S_{22}) - S_{12}S_{21}} I_2 \\
& + \frac{2\sqrt{Z_o}\,(1-S_{22})}{(1-S_{11})(1-S_{22}) - S_{12}S_{21}} \mathbf{b_{n1}} \\
& + \frac{2\sqrt{Z_o}\,S_{12}}{(1-S_{11})(1-S_{22}) - S_{12}S_{21}} \mathbf{b_{n2}}
\end{aligned} \tag{8.212}$$

$$\begin{aligned}
V_2 = & \; Z_o \frac{2S_{21}}{(1-S_{11})(1-S_{22}) - S_{12}S_{21}} I_1 \\
& + Z_o \frac{(1+S_{22})(1-S_{11}) + S_{12}S_{21}}{(1-S_{11})(1-S_{22}) - S_{12}S_{21}} I_2 \\
& + \frac{2\sqrt{Z_o}\,S_{21}}{(1-S_{11})(1-S_{22}) - S_{12}S_{21}} \mathbf{b_{n1}} \\
& + \frac{2\sqrt{Z_o}\,(1-S_{11})}{(1-S_{11})(1-S_{22}) - S_{12}S_{21}} \mathbf{b_{n2}}
\end{aligned} \tag{8.213}$$

These can be rewritten, using the relationship between S and Z parameters (see appendix C), as follows:

$$V_1 = Z_{11} I_1 + Z_{12} I_2 + \frac{2\sqrt{Z_o}(1 - S_{22})}{\Delta_5} \mathbf{b_{n1}} + \frac{2\sqrt{Z_o} S_{12}}{\Delta_5} \mathbf{b_{n2}} \tag{8.214}$$

$$V_2 = Z_{21} I_1 + Z_{22} I_2 + \frac{2\sqrt{Z_o} S_{21}}{\Delta_5} \mathbf{b_{n1}} + \frac{2\sqrt{Z_o}(1 - S_{11})}{\Delta_5} \mathbf{b_{n2}} \tag{8.215}$$

with $\Delta_5 = (1 - S_{11})(1 - S_{22}) - S_{12} S_{21}$ (see appendix C). Comparing eqns 8.204 and 8.205 with eqns 8.214 and 8.215, respectively, we conclude that

$$\begin{bmatrix} \mathbf{u_1} \\ \mathbf{u_2} \end{bmatrix} = \begin{bmatrix} \frac{2\sqrt{Z_o}(1-S_{22})}{\Delta_5} & \frac{2\sqrt{Z_o} S_{12}}{\Delta_5} \\ \frac{2\sqrt{Z_o} S_{21}}{\Delta_5} & \frac{2\sqrt{Z_o}(1-S_{11})}{\Delta_5} \end{bmatrix} \begin{bmatrix} \mathbf{b_{n1}} \\ \mathbf{b_{n2}} \end{bmatrix} \tag{8.216}$$

This can also be written, using the relationship between S and Z parameters (see appendix C), as follows:

$$\begin{bmatrix} \mathbf{u_1} \\ \mathbf{u_2} \end{bmatrix} = \begin{bmatrix} \frac{Z_o + Z_{11}}{\sqrt{Z_o}} & \frac{Z_{12}}{\sqrt{Z_o}} \\ \frac{Z_{21}}{\sqrt{Z_o}} & \frac{Z_o + Z_{22}}{\sqrt{Z_o}} \end{bmatrix} \begin{bmatrix} \mathbf{b_{n1}} \\ \mathbf{b_{n2}} \end{bmatrix} \tag{8.217}$$

We now summarise a systematic approach suitable for the application of the noise analysis method described above.

1. **Decompose the circuit to be analysed into its elementary two-port sub-circuits such as series impedances, shunt admittances, voltage- and current-controlled sources, etc.**

2. **Identify the types of interconnections between the various elementary two-port networks mentioned above (parallel, series, chain).**

3. **Characterise the noisy behaviour and the electrical response for each elementary two-port according to one of the two-port noise representations illustrated in figure 8.36. Use a noise representation for the elementary two-port network that takes into account the type of interconnection with the other elementary noisy networks.**

4. **Reconstruct the overall two-port circuit. Whenever appropriate use the transformation matrices shown in tables 8.1 and 8.2 to obtain the appropriate noise representations.**

The next example illustrates one application of these steps.

Example 8.5.3 Consider the common-emitter amplifier of figure 8.35 a). Determine the chain representation for the amplifier. The noise model for the BJT is described by eqns 8.125–8.128 (see also figure 8.24). For the BJT assume that $\beta = 200$, $C_\pi = 6$ pF, $C_\mu = 0.8$ pF, $r_x \simeq 0$ and that r_o can be neglected. The corner frequency associated with $\langle \mathbf{u_{nf}}\, \mathbf{u_{nf}^*} \rangle$ is $f_c = 200$ Hz. $I_C = 1$ mA.

Solution: Figure 8.41 a) shows the small-signal equivalent circuit of the amplifier including all the noise sources. It is assumed that R_L, R_E and R_B produce thermal noise only. Therefore, the spectral densities associated with these resistances are given by

$$\langle \mathbf{u_{nL}}\, \mathbf{u_{nL}^*} \rangle = 2\mathcal{K}\mathcal{T} R_L \qquad (8.218)$$
$$\langle \mathbf{u_{nE}}\, \mathbf{u_{nE}^*} \rangle = 2\mathcal{K}\mathcal{T} R_E \qquad (8.219)$$
$$\langle \mathbf{i_{nB}}\, \mathbf{i_{nB}^*} \rangle = 2\mathcal{K}\mathcal{T} R_B^{-1} \qquad (8.220)$$

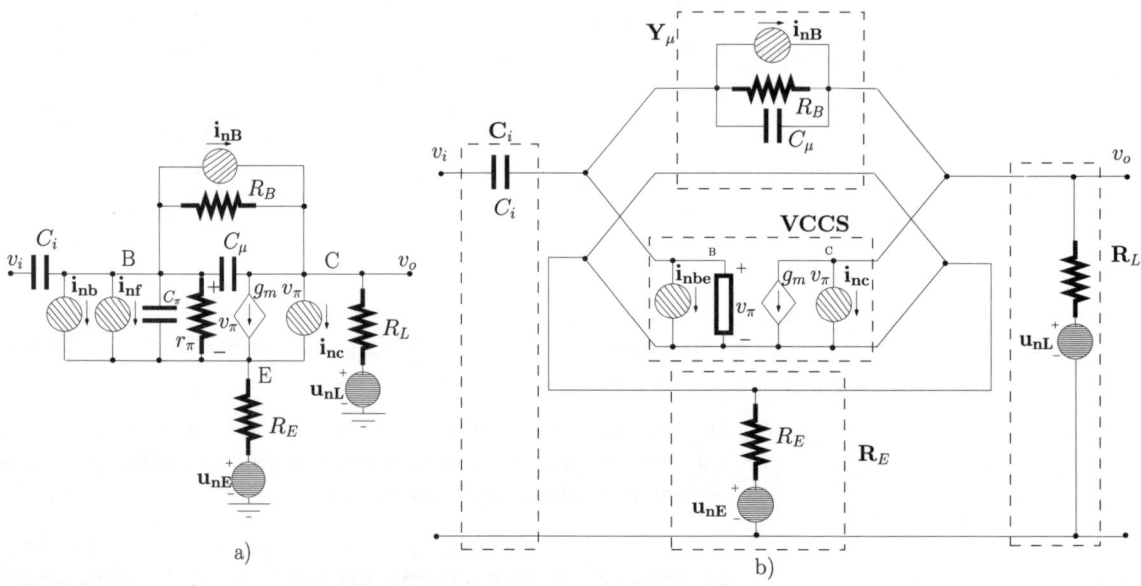

Figure 8.41: *Small-signal equivalent circuit for the common-emitter amplifier. b) The common-emitter amplifier as an interconnection of elementary two-port networks.*

Figure 8.41 b) shows the common-emitter amplifier as an interconnection of elementary two-port networks. $\mathbf{i_{nbe}}$ accounts for $\mathbf{i_{nb}}$ and $\mathbf{i_{nf}}$.

We start by characterising the transistor and the resistance R_B in terms of an interconnection of elementary two-port circuits. $g_m = I_C q/\mathcal{K}\mathcal{T} = 40$ mA/V, $r_\pi = \beta/g_m = 5$ kΩ.

Figure 8.42 a) shows the small-signal equivalent circuit for just the BJT and R_B including the noise sources. Figure 8.42 b) shows this equivalent circuit as an interconnection of elementary two-port circuits. From figure 8.42 b) we can see that the two-port circuit which characterises the BJT and R_B can be seen as the parallel connection of the two-port circuit which describes the voltage controlled current source, **VCCS**, with the two-port circuit which describes the base-collector admittance, $\mathbf{Y}_\mu$. $\mathbf{Y}_\mu$, in turn, is composed of the parallel connection of C_μ with R_B. Since **VCCS** is in parallel with $\mathbf{Y}_\mu$ it is appropriate to adopt admittance representations for these two networks. The admittance

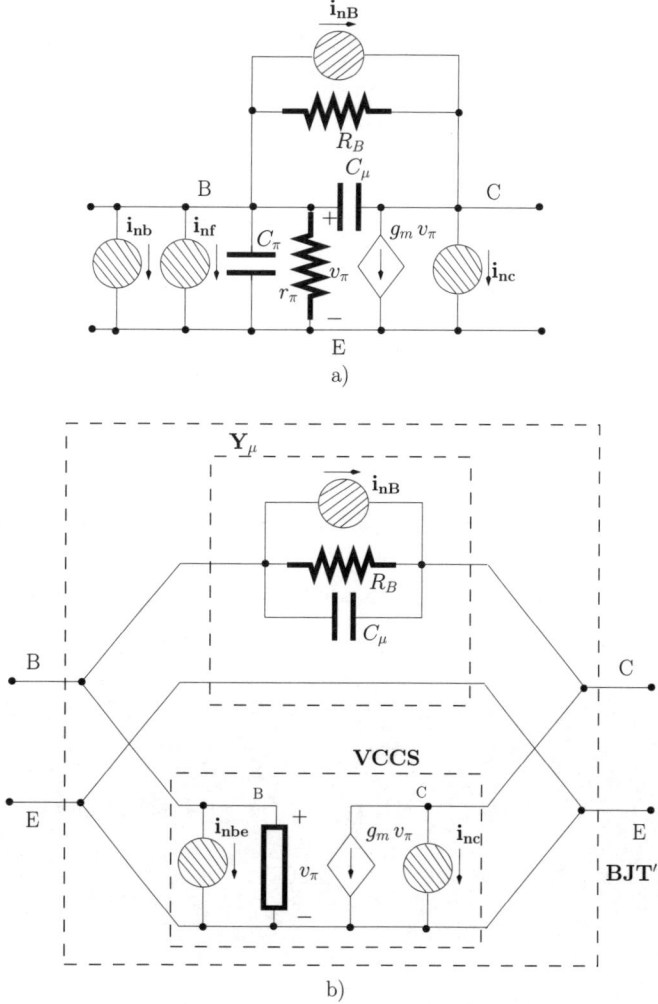

Figure 8.42: *The BJT and R_B as an interconnection of elementary two-port circuits.*

representation for $\mathbf{Y}_\mu$ is as follows (see also appendix C):

$$[\mathbf{Y}_{\mathbf{Y}_\mu}] = \begin{bmatrix} Y_\mu & -Y_\mu \\ -Y_\mu & Y_\mu \end{bmatrix} \quad (8.221)$$

where Y_μ corresponds to the admittance of the parallel connection of C_μ with R_B:

$$Y_\mu = \frac{1}{R_B} + j\omega C_\mu \qquad (8.222)$$

$$[\mathbf{C_{Y_\mu}}] = 2\mathcal{K}\mathcal{T} \begin{bmatrix} R_B^{-1} & -R_B^{-1} \\ -R_B^{-1} & R_B^{-1} \end{bmatrix} \qquad (8.223)$$

The admittance representation for **VCCS** is given by:

$$[\mathbf{Y_{VCCS}}] = \begin{bmatrix} Y_\pi & 0 \\ g_m & 0 \end{bmatrix} \qquad (8.224)$$

where Y_π corresponds to the parallel connection of r_π with C_π:

$$Y_\pi = \frac{1}{r_\pi} + j\omega C_\pi \qquad (8.225)$$

According to eqns 8.126–8.128 we can write:

$$[\mathbf{C_{Y_{VCCS}}}] = \begin{bmatrix} qI_B\left(1 + \frac{\omega_c}{\omega}\right) & 0 \\ 0 & qI_C \end{bmatrix} \qquad (8.226)$$

where $\omega_c = 2\pi f_c$ and $I_B = I_C/\beta$. The two-port circuit describing the parallel connection of **VCCS** with $\mathbf{Y}_\mu$ is represented here by **BJT'** and can now be characterised by an admittance representation given by (see also eqn 8.192):

$$[\mathbf{Y_{BJT'}}] = [\mathbf{Y_{C_\mu}}] + [\mathbf{Y_{VCCS}}] \qquad (8.227)$$

$$[\mathbf{C_{Y_{BJT'}}}] = [\mathbf{C_{Y_\mu}}] + [\mathbf{C_{Y_{VCCS}}}] \qquad (8.228)$$

that is

$$[\mathbf{Y_{BJT'}}] = \begin{bmatrix} Y_\pi + Y_\mu & -Y_\mu \\ g_m - Y_\mu & Y_\mu \end{bmatrix} \qquad (8.229)$$

$$[\mathbf{C_{Y_{BJT'}}}] = \begin{bmatrix} qI_B\left(1 + \frac{\omega_c}{\omega}\right) + 2\mathcal{K}\mathcal{T}R_B^{-1} & -2\mathcal{K}\mathcal{T}R_B^{-1} \\ -2\mathcal{K}\mathcal{T}R_B^{-1} & qI_C + 2\mathcal{K}\mathcal{T}R_B^{-1} \end{bmatrix} \qquad (8.230)$$

The two-port sub-circuit which characterises **BJT'** is in series with the two-port sub-circuit which characterises the resistance R_E, $\mathbf{R_E}$. Hence, it is appropriate to use impedance representations for both two-port sub-circuits;

$$[\mathbf{Z_{R_E}}] = \begin{bmatrix} R_E & R_E \\ R_E & R_E \end{bmatrix} \qquad (8.231)$$

8. Noise in electronic circuits

$$[\mathbf{C}_{\mathbf{Z}_{\mathbf{R}_E}}] = 2\mathcal{K}\mathcal{T}\,\text{Real}\,[\mathbf{Z}_{\mathbf{R}_E}] \tag{8.232}$$

and for **BJT'** we can write (see appendix C)

$$[\mathbf{Z}_{\mathbf{BJT'}}] = \begin{bmatrix} \frac{1}{g_m+Y_\pi} & \frac{1}{g_m+Y_\pi} \\ \frac{Y_\mu - g_m}{Y_\mu(g_m+Y_\pi)} & \frac{Y_\mu + Y_\pi}{Y_\mu(g_m+Y_\pi)} \end{bmatrix} \tag{8.233}$$

According to table 8.1 the transformation matrix from admittance to impedance is given by;

$$[\mathbf{T}_{(\mathbf{Y}\to\mathbf{Z})_{\mathbf{BJT'}}}] = \begin{bmatrix} \frac{1}{g_m+Y_\pi} & \frac{1}{g_m+Y_\pi} \\ \frac{Y_\mu - g_m}{Y_\mu(g_m+Y_\pi)} & \frac{Y_\mu + Y_\pi}{Y_\mu(g_m+Y_\pi)} \end{bmatrix} \tag{8.234}$$

Now the impedance correlation matrix for **BJT'** can be expressed, according to eqn 8.203, as follows:

$$[\mathbf{C}_{\mathbf{Z}_{\mathbf{BJT'}}}] = [\mathbf{T}_{(\mathbf{Y}\to\mathbf{Z})_{\mathbf{BJT'}}}]\,[\mathbf{C}_{\mathbf{Y}_{\mathbf{BJT'}}}]\,[\mathbf{T}_{(\mathbf{Y}\to\mathbf{Z})_{\mathbf{BJT'}}}]^+ \tag{8.235}$$

The two-port circuit composed by the series of **BJT'** with R_E, represented by **BJT''**, can be characterised by;

$$[\mathbf{C}_{\mathbf{Z}_{\mathbf{BJT''}}}] = [\mathbf{C}_{\mathbf{Z}_{\mathbf{BJT'}}}] + [\mathbf{C}_{\mathbf{Z}_{\mathbf{R}_E}}] \tag{8.236}$$

$$[\mathbf{Z}_{\mathbf{BJT''}}] = [\mathbf{Z}_{\mathbf{BJT'}}] + [\mathbf{Z}_{\mathbf{R}_E}] \tag{8.237}$$

$$[\mathbf{Z}_{\mathbf{BJT''}}] = \begin{bmatrix} \frac{1}{g_m+Y_\pi} + R_E & \frac{1}{g_m+Y_\pi} + R_E \\ \frac{Y_\mu - g_m}{Y_\mu(g_m+Y_\pi)} + R_E & \frac{Y_\mu + Y_\pi}{Y_\mu(g_m+Y_\pi)} + R_E \end{bmatrix} \tag{8.238}$$

From figure 8.43 a) we see that the two-port network representing the capacitor C_i is in a chain with **BJT''** which, in turn, is in a chain with the two-port sub-circuit representing R_L. Therefore, these three sub-circuits must be represented in chain representations and the overall amplifier will be also characterised by a chain representation. The chain representation for **BJT''** is given by:

$$[\mathbf{A}_{\mathbf{BJT''}}] = \begin{bmatrix} \frac{Y_\mu[1+R_E(g_m+Y_\pi)]}{Y_\mu[1+R_E(g_m+Y_\pi)]-g_m} & \frac{1+R_E(g_m+Y_\pi)}{Y_\mu[1+R_E(g_m+Y_\pi)]-g_m} \\ \frac{Y_\mu(g_m+Y_\pi)}{Y_\mu[1+R_E(g_m+Y_\pi)]-g_m} & \frac{Y_\pi+Y_\mu[1+R_E(g_m+Y_\pi)]}{Y_\mu[1+R_E(g_m+Y_\pi)]-g_m} \end{bmatrix} \tag{8.239}$$

and

$$[\mathbf{C}_{\mathbf{A}_{\mathbf{BJT''}}}] = [\mathbf{T}_{(\mathbf{Z}\to\mathbf{A})_{\mathbf{BJT''}}}]\,[\mathbf{C}_{\mathbf{Z}_{\mathbf{BJT''}}}]\,[\mathbf{T}_{(\mathbf{Z}\to\mathbf{A})_{\mathbf{BJT''}}}]^+ \tag{8.240}$$

with (see table 8.1)

$$[\mathbf{T}_{(\mathbf{Z}\to\mathbf{A})_{\mathbf{BJT''}}}] = \begin{bmatrix} 1 & -\frac{Y_\mu[1+R_E(g_m+Y_\pi)]}{Y_\mu[1+R_E(g_m+Y_\pi)]-g_m} \\ 0 & -\frac{Y_\mu(g_m+Y_\pi)}{Y_\mu[1+R_E(g_m+Y_\pi)]-g_m} \end{bmatrix} \tag{8.241}$$

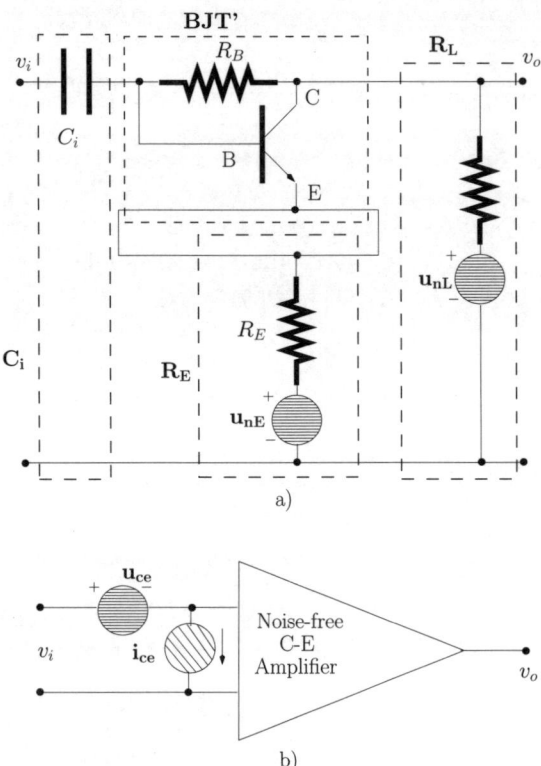

Figure 8.43: *a) The common-emitter amplifier as an interconnection of two-port sub-circuits. b) The equivalent amplifier model.*

The chain representation for $\mathbf{C_i}$ is given by:

$$[\boldsymbol{A}_{\mathbf{C_i}}] = \begin{bmatrix} 1 & (j\omega C_i)^{-1} \\ 0 & 1 \end{bmatrix} \quad (8.242)$$

We consider this capacitor as an ideal element. Hence $\left[\mathbf{C}_{\boldsymbol{A}_{C_i}}\right] = [\mathbf{0}]$ where $[\mathbf{0}]$ represents the null matrix.

The chain representation for $\mathbf{R_L}$ is given by (see example 8.5.1 and appendix C):

$$[\boldsymbol{A}_{\mathbf{R_L}}] = \begin{bmatrix} 1 & 0 \\ R_L^{-1} & 1 \end{bmatrix} \quad (8.243)$$

$$\left[\mathbf{C}_{\boldsymbol{A}_{\mathbf{R_L}}}\right] = 2\mathcal{KT} \begin{bmatrix} 0 & 0 \\ 0 & R_L^{-1} \end{bmatrix} \quad (8.244)$$

Hence the entire amplifier is characterised by

$$[\boldsymbol{A}_{\mathbf{ce}}] = [\boldsymbol{A}_{\mathbf{C_i}}][\boldsymbol{A}_{\mathbf{BJT''}}][\boldsymbol{A}_{\mathbf{R_L}}] \quad (8.245)$$

$$[\mathbf{C_{A_{aux}}}] = [\mathbf{A_{BJT''}}] \left[\mathbf{C_{A_{R_L}}}\right] [\mathbf{A_{BJT''}}]^+ + [\mathbf{C_{A_{BJT''}}}]$$
$$[\mathbf{C_{A_{ce}}}] = [\mathbf{A_{C_i}}] [\mathbf{C_{A_{aux}}}] [\mathbf{A_{C_i}}]^+ \quad (8.246)$$

It is now possible to obtain the voltage gain, A_v and the current gain, A_i, from eqn 8.245 as follows:

$$A_v = (A_{ce_{11}})^{-1} \quad (8.247)$$
$$A_i = (A_{ce_{22}})^{-1} \quad (8.248)$$

and the spectral densities associated with $\mathbf{u_{ce}}$ and to $\mathbf{i_{ce}}$ can be obtained from eqn 8.246 as follows:

$$\langle \mathbf{u_{ce}} \, \mathbf{u_{ce}^*} \rangle = C_{A_{ce\,11}} \quad (8.249)$$
$$\langle \mathbf{i_{ce}} \, \mathbf{i_{ce}^*} \rangle = C_{A_{ce\,22}} \quad (8.250)$$
$$\langle \mathbf{u_{ce}} \, \mathbf{i_{ce}^*} \rangle = C_{A_{ce\,12}} \quad (8.251)$$

Figure 8.44 shows the voltage gain, the current gain and the spectral densities $\langle \mathbf{u_{ce}} \, \mathbf{u_{ce}^*} \rangle$ and $\langle \mathbf{i_{ce}} \, \mathbf{i_{ce}^*} \rangle$. From figure 8.44 a) we observe that the voltage gain

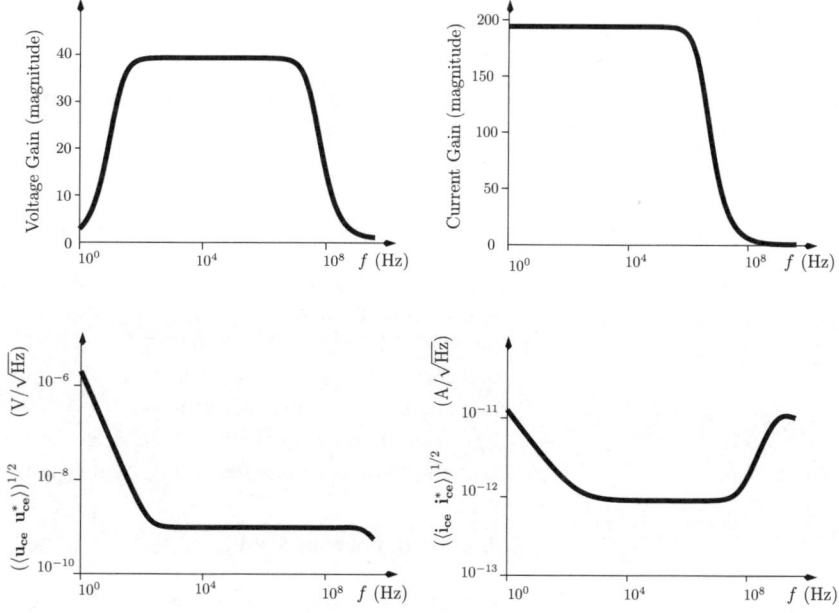

Figure 8.44: *a) Voltage gain. b) Current gain. c) Spectral density associated with* $\mathbf{u_{ce}}$*. d) Spectral density associated with* $\mathbf{i_{ce}}$*.*

in the medium frequency range is about 39.5 and it has low and high frequency cut-off frequencies of 13 Hz and 42 MHz, respectively. Figure 8.44 b) shows the current gain. In the mid range the current gain is about 194. The high cut-off frequency is 3.2 MHz while the low cut-off frequency is under 1 Hz. Figures 8.44 c) and 8.44 d) show $(\langle \mathbf{u_{ce}} \, \mathbf{u_{ce}} \rangle)^{1/2}$ and $(\langle \mathbf{i_{ce}} \, \mathbf{i_{ce}} \rangle)^{1/2}$. It can be seen that over most of the useful frequency range of the amplifier the equivalent RMS voltage noise is about 1 nV/$\sqrt{\text{Hz}}$ and the equivalent RMS current noise is about 1 pA/$\sqrt{\text{Hz}}$.

For a current amplifier, like that represented in figure 8.31, we can calculate the noise factor according to eqn 8.164, that is:

$$F = 1 + \frac{\langle i_n\, i_n^*\rangle + 2\,\text{Real}[\langle u_n\, i_n^*\rangle\, Y_s] + \langle u_n\, u_n^*\rangle |Y_s|^2}{2\mathcal{K}\mathcal{T}\,\text{Real}[Y_s]} \quad (8.252)$$

where we have dropped the dependency of Y_s on the angular frequency ω for simplicity. The last eqn can be written as:

$$F = 1 + \frac{[\mathbf{Y_s}][\mathbf{C_A}][\mathbf{Y_s}]^*}{2\mathcal{K}\mathcal{T}\,\text{Real}[Y_s]} \quad (8.253)$$

where

$$[\mathbf{Y_s}] = [Y_s\; 1] \quad (8.254)$$

and

$$[\mathbf{Y_s}]^* = \begin{bmatrix} Y_s^* \\ 1 \end{bmatrix} \quad (8.255)$$

$[\mathbf{C_A}]$ is the amplifier chain correlation matrix:

$$[\mathbf{C_A}] = \begin{bmatrix} \langle u_n\, u_n^*\rangle & \langle u_n\, i_n^*\rangle \\ \langle i_n\, u_n^*\rangle & \langle i_n\, i_n^*\rangle \end{bmatrix} \quad (8.256)$$

Figure 8.45 shows the noise figure of the common-emitter amplifier discussed in the previous example considering four different values for the source admittance Y_s: 10^{-2} S, 10^{-3} S, 10^{-4} S and 10^{-5} S. From this figure we observe that the noise figure exhibits a strong dependence on the admittance Y_s. There is an optimum source admittance, $Y_s = 10^{-3}$ S, which minimises the noise figure at about 0.94 dB in the frequency range 1 kHz–100 MHz. Note that this range includes most of the useful bandwidth of the amplifier.

A note on noise analysis

The plethora of different noise analysis and characterisation methods (equivalent input noise spectral density, noise figure, equivalent noise temperature, etc.) may look confusing in the sense of 'how, what and when to apply?'. It is important to stress that careful application of any of these methods must lead to the same result. The choice of a specific noise characterisation technique results from consideration of the circuit and its application. For example, in microwave systems where a known impedance termination is used (50 Ω) it is convenient to use the noise figure technique or to describe the system in terms of its noise temperature and/or resistance. However, for circuits where the input can be expressed as a single voltage or current source, it is normally more

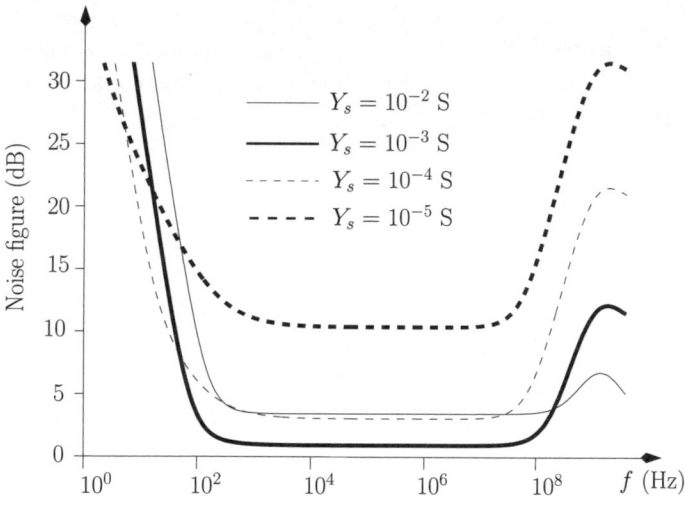

Figure 8.45: *Noise figure of the common-emitter amplifier.*

convenient to obtain an estimate of the input noise spectral density (either current or voltage, as appropriate). This allows a simple calculation of the input signal to noise ratio and consequently the error rate performance. Examples of such cases include optical receivers.

8.6 Bibliography

1. A. Papoulis, *Probability, Random Variables, and Stochastic Processes*, 1991 (McGraw-Hill International Editions), 3rd edition.

2. K.W. Cattermole and J.J. O'Reilly (Ed.), *Mathematical Topics in Telecommunications*, 1984 (Pentech Press London).

3. A.B. Carlson, P.B. Crilly, J.C. Rutledge, *Communication Systems: An Introduction to Signals and Noise in Electrical Communication*, 2001 (McGraw-Hill Series in Electrical Engineering), 4th edition.

4. P.J Fish, *Electronic Noise and Low Noise Design*, 1994, (McGraw-Hill).

5. S. B. Alexander, *Optical Communication Receiver Design*, 1997, (SPIE Optical Engineering Press, IEE).

6. H. Rothe and W. Dahlke, *Theory of Noisy Four-Poles*, Proc. of the IRE, pp. 811–818, 1956.

7. H. Hillbrand and P. Russer, *An Efficient Method for Computer Aided Noise Analysis of Linear Amplifier Networks*, IEEE Trans. on Circuits and Systems–I: Fundamental Theory and Applications, Vol. CAS–23, No. 4, pp. 235–238, Apr. 1976.

8. L. Moura, P. Monteiro, and I. Darwazeh, *Exact Noise Analysis of Distributed Preamplifiers for Optical Receivers*, Microwave and Optical Technology Letters (Wiley), Vol. 30, No. 6, pp. 416–418, Sep. 2001.

8.7 Problems

8.1 For a Gaussian random variable, X, with mean $\mu = -0.7$ and $\sigma = 2.2$. Determine the following probabilities

1. $P[X > 3]$
2. $P[X > -3]$
3. $P[-2 < X < 3]$
4. $P[X < -2]$

8.2 For a uniformly distributed random variable, X, with range $[-2, 3]$. Determine

1. The mean value of the distribution
2. The variance of the distribution
3. The third moment of the distribution

8.3 Consider the Poisson distribution with $\mu = 0.8$. Determine the characteristic function for this distribution. Using eqn 8.41 determine the mean of this distribution.

8.4 Consider a random variable Y resulting from the sum of three independent and uniformly distributed random variables with range $[-3, 3]$. Determine the following probabilities;

1. $P[Y > 4]$
2. $P[Y > 8.9]$

Consider now the Gaussian approximation to Y given by the central limit theorem. Calculate the probabilities mentioned above using the Gaussian approximation and compare the results with those obtained with the true distribution for Y.

8.5 A random variable X has a uniform distribution with range $[-1, 2]$. Consider the random process $a(t) = \exp(-3\,X\,t)$. Determine the expectation $E[a(t)]$ and and the autocorrelation function $R_a(t_1, t_2)$.

8.6 Consider a random process $v(t)$ with the autocorrelation function $R_v(\tau) = \sigma^2(1 - |\tau|/A)\,\text{rect}(t/2A)$ where $A > 0$. Determine the power spectral density of $v(t)$.

8.7 Consider the following transfer functions:

$$H_1(\omega) = \frac{\omega_n^2}{-\omega^2 + 2j\eta\omega_n\omega + \omega_n^2}$$

$$H_2(\omega) = \frac{2j\eta\omega_n\omega}{-\omega^2 + 2j\eta\omega_n\omega + \omega_n^2}$$

Derive an expression for the equivalent noise bandwidth of $H_1(\omega)$, $H_2(\omega)$ and $H_1(\omega) + H_2(\omega)$. Take $\eta < 1$.

8. Noise in electronic circuits

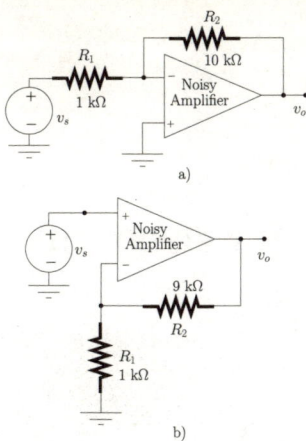

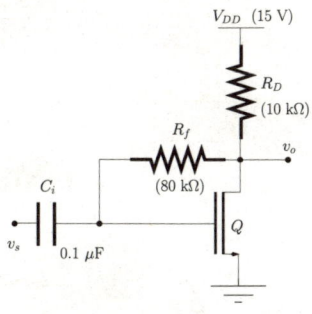

Figure 8.46: *Voltage amplifiers.*

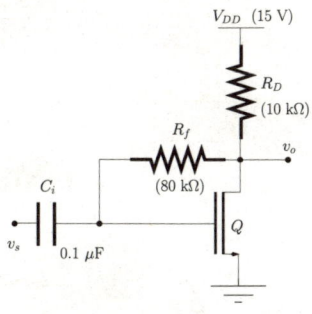

Figure 8.47: *Amplifier.*

8.8 Show that the equivalent current noise source for the amplifier of figure 8.31 b) can be expressed by eqn 8.151.

8.9 Show that the Power Spectral Density (PSD) of the equivalent voltage noise source for the amplifier of figure 8.33 b) can be expressed by eqn 8.153.

8.10 Consider the two voltage amplifiers of figure 8.46. Determine the equivalent input voltage noise source for each amplifier. The operational amplifiers are characterised by a differential voltage gain of 5000, $R_{in} = 10$ MΩ and $R_o = 50$ Ω. Assume that the noise of each op-amp is described by equivalent input noise sources such that;

$$\langle \mathbf{u_n\ u_n^*} \rangle^{\frac{1}{2}} = 0.6\ \text{nV}/\sqrt{\text{Hz}}$$
$$\langle \mathbf{i_n\ i_n^*} \rangle^{\frac{1}{2}} = 2\ \text{pA}/\sqrt{\text{Hz}}$$
$$\langle \mathbf{u_n\ i_n^*} \rangle \simeq 0$$

8.11 Show that R_n and g_n, defined by eqns 8.173 and 8.175, respectively, can be related by $Y_{s_{opt}}$ as follows: $g_n = R_n\ |Y_{s_{opt}}|^2$.

8.12 Show that the chain correlation matrix of a two-port network for which R_n, $Y_{s_{opt}}$ and F_{min} have been measured can be expressed as eqn 8.185.

8.13 Consider the amplifier of figure 8.47. Determine the PSD of the equivalent input current and voltage noise sources. $f_c = 1$ kHz, $I_G = 10$ pA, $V_{Th} = 1.5$ V, $k_n\ W/L = 0.25$ mA/V^2, $V_A = 50$ V, $C_{gd} = 4$ pF and $C_{gs} = 16$ pF. Determine the noise figure assuming a 2 kΩ source load.

A Mathematical formulae for electrical engineering

Function definition

Error function (Gaussian probability)	$Q(x) = \dfrac{1}{\sqrt{2\pi}} \displaystyle\int_x^\infty e^{-\lambda^2/2}\, d\lambda$	
Error function (2)	$\mathrm{erf}(x) = \dfrac{2}{\sqrt{\pi}} \displaystyle\int_0^x e^{-\lambda^2}\, d\lambda$	
Error function (3)	$\mathrm{erfc}(x) = \dfrac{2}{\sqrt{\pi}} \displaystyle\int_x^\infty e^{-\lambda^2}\, d\lambda$	
Sinc	$\mathrm{sinc}(t) = \dfrac{\sin(\pi t)}{\pi t}$	
Sign	$\mathrm{sign}(t) = \begin{cases} 1, & t > 0 \\ -1, & t < 0 \end{cases}$	
Step	$u(t) = \begin{cases} 1, & t > 0 \\ 0, & t < 0 \end{cases}$	
Rectangle	$\mathrm{rect}\left(\dfrac{t}{\tau}\right) = \begin{cases} 1, & |t| < \frac{\tau}{2} \\ 0, & |t| > \frac{\tau}{2} \end{cases}$	
Triangle	$\mathrm{triang}\left(\dfrac{t}{\tau}\right) = \begin{cases} 1 - \dfrac{|t|}{\tau}, & |t| < \tau \\ 0, & |t| > \tau \end{cases}$	

Fourier transform

Theorems[1]

Operation	Function	Transform
Superposition	$a_1 x_1(t) + a_2 x_2(t)$	$a_1 X_1(f) + a_2 X_2(f)$
Delay	$x(t-a)$	$X(f) e^{-j2\pi f a}$
Scale factor (time)	$x(\alpha t)$	$\dfrac{1}{|\alpha|} X\left(\dfrac{f}{\alpha}\right)$
Conjugate	$x^*(t)$	$X^*(-f)$
Duality	$X(t)$	$x(-f)$
Frequency translation	$x(t) e^{j2\pi f_c t}$	$X(f - f_c)$
Convolution	$x(t) * y(t)$	$X(f) Y(f)$
Multiplication	$x(t) y(t)$	$X(f) * Y(f)$
Multiplication by t^n	$t^n x(t)$	$(-j2\pi)^{-n} \dfrac{d^n X(f)}{df^n}$
Poisson's	$\displaystyle\sum_{n=-\infty}^{\infty} e^{\pm j2\pi n \lambda T}$	$\dfrac{1}{T} \displaystyle\sum_{m=-\infty}^{\infty} \delta\left(\lambda - \dfrac{m}{T}\right)$
Integral	$\displaystyle\int_{-\infty}^{\infty} x(t) y^*(t)\, dt$	$\displaystyle\int_{-\infty}^{\infty} X(f) Y^*(f)\, df$

---

[1] See also sections 4.3.1 and 4.3.2.

Fourier transforms

Function	$x(t)$	$X(f)$
Constant	1	$\delta(f)$
Impulse	$\delta(t-a)$	$e^{-j2\pi fa}$
Step	$u(t)$	$\dfrac{1}{j2\pi f} + \dfrac{1}{2}\delta(f)$
Rectangle	$\operatorname{rect}\left(\dfrac{t}{\tau}\right)$	$\tau\operatorname{sinc}(f\tau)$
Triangle	$\operatorname{triang}\left(\dfrac{t}{\tau}\right)$	$\tau\operatorname{sinc}^2(f\tau)$
Exponential (causal)	$e^{-\beta t}u(t)$	$\dfrac{1}{\beta+j2\pi f}$
Symmetrical exponential	$e^{-\beta|t|}u(t)$	$\dfrac{2\beta}{\beta^2+(2\pi f)^2}$
Phasor	$e^{-j(2\pi f_c t+\phi)}$	$e^{j\phi}\delta(f-f_c)$
Sine	$\sin(2\pi f_c t+\phi)$	$\dfrac{1}{2j}\left[e^{j\phi}\delta(f-f_c)-e^{-j\phi}\delta(f+f_c)\right]$
Cosine	$\cos(2\pi f_c t+\phi)$	$\dfrac{1}{2}\left[e^{j\phi}\delta(f-f_c)+e^{-j\phi}\delta(f+f_c)\right]$
Sign	$\operatorname{sign}(t)$	$\dfrac{1}{j\pi f}$
Sampling	$\displaystyle\sum_{k=-\infty}^{\infty}\delta(t-kT_s)$	$\dfrac{1}{T_s}\displaystyle\sum_{n=-\infty}^{\infty}\delta\left(t-\dfrac{n}{T_s}\right)$

A. Mathematical formulae for electrical engineering

Laplace transforms

$x(t)$	$X(s)$
$u(t)$	$\dfrac{1}{s}$
$t\,u(t)$	$\dfrac{1}{s^2}$
$t^n\,u(t)$	$\dfrac{n!}{s^{n+1}}$
$\dfrac{u(t)}{\sqrt{t}}$	$\sqrt{\dfrac{\pi}{s}}$
$e^{at}\,u(t)$	$\dfrac{1}{s-a}$
$t^n\,e^{at}\,u(t)$	$\dfrac{n!}{(s-a)^{n+1}}$
$\dfrac{1}{a-b}\left(e^{at}-e^{bt}\right)u(t)$	$\dfrac{1}{(s-b)(s-a)}$
$\dfrac{1}{a-b}\left(a\,e^{at}-b\,e^{bt}\right)u(t)$	$\dfrac{s}{(s-b)(s-a)}$
$\dfrac{(c-b)e^{at}+(a-c)e^{bt}}{(a-b)(b-c)(c-a)}u(t)$ $+\dfrac{(b-a)e^{ct}}{(a-b)(b-c)(c-a)}u(t)$	$\dfrac{1}{(s-c)(s-b)(s-a)}$
$\sin(a t)\,u(t)$	$\dfrac{a}{s^2+a^2}$
$\cos(a t)\,u(t)$	$\dfrac{s}{s^2+a^2}$
$t\sin(a t)\,u(t)$	$\dfrac{2as}{(s^2+a^2)^2}$
$t\cos(a t)\,u(t)$	$\dfrac{(s^2-a^2)}{(s^2+a^2)^2}$
$\dfrac{\cos(a t)-\cos(b t)}{b^2-a^2}u(t)$	$\dfrac{s}{(s^2+a^2)(s^2+b^2)}$

A. Mathematical formulae for electrical engineering

$x(t)$	$X(s)$
$e^{at}\sin(bt)\,u(t)$	$\dfrac{b}{(s-a)^2+b^2}$
$e^{at}\cos(bt)\,u(t)$	$\dfrac{s-a}{(s-a)^2+b^2}$
$\sinh(at)\,u(t)$	$\dfrac{a}{s^2-a^2}$
$\cosh(at)\,u(t)$	$\dfrac{s}{s^2-a^2}$
$\sin(at)\sinh(at)\,u(t)$	$\dfrac{2a^2 s}{s^4+4a^4}$
$\dfrac{e^{at}(1+2at)}{\sqrt{\pi t}}\,u(t)$	$\dfrac{s}{(s-a)^{\frac{3}{2}}}$
$\dfrac{1}{t}\sin(at)\,u(t)$	$\tan^{-1}\left(\dfrac{a}{s}\right)$
$\dfrac{2}{t}[1-\cos(at)]\,u(t)$	$\ln\left(\dfrac{s^2+a^2}{s^2}\right)$
$\dfrac{2}{t}[1-\cosh(at)]\,u(t)$	$\ln\left(\dfrac{s^2-a^2}{s^2}\right)$
$\operatorname{erfc}\left(\dfrac{a}{2\sqrt{t}}\right)u(t)$	$\dfrac{1}{s}e^{-a\sqrt{s}}$
$\dfrac{1}{\sqrt{\pi t}}e^{\frac{-a^2}{4t}}\,u(t)$	$\dfrac{1}{\sqrt{s}}e^{-a\sqrt{s}}$
$\dfrac{1}{\sqrt{\pi(t+a)}}\,u(t)$	$\dfrac{1}{\sqrt{s}}e^{as}\operatorname{erfc}(\sqrt{as})$
$\dfrac{1}{\pi t}\sin(2a\sqrt{t})\,u(t)$	$\operatorname{erfc}\left(\dfrac{a}{\sqrt{s}}\right)$
$\delta(t-a)$	e^{-as}
$a^{t/T}\,u(t)$	$\dfrac{1}{s-\frac{1}{T}\ln(a)}$
$\dfrac{t^{n-1}}{(n-1)!}e^{-at}$	$\dfrac{1}{(s+a)^n}$

Trigonometric identities

$$e^{\pm j\theta} = \cos(\theta) \pm j\sin(\theta)$$
$$e^{j2\alpha} + e^{j2\beta} = 2\cos(\alpha - \beta)e^{j(\alpha+\beta)}$$
$$e^{j2\alpha} - e^{j2\beta} = j2\sin(\alpha - \beta)e^{j(\alpha+\beta)}$$
$$\cos(\theta) = \frac{e^{j\theta} + e^{-j\theta}}{2}$$
$$\sin(\theta) = \frac{e^{j\theta} - e^{-j\theta}}{2j}$$
$$\cos^2(\theta) - \sin^2(\theta) = \cos(2\theta)$$
$$\cos^2(\theta) = \frac{1}{2}[1 + \cos(2\theta)]$$
$$\cos^3(\theta) = \frac{1}{4}[3\cos(\theta) + \cos(3\theta)]$$
$$\sin^2(\theta) = \frac{1}{2}[1 - \cos(2\theta)]$$
$$\sin^3(\theta) = \frac{1}{4}[3\sin(\theta) - \sin(3\theta)]$$
$$\sin(\alpha \pm \beta) = \sin(\alpha)\cos(\beta) \pm \sin(\beta)\cos(\alpha)$$
$$\cos(\alpha \pm \beta) = \cos(\alpha)\cos(\beta) \mp \sin(\beta)\sin(\alpha)$$
$$\tan(\alpha \pm \beta) = \frac{\tan(\alpha) \pm \tan(\beta)}{1 \mp \tan(\alpha)\tan(\beta)}$$
$$\sin(\beta)\sin(\alpha) = \frac{1}{2}\cos(\alpha - \beta) - \frac{1}{2}\cos(\alpha + \beta)$$
$$\cos(\beta)\cos(\alpha) = \frac{1}{2}\cos(\alpha - \beta) + \frac{1}{2}\cos(\alpha + \beta)$$
$$\cos(\beta)\sin(\alpha) = \frac{1}{2}\sin(\alpha - \beta) + \frac{1}{2}\sin(\alpha + \beta)$$

$$x\sin(\alpha) + y\cos(\alpha) = R\sin(\alpha + \beta)$$
$$x\cos(\alpha) - y\sin(\alpha) = R\cos(\alpha + \beta)$$

with

$$R = \sqrt{x^2 + y^2}$$
$$\beta = \tan^{-1}\left(\frac{y}{x}\right)$$

Series

Arithmetic

$$a + (a + r) + (a + 2r) + \ldots + [a + (n-1)r] = \frac{n}{2}[2a + (n-1)r]$$

Geometric

$$1 + r + r^2 + \ldots + r^{n-1} = \frac{1 - r^n}{1 - r}, \quad r \neq 1$$

Infinite geometric

$$1 + r + r^2 + \ldots + r^{n-1} + \ldots = \sum_{k=0}^{\infty} r^k$$
$$= \frac{1}{1 - r}, \quad |r| < 1$$

Binomial

$$(1 + x)^n = 1 + nx + \frac{n(n-1)}{2!} x^2 + \ldots + \frac{n!}{(n-k)!k!} x^k + \ldots, \quad |x| < 1$$

Taylor

$$\begin{aligned} f(x) &= f(a) + (x - a)f'(a) + \frac{(x - a)^2}{2!} f''(a) \\ &\quad + \ldots + \frac{(x - a)^n}{n!} f^{(n)}(a) + \ldots \end{aligned}$$

Maclaurin

$$f(x) = f(0) + xf'(0) + \frac{x^2}{2!} f''(0) + \ldots + \frac{x^n}{n!} f^{(n)}(0) + \ldots$$

Maclaurin series expansions of some functions

$$\cos(x) = \sum_{n=0}^{\infty} (-1)^n \frac{x^{2n}}{(2n)!}$$

$$\sin(x) = \sum_{n=0}^{\infty} (-1)^n \frac{x^{2n+1}}{(2n+1)!}$$

$$\exp(x) = \sum_{n=0}^{\infty} \frac{x^n}{n!}$$

$$\frac{1}{1 - x} = \sum_{n=0}^{\infty} x^n, \quad \text{for } -1 < x < 1$$

$$\ln(1 + x) = \sum_{n=1}^{\infty} (-1)^{n+1} \frac{x^n}{n}, \quad \text{for } -1 < x \leq 1$$

A. Mathematical formulae for electrical engineering

Error functions

$$\operatorname{erf}(x) = 1 - 2Q\left(\sqrt{2}\,x\right)$$
$$\operatorname{erfc}(x) = 2Q\left(\sqrt{2}\,x\right)$$

For x less than zero we have:

$$Q(-|x|) = 1 - Q(|x|)$$

For values of $x > 3$ it is possible to show that:

$$Q(x) \simeq \frac{1}{\sqrt{2\pi}\,x}\, e^{-x^2/2}$$

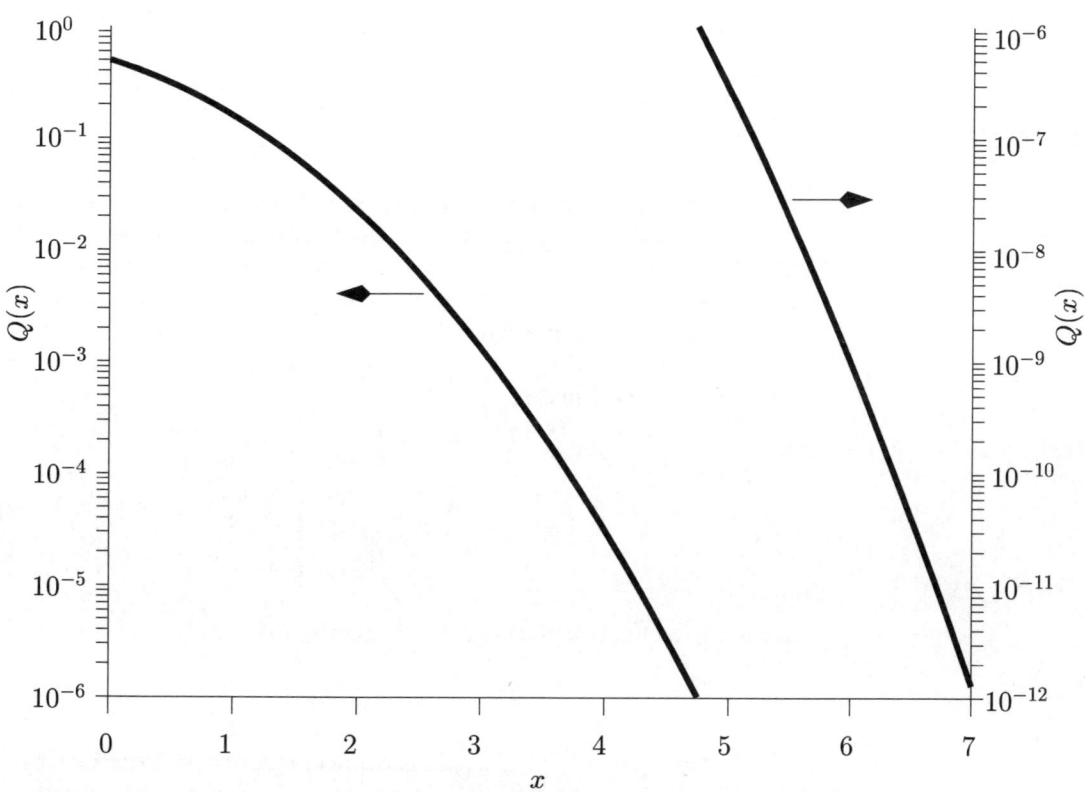

B Elementary matrix algebra

Definitions

A **matrix** is a rectangular array of elements and it is usually represented as follows:

$$[X] = \begin{bmatrix} x_{11} & x_{12} & \cdots & x_{1m} \\ x_{21} & x_{22} & \cdots & x_{2m} \\ \vdots & \vdots & \ddots & \vdots \\ x_{n1} & x_{n2} & \cdots & x_{nm} \end{bmatrix} \tag{B.1}$$

Each element x_{kl} can represent a real or complex number or functions of time or frequency. The matrix shown above has n rows and m columns and therefore is said to be an $n \times m$ matrix. The subscript $_{kl}$ identifies the position of each element in the matrix. Hence, the element x_{kl} is located at the intersection of the kth row with the lth column. If $n = m$ the matrix is called a square matrix.

A **vector** is a single column or single row matrix. A column vector is defined as an $n \times 1$ matrix;

$$[V] = \begin{bmatrix} v_1 \\ v_2 \\ \vdots \\ v_n \end{bmatrix} \tag{B.2}$$

and a row vector is defined as a $1 \times m$ matrix;

$$[U] = [u_1 \, u_2 \, \ldots \, u_m]$$

The **transpose** of a matrix $[X]$ is written as $[X]^T$ and is obtained by interchanging the rows and columns of the matrix $[X]$. For example, the transpose of a column vector $[V]$ yields a row vector $[V]^T$.

The *addition* and *subtraction* of matrices applies only to matrices of the same order. Hence, the sum or difference of two matrices $[X]$ and $[Y]$, both of order $n \times m$, produces a matrix $[Z]$ also of order $n \times m$. Each element z_{kl} of $[Z] = [X] \pm [Y]$ is given by

$$z_{kl} = x_{kl} \pm y_{kl}$$

B. *Elementary matrix algebra*

where x_{kl} and y_{kl} are the elements of $[X]$ and $[Y]$, respectively. These two operations are commutative and associative, that is:

$$[X] \pm [Y] = [Y] \pm [X]$$
$$([X] \pm [Y]) \pm [W] = [X] \pm ([Y] \pm [W])$$

The **Multiplication** of a **matrix** $[X]$, of order $n \times m$, by a **scalar** s produces a matrix $[Z]$ of order $n \times m$. Each element z_{kl} of $[Z] = s \times [X]$ is given by

$$z_{kl} = s \times x_{kl}$$

The **Multiplication of two matrices** $[X] \times [Y]$ can only be performed if they *conform*. The matrices $[X]$ of order $n \times m$ and $[Y]$ of order $p \times q$ are said to conform if $m = p$, that is, the number of columns of $[X]$ must be equal to the number of rows of $[Y]$. Under this condition $[X] \times [Y]$ produces a matrix $[Z]$ of order $n \times q$. Each element of $[Z]$ is given by

$$z_{kl} = \sum_{i=1}^{m} x_{ki} \times y_{il}$$

Note that when $[X]$ is of order $n \times m$ and $[Y]$ is of order $m \times n$ each of the products $[X] \times [Y]$ and $[Y] \times [X]$ conform. However, in general, we have that

$$[X] \times [Y] \neq [Y] \times [X]$$

that is the product of $[X]$ and $[Y]$ is non-commutable.

A **unity** or identity matrix, denoted here by[1] $[1]$, is a square matrix in which the elements on the principal diagonal are unity and the rest are zeros. An example of a unity matrix is given below:

$$[1] = \begin{bmatrix} 1 & 0 & 0 & 0 \\ 0 & 1 & 0 & 0 \\ 0 & 0 & 1 & 0 \\ 0 & 0 & 0 & 1 \end{bmatrix} \qquad (B.3)$$

The multiplication of a square matrix $[X]$ by the unit matrix of identical order does not change $[X]$, that is

$$[X] \times [1] = [1] \times [X] = [X]$$

The **inverse** of a square matrix $[X]$, represented by $[X]^{-1}$, is such that

$$[X] \times [X]^{-1} = [X]^{-1} \times [X] = [1]$$

Here we define $[Y]/[X]$ as follows:

$$\frac{[Y]}{[X]} = [X]^{-1} \times [Y]$$

---

[1]The unity matrix is usually denoted by $[I]$. In this book, we do not use this symbol in order to avoid confusion with the symbol which represents the electrical current.

C Two-port electrical parameters

Conversion between electrical parameters

Parameters	Admittance		Impedance	
Admittance	Y_{11} Y_{12}		$\frac{Z_{22}}{|Z|}$ $\frac{-Z_{12}}{|Z|}$	
	Y_{21} Y_{22}		$\frac{-Z_{21}}{|Z|}$ $\frac{Z_{11}}{|Z|}$	
Impedance	$\frac{Y_{22}}{|Y|}$ $\frac{-Y_{12}}{|Y|}$		Z_{11} Z_{12}	
	$\frac{-Y_{21}}{|Y|}$ $\frac{Y_{11}}{|Y|}$		Z_{21} Z_{22}	
Chain	$\frac{-Y_{22}}{Y_{21}}$ $\frac{-1}{Y_{21}}$		$\frac{Z_{11}}{Z_{21}}$ $\frac{|Z|}{Z_{21}}$	
	$\frac{-|Y|}{Y_{21}}$ $\frac{-Y_{11}}{Y_{21}}$		$\frac{1}{Z_{21}}$ $\frac{Z_{22}}{Z_{21}}$	
Scattering	$S_{11} = \frac{(1-y'_{11})(1+y'_{22})+y'_{12}y'_{21}}{\Delta_1}$ $S_{12} = \frac{-2y'_{12}}{\Delta_1}$ $S_{21} = \frac{-2y'_{21}}{\Delta_1}$ $S_{22} = \frac{(1+y'_{11})(1-y'_{22})+y'_{12}y'_{21}}{\Delta_1}$		$S_{11} = \frac{(z'_{11}-1)(z'_{22}+1)-z'_{12}z'_{21}}{\Delta_2}$ $S_{12} = \frac{2z'_{12}}{\Delta_2}$ $S_{21} = \frac{2z'_{21}}{\Delta_2}$ $S_{22} = \frac{(z'_{11}+1)(z'_{22}-1)-z'_{12}z'_{21}}{\Delta_2}$	

Table C.1: Transformations between electrical parameters.

$$\Delta_1 = (y'_{11}+1)(y'_{22}+1) - y'_{12}y'_{21} \qquad y'_{ij} = Y_{ij}Z_o \quad i,j = 1,2$$
$$\Delta_2 = (z'_{11}+1)(z'_{22}+1) - z'_{12}z'_{21} \qquad z'_{ij} = \frac{Z_{ij}}{Z_o} \quad i,j = 1,2$$

C. Two-port electrical parameters

Parameters	Chain	Scattering
Admittance	$\dfrac{A_{22}}{A_{12}} \quad \dfrac{-\|A\|}{A_{12}}$ $\dfrac{-1}{A_{12}} \quad \dfrac{A_{11}}{A_{12}}$	$Y_{11} = \dfrac{(1+S_{22})(1-S_{11})+S_{12}S_{21}}{\Delta_4 Z_o}$ $Y_{12} = \dfrac{-2S_{12}}{\Delta_4 Z_o}$ $Y_{21} = \dfrac{-2S_{21}}{\Delta_4 Z_o}$ $Y_{22} = \dfrac{(1-S_{22})(1+S_{11})+S_{12}S_{21}}{\Delta_4 Z_o}$
Impedance	$\dfrac{A_{11}}{A_{21}} \quad \dfrac{\|A\|}{A_{21}}$ $\dfrac{1}{A_{21}} \quad \dfrac{A_{22}}{A_{21}}$	$Z_{11} = Z_o \dfrac{(1-S_{22})(1+S_{11})+S_{12}S_{21}}{\Delta_5}$ $Z_{12} = Z_o \dfrac{2S_{12}}{\Delta_5}$ $Z_{21} = Z_o \dfrac{2S_{21}}{\Delta_5}$ $Z_{22} = Z_o \dfrac{(1+S_{22})(1-S_{11})+S_{12}S_{21}}{\Delta_5}$
Chain	$A_{11} \quad A_{12}$ $A_{21} \quad A_{22}$	$A_{11} = \dfrac{(1-S_{22})(1+S_{11})+S_{12}S_{21}}{2S_{21}}$ $A_{12} = Z_o \dfrac{(1+S_{22})(1+S_{11})-S_{12}S_{21}}{2S_{21}}$ $A_{21} = \dfrac{(1-S_{22})(1-S_{11})-S_{12}S_{21}}{2Z_o S_{21}}$ $A_{22} = \dfrac{(1+S_{22})(1-S_{11})+S_{12}S_{21}}{2S_{21}}$
Scattering	$S_{11} = \dfrac{a'_{11}+a'_{12}-a'_{21}-a'_{22}}{\Delta_3}$ $S_{12} = \dfrac{2(a'_{11}a'_{22}-a'_{12}a'_{21})}{\Delta_3}$ $S_{21} = \dfrac{2}{\Delta_3}$ $S_{22} = \dfrac{-a'_{11}+a'_{12}-a'_{21}+a'_{22}}{\Delta_3}$	$S_{11} \quad S_{12}$ $S_{21} \quad S_{22}$

Table C.2: Transformations between electrical parameters. *(Cont.)*

$$\Delta_3 = a'_{11} + a'_{12} + a'_{21} + a'_{22}$$
$$|A| = A_{11}A_{22} - A_{12}A_{21}$$
$$\Delta_4 = (1+S_{11})(1+S_{22}) - S_{12}S_{21}$$
$$\Delta_5 = (1-S_{11})(1-S_{22}) - S_{12}S_{21}$$

$$a'_{11} = A_{11}, \ a'_{12} = A_{12}/Z_o,$$
$$a'_{21} = A_{21}Z_o, \ a'_{22} = A_{22}$$
$$|Y| = Y_{11}Y_{22} - Y_{12}Y_{21}$$
$$|Z| = Z_{11}Z_{22} - Z_{12}Z_{21}$$

C Two-port electrical parameters

Conversion between electrical parameters

Parameters	Admittance		Impedance	
Admittance	Y_{11} Y_{12}		$\frac{Z_{22}}{|Z|}$ $\frac{-Z_{12}}{|Z|}$	
	Y_{21} Y_{22}		$\frac{-Z_{21}}{|Z|}$ $\frac{Z_{11}}{|Z|}$	
Impedance	$\frac{Y_{22}}{|Y|}$ $\frac{-Y_{12}}{|Y|}$		Z_{11} Z_{12}	
	$\frac{-Y_{21}}{|Y|}$ $\frac{Y_{11}}{|Y|}$		Z_{21} Z_{22}	
Chain	$\frac{-Y_{22}}{Y_{21}}$ $\frac{-1}{Y_{21}}$		$\frac{Z_{11}}{Z_{21}}$ $\frac{|Z|}{Z_{21}}$	
	$\frac{-|Y|}{Y_{21}}$ $\frac{-Y_{11}}{Y_{21}}$		$\frac{1}{Z_{21}}$ $\frac{Z_{22}}{Z_{21}}$	
Scattering	$S_{11} = \frac{(1-y'_{11})(1+y'_{22})+y'_{12}y'_{21}}{\Delta_1}$ $S_{12} = \frac{-2y'_{12}}{\Delta_1}$ $S_{21} = \frac{-2y'_{21}}{\Delta_1}$ $S_{22} = \frac{(1+y'_{11})(1-y'_{22})+y'_{12}y'_{21}}{\Delta_1}$		$S_{11} = \frac{(z'_{11}-1)(z'_{22}+1)-z'_{12}z'_{21}}{\Delta_2}$ $S_{12} = \frac{2z'_{12}}{\Delta_2}$ $S_{21} = \frac{2z'_{21}}{\Delta_2}$ $S_{22} = \frac{(z'_{11}+1)(z'_{22}-1)-z'_{12}z'_{21}}{\Delta_2}$	

Table C.1: Transformations between electrical parameters.

$$\Delta_1 = (y'_{11}+1)(y'_{22}+1) - y'_{12}y'_{21} \qquad y'_{ij} = Y_{ij}Z_o \quad i,j = 1,2$$
$$\Delta_2 = (z'_{11}+1)(z'_{22}+1) - z'_{12}z'_{21} \qquad z'_{ij} = \frac{Z_{ij}}{Z_o} \quad i,j = 1,2$$

C. Two-port electrical parameters

		IMPEDANCE		ADMITTANCE		CASCADE	
Series Impedance (impedance Z)		NON-EXISTENT		Y	$-Y$	1	Z
				$-Y$	Y	0	1
Shunt Admittance (admittance Y)		Z	Z	NON-EXISTENT		1	0
		Z	Z			Y	1
T Circuit (Z_1, Z_2, Z_3)		$Z_1 + Z_3$	Z_3	$\frac{Y_1(Y_2+Y_3)}{Y_1+Y_2+Y_3}$	$\frac{-Y_1 Y_2}{Y_1+Y_2+Y_3}$	$\frac{Z_3+Z_1}{Z_3}$	$\frac{(Z_1+Z_2)Z_3+Z_1 Z_2}{Z_3}$
		Z_3	$Z_2 + Z_3$	$\frac{-Y_1 Y_2}{Y_1+Y_2+Y_3}$	$\frac{Y_2(Y_1+Y_3)}{Y_1+Y_2+Y_3}$	$\frac{1}{Z_3}$	$\frac{Z_3+Z_2}{Z_3}$
PI Circuit (Y_1, Y_2, Y_3)		$\frac{Z_1(Z_2+Z_3)}{Z_1+Z_2+Z_3}$	$\frac{Z_1 Z_2}{Z_1+Z_2+Z_3}$	$Y_1 + Y_3$	$-Y_3$	$\frac{Y_3+Y_2}{Y_3}$	$\frac{1}{Y_3}$
		$\frac{Z_1 Z_2}{Z_1+Z_2+Z_3}$	$\frac{Z_2(Z_1+Z_3)}{Z_1+Z_2+Z_3}$	$-Y_3$	$Y_2 + Y_3$	$\frac{(Y_1+Y_2)Y_3+Y_1 Y_2}{Y_3}$	$\frac{Y_3+Y_1}{Y_3}$
General Tx Line (Z_o, γ, length l)		$Z_o \coth(\gamma l)$	$Z_o \csch(\gamma l)$	$\frac{\coth(\gamma l)}{Z_o}$	$-\frac{\csch(\gamma l)}{Z_o}$	$\cosh(\gamma l)$	$Z_o \sinh(\gamma l)$
		$Z_o \csch(\gamma l)$	$Z_o \coth(\gamma l)$	$-\frac{\csch(\gamma l)}{Z_o}$	$\frac{\coth(\gamma l)}{Z_o}$	$\frac{\sinh(\gamma l)}{Z_o}$	$\cosh(\gamma l)$
Lossless Tx Line ($Z_o, j\beta$, length l)		$-jZ_o \cot(\beta l)$	$-jZ_o \csc(\beta l)$	$-\frac{j\cot(\beta l)}{Z_o}$	$\frac{j\csc(\beta l)}{Z_o}$	$\cos(\beta l)$	$jZ_o \sin(\beta l)$
		$-jZ_o \csc(\beta l)$	$-jZ_o \cot(\beta l)$	$\frac{j\csc(\beta l)}{Z_o}$	$-\frac{j\cot(\beta l)}{Z_o}$	$\frac{j\sin(\beta l)}{Z_o}$	$\cos(\beta l)$
Voltage controlled Voltage Source ($A \cdot v$)		NON-EXISTENT		NON-EXISTENT		$\frac{1}{A}$	0
						0	0
Voltage controlled Current Source ($G \cdot v$)		NON-EXISTENT		0	0	0	$-\frac{1}{G}$
				G	0	0	0
Current controlled Current Source ($A \cdot i$)		NON-EXISTENT		NON-EXISTENT		0	0
						0	$-\frac{1}{A}$
Current controlled Voltage Source ($R \cdot i$)		0	0	NON-EXISTENT		0	0
		R	0			$\frac{1}{R}$	0

Index

ABCD parameters, 157
acceptor, 183
active devices, 183
admittance, 59
admittance parameters, 153
alternating current (AC), 3, 32, 48
ampere, 2
amplifier
    adder, 181
    common-base, 208
    common-collector, 211
    common-emitter, 187
    difference, 181
    differential pair, 215
    differentiator, 180
    instrumentation, 182
    integrator, 180
    inverting, 179
    noisy, 310
    non-inverting, 178
angle, 39, 48
Argand diagram, 32, 57
asymptotes, 80, 81, 83
autocorrelation function, 295–298
average power, 4, 5, 7
    random processes, 297
average value, 66
    random variable, 281
    random variables, 285

bandwidth
    amplifier, 169, 171
    circuit, 73, 76, 78, 82
    noise, 298
    random processes, 297, 298
    signal, 73–75
    system, 82
    unity-gain, 196
Boltzmann's constant, 183

capacitance, 8, 51, 55, 115, 129
capacitor, 8

AC-coupling, 170
DC-blocking, 170, 200
noise model, 308
causal exponential, 97
causal signal, 97, 109
central limit theorem, 290
chain parameters, 157
chain/cascade connection, 158
channel length, 193
channel width, 193
channel-length modulation, 195
characteristic function, 288, 289
characteristic impedance, 230
charge (electron), 2
Chebyshev's inequality, 286
clipping, 188
common-base
    DC-analysis, 208
    high-frequency analysis, 211
    mid-frequency analysis, 209
common-collector
    DC analysis, 212
    mid-frequency analysis, 213
common-emitter
    DC analysis, 197
    high-frequency, 204
    low-frequency analysis, 198
    mid-frequency analysis, 204
    noise analysis, 323
    voltage gain, 188, 198
complex exponentials, 54, 66, 73, 77
complex numbers, 32
    addition, 34
    angle, 39
    argument, 39
    Cartesian representation, 33
    complex conjugate, 36
    division, 37, 40
    equality, 33
    equations, 38
    exponential form, 41
    imaginary numbers, 33

INDEX

357

magnitude, 39
modulus, 39
multiplication, 35, 40
number j, 33
phasor representation, 42
polar coordinates, 39
polar representation, 39
powers, 43
rectangular representation, 33
roots, 43
subtraction, 34
complex plane, 32
computer-aided
electrical analysis, 163
noise analysis, 323
noise modelling, 323
conductance, 5, 6
controlled sources, 21
convolution operation, 100, 101, 110, 112, 289
correlation, 295
matrix, 328
admittance, 324
chain, 325
impedance, 325
scattering, 325
time, 300
transformation matrices, 328
coulomb, 2, 8
CR circuit, 83
transient response, 138
critically damped, 134, 140
cross-correlation function, 301
current, 1
amplifier, 176
current divider
resistive, 20
current gain, 98, 157
current mirror, 220
output resistance, 220
current source, 3, 220
current-controlled, 21
DC, 3
output resistance, 4
voltage-controlled, 21
cut-off frequency, 76

damping factor, 124
De Moivre's theorem, 43
decibel (dB), 80
delay time, 137
dependent sources, 21

differential pair
common-mode, 215, 216
differential mode, 216, 217
diode, 183
p–n junction, 183
effect, 185
geometry, 183
saturation current, 183
symbol, 183
Dirac delta function, 92, 98, 281
direct current (DC), 3
distortion
linear, 79, 171, 243
non-linear, 79
clipping, 188
donor, 183
duty-cycle, 73

Early effect, 192
Ebers-Moll model, 184
effective value
sinusoidal waveform, 50
triangular waveform, 50
effective values, 49
electrical length, 234
electrical parameters
ABCD, 157
admittance, 153
chain, 157
conversion between, 159, 352
impedance, 150
scattering, 248
electro-motive force (emf), 131
electron charge, 2
electronic amplifier, 169
energy stored
capacitor, 9
inductor, 9
enhancement, 193
ensemble average, 295
equivalent conductance, 12
equivalent resistance, 11, 13, 17
error function, 282, 349
Euler's formula, 42, 54
expectation operator, 285
multivariate, 287

fall-time, 137
farad, 8
Faraday's law, 9
feedback, 178
figures of merit, 169

filter
- band-pass, 86
- high-pass, 84
- low-pass, 76, 77

forced response, 109, 111
forcing signal, 111
Fourier integral, 91
Fourier series, 65
- average value, 66
- coefficients, 66, 87
- harmonics, 66, 77, 87
- time delay, 71

Fourier transform, 88
- from Laplace transform, 126
- convolution theorems, 106, 289
- DC signal, 93
- duality, 91
- generalised transform, 94
- periodic functions, 95
- table, 344
- time delay, 92
- transient analysis, 113
  - differentiation theorem, 113
  - integration theorem, 114

frequency
- angular, 48
- linear, 48

frequency domain, 65
frequency response
- high-frequency, 171
- low-frequency, 170
- mid-frequency, 171

frequency selectivity, 86
function
- s-functions, 126
- Gaussian probability, 342
- parabolic, 286
- rect, 73, 342
- signum, 94, 95, 342
- sinc, 90
- table of, 342
- triang, 105, 342
- unit-step, 94, 118, 342

gain
- amplifier, 169

Gaussian distribution, 281
- bivariate, 293

geometric series, 348
ground terminal, 18

harmonic oscillator, 124

henry, 9
histogram, 280
hybrid-π
- BJT, 190
- IGFET, 194

hydraulic analogue
- capacitor, 8
- inductor, 9
- LC circuit, 133
- oscillator, 133
- resistance, 2
- RLC circuit, 135

imaginary axis, 33
impedance
- capacitive, 55
- generalised, 56, 58
- generalised (s-domain), 130
- inductive, 56
- parallel connection, 59
- series connection, 59

impedance matching, 272, 275
impedance parameters, 150
impulse response, 98
- RC circuit, 98

incident wave, 232
inductance, 9, 52, 56, 115, 129
inductor, 9
- noise model, 308

initial conditions, 110
insertion loss, 234
instantaneous power, 4, 5, 7
integration (by parts), 113
International System of Units, 1

joule, 9

Kirchhoff's laws, 10, 18, 52
- current law, 10, 19
- voltage law, 10

L-sections, 272
Laplace transform
- s-domain differentiation, 121
- convolution, 121
- definition, 117
- inverse transform, 119
- linearity, 119
- partial-fraction, 123
- region of convergence, 118, 126
- solving differential equations, 127
- table, 345

INDEX

time delay, 119
time differentiation, 120
time integration, 120
time periodicity, 122
time scaling, 121
transient analysis, 127
LC circuit
    natural response, 133
load matching, 272, 275
    $\lambda/4$ transformer, 240
load-line, 188
logarithmic scale, 80
LR circuit
    transient response, 139

Maclaurin series, 348
magnetic flux, 9
matrix algebra, 350
mean
    random process, 295, 296
        time averaging, 296
    random variable, 280, 281, 288, 291
    random variables, 285
mean-square, 286
microstrip lines, 247
    attenuation, 248
    characteristic impedance, 247
    dielectric permittivity, 247
    electrical length, 254
    geometry, 247
    loss tangent, 248
    propagation constant, 248
microwaves
    frequency range, 224
    wavelength, 224
Miller's theorem, 161
    common-emitter, 205, 206
    forward voltage gain, 162
    high-frequency response, 163
    input impedance, 162
moments
    random variables, 285, 288

natural frequency, 124
natural response, 109, 111
neper, 243
nodal analysis, 18, 56
node zero, 18
noise, 279, 295
    $1/f$, 306
    admittance representation, 324
    analysis, 299, 338
    analysis method, 310, 331
    bandwidth, 295, 298
    bipolar transistor, 309
    capacitor, 308
    chain representation, 325
    characterisation, 338
    common-emitter, 318, 323, 332
    computer-aided, 323
    cross-power spectral density, 301
    current spectral density (RMS), 317
    equivalent bandwidth, 304
    equivalent current spectral density, 317
    equivalent input sources, 310
    equivalent input spectral density, 338
    equivalent resistance, 322
    equivalent temperature, 323, 338
    equivalent voltage spectral density, 319
    factor, 320
        minimum, 321
    field-effect transistor, 308
    figure, 319, 338
        chain of amplifiers, 321
    impedance representation, 324
    inductor, 308
    Johnson, 305
    mean, 297
    optimum output admittance, 321
    power, 295, 297, 300, 320
    power spectral density, 298, 299
    power spectrum, 298
    resistor, 307
    RMS, 297
    scattering representation, 325
    shot, 306
    sources, 279
    thermal, 305
    transformation matrices, 328
    two-port, 323
    voltage spectral density (RMS), 317
    white, 303
Norton equivalent
    $1/f$ noise, 306
    AC circuit, 62
    DC circuit, 24
    noisy amplifier, 318
    shot noise, 306
    thermal noise, 305

Norton's theorem, 23, 24, 62

ohm, 2
Ohm's law, 2, 17, 50, 58
op-amp
    ideal, 177
open-circuit, 22
open-circuit time constants, 207
open-loop, 177
    gain, 178
operating point, 188
operational amplifiers, 177
oscillator, 124, 133
overdamped, 134, 140
overshoot, 140

parallel connection
    capacitor, 14
    inductor, 16
    resistor, 12
    two-port circuits, 156
Parseval's theorem, 71
peak overshoot, 140
periodic waveforms, 66
permeability, 248
phase
    difference, 49, 52, 55, 56
    instantaneous, 48
phasor, 32, 55, 66, 72, 89, 224
    density, 91
    rotating, 57, 72
    static, 58
    stationary, 58, 72
    travelling wave, 225
phasor analysis, 54
Poisson distribution, 284
poles and zeros, 83
power, 4, 297
    available, 255
    average, 4, 5, 7, 62
    average (normalised), 70
    dissipation, 6, 7, 62, 70
    instantaneous, 4, 5, 7
    instantaneous (normalised), 70
    maximum transfer, 64, 272
    noise, 300
    normalised, 70
    Parseval's theorem, 71
    transducer gain, 257
    waves, 254
power spectrum, 298
probability, 281

probability density function, 281
probability density functions
    joint, 293
propagation
    constant, 227, 230
    speed, 225
pulse
    rectangular, 89, 92
    square, 100
Pythagoras's theorem, 39

Quality Factor, 86
quiescent, 188

radian, 48
radio-frequency, 224
random processes, 295
    autocorrelation, 296
    average, 295
    bandwidth, 297, 298
    ensemble averages, 295
    ergodic, 296
        autocorrelation, 296
        mean, 296
    filtered, 303
        autocorrelation, 303
        mean, 303
    Gaussian, 302
    mean, 295
    sample function, 295
    stationary, 296
random signals, 303
random variables, 279
    average, 280
    average value, 285
    central limit theorem, 290
    characteristic function, 288
    continuous, 284
    covariance, 293
    discrete, 284
    expectation, 285
    Gaussian, 293
    Gaussian distribution, 281
    joint PDF, 293
    mean, 280, 285
    mean-square, 286
    moments, 285, 288
    multiple, 287
    Poisson, 284
    probability density function, 281
    standard deviation, 281
    statistically dependent, 293

INDEX

sums of, 289, 290
uniform, 284
variance, 281, 286
Rayleigh's energy theorem, 96
RC circuit, 73, 100
transient response, 136
reactance
capacitive, 51, 55
inductive, 52, 56
real axis, 33
rectangular waveform, 66, 67
reference node, 18
reflected wave, 232
reflection coefficient, 233
relative permittivity, 247
resistance, 2, 5, 50, 55, 115, 128
dynamic, 191
resistor, 5
noise model, 307
rise-time, 137
RL circuit, 53
natural response, 130
transient response, 138
RLC circuit, 85, 135
natural response, 133
critically damped, 134
overdamped, 134
underdamped, 134
transient analysis
critically damped, 140
overdamped, 140
underdamped, 140
transient response, 139
RMS, 50, 51
noise, 297
root-mean-square, 50, 297

sampling property, 93
scattering parameters, 248
generalised, 258
reference planes, 254
second-order, 124
series, 347
binomial, 348
Maclaurin, 348
series connection
capacitor, 13
inductor, 15
resistor, 11
two-port circuits, 153
settling time, 141
short-circuit, 18
virtual, 178, 179
short-circuit time constants, 200
siemen, 6
signal filtering, 85
signal shaping, 85
signal-to-noise ratio, 320
small-signal
amplifier, 175
Smith chart, 264
admittance representation, 267
impedance representation, 266
spectral density, 89
spectrum
continuous, 89
line, 72
speed of light, 248
standard deviation, 281
steady-state, 48, 51, 109, 111
stochastic processes, 295
stub-tuning, 276
superposition theorem, 25, 76, 86

Thévenin equivalent
AC circuit, 62
DC circuit, 24
noisy amplifier, 318
noisy shunt admittance, 326
thermal noise, 305
Thévenin's theorem, 22, 24, 62
thermal voltage, 183
threshold voltage, 193
time constant, 75, 77
time domain reflectometry, 239
transconductance
BJT, 191
IGFET, 194
transconductance gain, 98, 154, 157
transducer power gain, 257
transfer function, 75, 80
admittance, 98, 154
amplifier, 170
angle, 82
CR circuit (high-pass), 83
current, 98
impedance, 98, 151
magnitude, 82, 84
phase, 82, 84
poles and zeros, 83
RC circuit(low-pass), 80
RLC circuit (band-pass), 85
voltage, 98
transient analysis

transmission line, 237, 240
    using Fourier transform, 113
    using Laplace transforms, 127
transient response, 111
transimpedance gain, 98, 151, 157
transistor
    bipolar (BJT), 183
        hybrid-π, 190
        noise model, 309
        transconductance, 191
    current gain, 187
    effect, 185
    field-effect, 192
        n-channel, 193
        p-channel, 193
        large signal model, 193
        noise model, 308
    hetero-junction bipolar, 309
    high-frequency models, 195
    IGFET, 193
        transconductance, 194
    internal capacitances, 171, 195
    mobility, 193
    MOSFET, 193
    NPN
        geometry, 184
        symbol, 184
    PNP
        geometry, 184
        symbol, 184
transmission coefficient, 234
transmission line, 227
    $\lambda/4$ transformer, 239
    impedance matching, 275
    load matching
        transient analysis, 240
    lossless
        characteristic impedance, 230
        input impedance, 234
        matched, 230, 235
        model, 228
        open-circuit, 235
        propagation constant, 230
        short-circuited, 237
    lossy, 242
        attenuation constant, 243
        characteristic impedance, 243
        input impedance, 247
    microstrip, 247
    reflection coefficient, 233
    transient analysis, 237
triangular waveform, 66

trigonometric identities, 347
two-port circuit, 150
    S-parameters, 250
    Y-parameters, 153
    Z-parameters, 150
    ABCD (chain)-parameters, 157
    noise analysis, 323
    phasor analysis, 150
    table, 352
    unilateral circuit, 161

underdamped, 134, 140
uniform distribution, 284
unilateral circuit, 161
unity-gain bandwidth, 196

variance, 281, 286, 291
volt, 2
voltage, 2
    amplifier, 176
    polarity, 52
    propagating, 231
voltage divider
    resistive, 20
    with impedances, 64, 85, 174
voltage gain, 98, 157
voltage source, 2
    AC, 3
    current-controlled, 21
    DC, 3
    output resistance, 3
    voltage-controlled, 21
voltage standing wave ratio (VSWR), 234

watt, 4
waveform
    rectangular, 66, 67
    triangular, 66
wavelength, 225
white noise, 303
    filtered, 304
white spectrum, 303
Wiener-Kinchine theorem, 298

zeros and poles, 83